Representing Place

Representing Place

LANDSCAPE PAINTING AND MAPS

Edward S. Casey

University of Minnesota Press
Minneapolis / London

The University of Minnesota Press acknowledges the work of Edward Dimendberg, editorial consultant, on this project.

Every effort was made to obtain permission to reproduce the illustrations in this book. If any proper acknowledgment has not been made, we encourage copyright holders to notify the publisher.

Published by the University of Minnesota Press
111 Third Avenue South, Suite 290
Minneapolis, MN 55401-2520
http://www.upress.umn.edu

Library of Congress Cataloging-in-Publication Data

Casey, Edward S., 1939–
    Representing place : landscape painting and maps / Edward S. Casey.
        p.   cm.
    Includes index.
    ISBN 0-8166-3714-8 (HC : alk. paper)—ISBN 0-8166-3715-6 (PB : alk. paper)
    1. Landscape painting.   2. Map drawing.   I. Title.
    ND1340 .C38 2002
    758'.1—dc21

                                            2001004577

Printed in the United States of America on acid-free paper

The University of Minnesota is an equal-opportunity educator and employer.

12 11 10 09 08 07 06 05 04 03 02   10 9 8 7 6 5 4 3 2 1

Dedicated to Dan Rice
  Master of Landscape Painting
  Intrepid Mapper of the Place-World

Contents

List of Illustrations

PLATES

What Does It Mean to Represent Landscape?

The problem is to understand these strange relationships which are woven between the parts of the landscape, or between it and me as an incarnate subject, and through which an object perceived can concentrate in itself a whole scene or become the image of a whole segment of life.

—Maurice Merleau-Ponty, *Phenomenology of Perception*

Let us set aside the apparently straightforward (but in fact quite complicated) question: What does it mean to *represent* landscape, that is, a portion of the perceived earth that lies before and around us? Better to begin instead with a still more preliminary question: Why should we *want to do so* in the first place? Why bother to do so if it is the case that a given landscape—say, that of the Connecticut River viewed from Mount Holyoke, or Yarlung Valley in Tibet seen from Yumbu Lhakhang—suffices unto itself as a scene of perception, as something to behold on its own terms? Does not its mere presence, its massive moving plenitude, take care of itself? It certainly does not need *us* to represent *it*! Why take the trouble to represent what we already possess—what we now see with our eyes and now feel under our feet? Not only does representing landscape in paintings and photographs seem cognitively greedy (as if such representation were more important or more satisfying than experiential immersion), but it also seems to be unnecessary: What need is there to represent a landscape vista that is enticing or entrancing in itself, engaging us fully? Does not the vast implacable river below me—"sullen, untamed, and intractable"[1]—speak for itself? Is this not enough already? Why move to representation when the experience of landscape is dense enough, and frequently pleasing enough, in its own way? Why seek *other* ways, particularly representational ones that appear to signify a secondary status and that only complicate matters further? Why re-present what is already presented so effectively and thoroughly in ordinary direct experience?

And yet human beings go on representing landscape at every turn—and have done so for at least nine thousand years: witness the great map-painting on a crumbling wall at Çatal Hüyük in ancient Anatolia (6200 B.C.), or the enduring petroglyphic maps inscribed in the late Iron Age on boulders high in the Valcamonica region of northern Italy. (See Figures 7.3, 7.4, 7.5.) John Muir, that indefatigable explorer of the High Sierras and Glacier Bay, represented landscape in words and images fashioned on the spot, *sur place,* as the French so tellingly say. Constable and Cézanne and Monet, all of them *plein-air* painters, refashioned their immediately perceived worlds in their painted works. Such overt actions of representing the circumambient landscape continue, unabatedly and irrepressibly, in many times and places, ranging from sketches on the walls of caves and cities to digital videos with the help of the most advanced equipment. Far from being a luxury and superfluous—merely abetting or repeating primary perceptual experience—such continual and diversified representing is ingredient in the experience of landscape itself: the brute being of every landscape calls for—indeed, already includes, if only by anticipation—an insistent represented being as part of its very identity. The coimplication of *l'être sauvage* and *l'être représenté,* their literal com-plication, is pervasively present in landscape painting. As Merleau-Ponty puts it: "Everything is cultural in us (our *Lebenswelt* is 'subjective') (our perception is cultural-historical) and everything is natural in us (even the cultural rests on the polymorphism of wild being)."[2]

Representing landscape occurs in line and in paint, in gesture and in photograph, in image and in word. Of all these ways, whose complete cataloging would require a separate study, two in particular form the focus of this book: maps and landscape paintings. Although I have not chosen these two kinds of representation for the sake of symmetry, they are in fact quite complementary to each other. Whereas maps orient us in the practical world, landscape paintings possess the decidedly nonpractical function of helping us to appreciate the natural world's inherent beauty and sublimity. Maps facilitate our access to the life-world of action (by literally guiding this action), whereas landscape paintings aid in contemplating the surrounding world (contemplation is itself an action, but one that lacks a practical intentionality). Paintings call mainly upon darting eye movements and slight shifts of stance, if only on the part of the viewer's imaginative body[3]—in contradistinction to maps, which project and often induce the movement of their user's actual body. Once maps are seen (unless they are seen for aesthetic values—a situation we shall encounter later in this book), they direct the movement of the whole body as it makes its way to the place on the map that is its destination. Once landscape paintings are seen, on the other hand, they may call for further seeing, but rarely for any particular actions in relation to the depicted landscape: such actions are adventitious to the experience itself of these works.

Maps and landscape paintings are also to be contrasted in terms of their respective aims as forms of representation. Maps aim (at least officially) at representing the exact contours of land or sea masses and the precise distances between them—a cartographic concern—whereas landscape paintings attempt to convey the sensuous aspects of the environing place-world. The difference is that between Buckminster Fuller's Dymaxion Sky-Ocean World Map (a football field–sized map that accurately represents the entirety of the earth's surface as a single world-island in a single world-sea) and an imaginative painting of the sea by J. M. W. Turner, who tied himself to the mast of a moving ship in order to paint the sea evocatively as he experienced it firsthand.[4] As Svetlana Alpers has written, "maps give us the measure of a place and the relationship between places, quantifiable data, while landscape pictures are evocative, and aim rather to give us the quality of a place or of the viewer's sense of it. One is closer to science, the other to art."[5]

This is not to say that the two modes of representation are altogether antithetical.

"Topographical" painting of the sort practiced by Turner early in his career—or in realistic depictions of, say, colonial Bhutan[6]—brings together the cartographic and the painterly in one coherent representational project. The same combination holds for Jasper Johns's otherwise audaciously different *Map* (1961), a *painting* of a *map* of the United States, both at once. The two modes of representation can certainly coexist and may even enhance each other's presence, as we shall see in the intriguing instance of seventeenth-century Dutch art and cartography. Nevertheless, I shall start by considering painting and mapping in comparative isolation from each other, bringing them into close proximity only later on. Each form of representation is intricate enough to merit its own discussion, and each has its own history. But each is capable of speaking to the other in ways that are as illuminating as they are rarely discussed.

In this book I open up a dialogue between maps and landscape paintings understood as alternative but highly compatible ways of representing landscape. I do so by a close look at how land (and sea) come to be represented in terms of the concrete *places* that constitute them. Places I take to be the constituent units of every landscape, its main modules, its prime numbers, as it were. Close attention will be paid to the phenomenon of *re-implacement,* that is to say, the ways in which places are altered and transmuted even as they are reinstated in paintings and maps. Re-implacement also points to the vicissitudes of places as they come to be represented in diverse media: their changing representational status in being mapped or painted in various ways. In exploring these vicissitudes, we shall learn significant things about place and representation alike, especially with respect to their close collusion.

Places, like the landscapes they collectively compose, are bound up with representation, just as representation in turn calls for places as the bounded particulars of any given landscape domain. The truth is that *representation is not a contingent matter, something merely secondary; it is integral to the perception of landscape itself—indeed, part of its being and essential to its manifestation.* Rather than attempting to find a definitive reason for why we represent landscape—a reason that is notoriously elusive and in any case multiplex (i.e., historical, psychological, aesthetical, social, etc.)—we should ask ourselves instead if there is any such thing as *unrepresented* landscape: any landscape that is *not* part of an actual or potential painting or integral to some existing or possible map. (There is certainly unrepresented *space,* including large tracts of the earth; but to be a landscape is to be a *place* already on the road to representation: at the very least, it is to be the more or less coherent setting of an embodied point of view.) And even if there is no simple answer to the question of why we should *want* to represent landscape—the desire to do so being personally and culturally complicated—it is nonetheless evident that we have no choice but to do so.[7] To be a landscape at all, to be an integral part of a sensuously qualified place-world, is already to have entered the encompassing embrace of the representational enterprise.

A word is in order concerning my approach to the matters treated in this book. I am a phenomenologist by training, but I do not adhere to any single school or doctrine. My effort is to approach philosophical issues by means of a close description of those things that are at stake in these issues. My working premise is that such description, even if not solving philosophical problems outright, illuminates them in ways that contribute to eventual solutions. Moreover, it is valuable in its own right, given that *every* phenomenon shows itself upon detailed examination to be far more intricate (and often far more fascinating) than appears at first blush. There is thus a definite gain in a phenomenological approach, which at the least (and if skillfully done) provides descriptive detail and at the most offers a suggested resolution of certain traditional philosophical quandaries.

One of these quandaries is precisely that of representation, a problematic term that has been a central part of philosophical discourse since Descartes, Leibniz, and Kant first explored its intricacies—in other words, since the advent of modernity in philosophy. What is problematic about representation is not found in the fact *that* it happens (it happens continually and in myriad ways, in addition to those that characterize mapping and painting) but in *how* it happens. How can something so seemingly insubstantial as a mere idea or image—that is, the prototypes of representation—relate to perceived or felt or remembered worlds in any reliable way? What guarantees the truth of representation? *Is* there any such truth, or is it only (as Nietzsche put it indelicately) a "lie"? If it cannot provide truth on any rigorous epistemic basis, is representation to be absorbed into semiology (as Peirce and de Saussure both presumed)? Is it to be seen as an expression or product of historical or political forces? Is representation mainly a matter of "Power"—in Foucault's tag for culturally and socially encoded institutional practices?

This last line of thought is particularly tempting when it comes to analyzing paintings and maps as forms of representation. Both are undeniably cultural in status (indeed, they are among the most highly prized products of virtually any known world culture) and equally undeniably social in formation: they reflect the class origins of their creators on their very faces. Moreover, they are inextricably intertwined with the power relations of the society from which they stem, embodying (albeit often in disguised formats) its political beliefs and forces—in particular, its "social imaginary," in Castoriadis's term. All of this is incontrovertibly the case and highly pertinent to the full understanding of any given map or landscape painting. Many studies of the past three decades, especially in Europe and America, have testified eloquently to this unforeclosable facticity.[8]

Even so, and just so (that is, just because of this increasingly fashionable approach to all cultural products), I take another course in this book, where I ask different questions and pursue different answers. I am concerned less with how representation works (though I do address this question in chapter 12) or with its cultural content or historical position or political purport (these matters I leave to other-minded contemporaries of mine) than with how the representation of landscape places affects us, the viewers—affects us not socially or politically as such but *experientially.* Not that human experience is ever *not* cultural/social/political—of course, it is also all of these things!—but there is an identifiable stratum of that experience that has its own modes of receptivity and understanding. And its own modes of representation! Most of these modes are bodily rather than mental: it is not just a matter of "eye and mind" (as Merleau-Ponty felicitously put it) but of the eye—and hand and foot and back and neck, each of these in actual as well as virtual forms of realization.

This is all the more the case when it comes to the representation of the places of the world's landscapes. As I have argued elsewhere, wherever place is at stake in human experience there body will be found as well.[9] The same rule obtains when it comes to *representing* places. Not, again, that the representing body is unpolitical or unsocial, much less acultural. Far from it! I subscribe to the view that the body itself is ineluctably natural *and* cultural, both at once (where "cultural" entails being socially and politically imbued).[10] But this does not mean that the body, and a fortiori the representing body, is not *also* capable of alliances with landscape that call for their own analysis, in their own quite different terms. Among these alliances are those found in the creation and appreciation of maps and landscape paintings. And it is to the detailed description of such alliances that this book is devoted.

Representing Places proceeds in three parts. In "Painting the Land" I take up the curious evolution of landscape painting out of its origins in decoration—as mere backdrop and background, what was once called "landskip." I trace out the emerging importance of giving representation to such elemental things as earth and sky and water, focusing on several sorts of sublimity achieved in English and American landscape painting of the nineteenth century—a pivotal period in the history of the subject. I also take a concerted look at Northern Sung landscape art of the tenth to twelfth centuries A.D. Throughout, the primary issue is that of the status and fate of representation, and more specifically of re-implacement, in landscape painting. My concern is with the representation of *place*— how various particular places come to be represented in such paintings—rather than with the cultural politics of such representation.

In "Mapping the Land" I consider the same issue with regard to mapping. How do maps represent and re-implace land and sea? Here I distinguish between cartography and chorography—roughly, the exact mapping of sites versus the qualitative mapping of regions. I discuss portolan charts (i.e., late-medieval and Renaissance navigational maps) as well as certain Chinese and Japanese maps as instances of the creative combination of mapping and painting practices—a commingling of representational activities that are kept rigorously apart in the early American land surveys that I analyze at the end of this part.

In the final part, "Re-implacement in Mapping and Painting," I bring together what has been for the most part treated separately until this late point. I do so mainly in terms of the basic models of representation at stake in landscape painting and mapping, and I here distinguish between representation in general and the peculiar but powerful *re-presentation* that characterizes many landscape paintings and certain maps. An Epilogue draws conclusions from the book as a whole, with particular reference to the character of history in comparison with mapping and painting.

This text is the newest step in my continuing quest to rehabilitate the neglected significance of place in contemporary thought. It is the third in a series of books whose first two members are *Getting Back into Place* and *The Fate of Place*.[11] Where the first volume takes up the *experience* of place—for example, in architecture and in wilderness— the second traces out the history of the *idea* of place vis-à-vis space. *Representing Places,* in contrast, considers how *representation* bears on place, calls for it, and thrives in its multifarious presence. And the reverse: place elicits representation and is enhanced by it. It is my conviction that an appreciation of the intimate, coeval relationship between place and its representation in paintings and maps will augment our overall sensitivity to place while contributing to a better understanding of the operation and scope of human representational powers. It will also instruct us as to the remarkable ways in which painting and mapping realize these powers—how they differ even as they converge in their conjoint work of re-implacement.

Acknowledgments

This book has been a long time in the making. I started writing it before I had undertaken other writings on place, and now I am bringing it out after these latter have already been published. The places in which I composed the manuscript were almost as diverse as the places discussed therein, except that they were all in the United States: Sound Beach, Quogue, Guilford, Amherst, Blue Hill, Penobscot, Chicago, Vallejo. Even more important than these places of composition were the many people who contributed to the venture of this volume. Probably the first, and certainly the most skeptical, reader was Eve Ingalls, who took me to task for my circumlocutions and neologisms and who generously lent me her discerning eye and mind when it came to thorny theoretical matters. Janet Gyatso provided a nuanced critical reading of the text at several significant moments. Indeed, all of my work on place in the last decade and a half has been done under her guiding genius. Without her example, moreover, I would never have attempted to address non-Western painting and mapping. Samuel C. Morse, her colleague and eminent historian of art, furnished invaluable knowledge of Northern Sung landscape painting.

Nor could this book have been completed without the generous assistance of members of a younger generation. Eric Casey gallantly and tirelessly clarified issues and questions bearing on the ancient world of the West. Ann Cahill encouraged me to deepen and develop the manuscript, making many specific suggestions for improvement. Gaile Pohlhaus proved to be a super sleuth at the time when a number of references eluded me; she also secured permissions to reproduce many of this book's illustrations, sometimes having to seek them in the far corners of the earth. She and Talia Welsh composed the index to the book. Senior colleagues at Stony Brook offered ongoing inspiration, most notably Mary Rawlinson, David Allison, Hugh Silverman, Robert Crease, and Kelly Oliver. Virginia Massaro was indefatigably and unforgettably present throughout.

Véronique Fóti gave astute counsel out of her intimate acquaintance with painting, both theoretical and practical. Irene Klaver shared with me most insightfully her passion for landscape and landscape painting.

I wish to thank Pieter Martin of the University of Minnesota Press for his persistent efforts in the early stages of production, as well as David Thorstad, who did expert copyediting, and Adam Grafa in later stages. Special thanks to Edward Dimendberg for his lively support of my work on place, and to Doug Armato, director of the Press, to whom I am indebted for his enthusiastic support of the project.

A number of persons from Connecticut were of crucial importance: James Hillman,

xix

friend and mentor who continues to foster my newest work in challenging ways; Margot Mclean, whose painting and talks have been especially helpful; Brenda Casey, who steered me in the right direction during moments of doubt and who provided timely assistance on questions of linguistic felicity; Kent Bloomer, who led me to extend my conception of choric space in new directions. I dedicate this book to Dan Rice for his inimitable guidance in matters of painting and for his exemplary personal courage. Around him there is a circle of aspiring artists to which I belong myself, and to them I wish to signal my gratitude for supporting this work in diverse ways: Virginia Foster, Ed Horan, Parviz Mohassel, Christina Maile, Gwen Gunn, Alice Hayden, Ellen Lowe, Marja Watson, Molly McDonald, McCrady Axton, Bernie Braverman, Catherine Ferguson, Scott Paterson, Nancy Pedraza, Rex Walden. They, too, were indispensable to the completion of this long-pursued book. May its appearance, at last, justify the confidence they and others have placed in it.

PART I

Painting the Land

The real voyage of discovery consists not in seeking new landscapes, but in having new eyes.

—Marcel Proust, "The Captive," *Remembrance of Things Past*

From Landskip to Landscape

Landskips how gay! arise in every Light,
And fresh Creations rush upon the Sight.
Through fairy Scenes the roving Fancy strays,
Lost in the endless visionary Maze.

—Mather Byles, "To Mr. Smibert on the Sight of His Pictures"
(*London Daily Courant,* April 14, 1730)

THE PROBLEM OF REPRESENTING LANDSCAPE

It is a remarkable fact that what we now call routinely "landscape painting" was unknown in the ancient world of the West. Nothing like the broad vistas, the commodious scenes that we consider to be the sine qua non of landscape painting, is to be found in the art of earlier times. At the most, this art included a schematic landscape vista that served as a literal background for the myth or story that was the subject matter and the primary focus of the scene. We see this, for example, in the recently restored frescoes from Boscotrecase near Pompeii. In these works of the first century B.C., "landscape vignettes" (as we might call such fragmentary views) either accompany the main action (e.g., a mythical motif such as "Polyphemus and Galatea" or "Perseus Saving Andromeda") or they float unattached and suspended in a featureless black void that is bounded by a contrivance such as depicted columns that serve to frame the void itself.

Landscape as place-setting also characterizes much medieval and Renaissance painting. When landscape figures at all in the art of these periods (most notably, in the form of a garden), it enters mainly as a decorative enframing element—for instance, in the late-medieval theme of the Hortus Conclusus, in which the Virgin and Child sit enclosed in a protective bower. Overall, landscape is a mere place-marker, a reminder of an outside world that is only glimpsed through the windows and doors of a scene that is located securely indoors, within the security of domestic or ecclesiastical architecture. Even Giotto, who is exceptional in presenting certain out-of-doors settings on their own terms (e.g., in several of the frescoes depicting the life of Saint Francis in the cathedral at Assisi), almost always ensconces the focal scene of action within a building, often a domed aedicula, designed exclusively to contain this scene. Typically, we (though not the primary figures in the painting) are only permitted to peek through apertures in the architecture—for example, in Fra Filippo Lippi's mid-fifteenth-century *Annunciation,* in which the annunciating angel and the Virgin are wholly indifferent to an outside world that, though depicted in the center of the painting, is starkly severed by a standing column placed in front of the very door through which it is seen. Or, in an equally evasive maneuver, a significant stretch of landscape is allowed in the outer wings of a triptych but denied in its central panel (as, for instance, in the *Lamentation* triptych of the *Follower of the Master of the Virgin among Virgins,* a late-fifteenth-century Dutch painting now in the Metropolitan Museum).

It was not until the late fifteenth century that landscape began to be presented or more exactly, *represented* as a presence of its own, that is, as something more than a mere backdrop. In Hans Memling's *Virgin and Child with Saints Catherine of Alexandria and Barbara* (see Plate 1), the setting is outdoors to begin with: only a breezy grape arbor (which may have been a later addition to the work) encloses the Virgin and Child, while the figures of Saint Catherine, Saint Barbara, a young monk or patron, and two angels are situated in the landscape around the sacred couple. The landscape itself, composed of rolling hills and towers in the middle distance, sweeps *around* and *under* these latter figures. The brilliant green of the underlying grass extends from the foreground into the middle and far distances, tying the entire composition together in a single shared space; this in contrast with many other paintings of the same period that act to exclude the landscape from the lower part of the pictorial space (e.g., the *Madonna and Child* [c. 1470] of the Workshop of Verrocchio, in which the landscape is abruptly cut off by a marble wall behind the two central figures). From having been confined to the background, the landscape becomes, in Memling's masterwork, the very ground of the

Figure 1.1. Fra Filippo Lippi, *Annunciation,* c. 1450. The Metropolitan Museum, New York, Maitland F. Griggs Collection, Bequest of Maitland F. Griggs, 1943 (43.98.2). Reprinted with permission.

action, an integral part of the representation. Mere "scenery" is here on the way to becoming full-fledged landscape.

In landscape painting proper, a natural scene is the primary subject matter. Human (or divine) actions are presented in *its* terms: they become the foreground event, the precipitating cause of the larger action embodied in the landscape itself. Whenever we may decide to date the origin of landscape painting in this more complete or "modern" sense—whether we set it at 1414–17 (when the *Hours of Turin* was composed by Hubert or Jan van Eyck) or at 1520 (as does John Constable in commenting on Titian's *St. Peter Martyr*)[1] or in the mid-seventeenth century (as do many art historians)—it is an indisputable fact that landscape painting as a genre of its own starts late in the history of Western art, later than portraiture or still life and much later than allegorical or dramatic painting. As Kenneth Clark remarks:

> In the West, landscape painting has had a short and fitful history. In the greatest ages
> of European art, the age of the Parthenon and the age of Chartres Cathedral, landscape did not and could not exist; to Giotto and Michelangelo it was an impertinence. It is only in the seventeenth century that great artists take up landscape painting for its own sake, and try to systematize the rules. Only in the nineteenth century
> does it become the dominant art, and create a new aesthetic of its own.[2]

We ought to feel a certain astonishment about the deferred development of landscape painting. Why did it take so long for such painting to come into its own? Why, in Constable's words, is Western landscape painting "the child of history"[3]—a quite belated child? It was certainly not a matter of a lack of technical means: the painters of Boscotrecase could represent landscape perfectly well, as we know from their partial (yet still convincing) efforts; and the same is certainly true of medieval and early Renaissance artists. Moreover, in other parts of the world (e.g., in China), landscape painting was an advanced art by the time of the Dark Ages in Europe.[4] What, then, lies behind such considerable deferment in the West?

It would not be an adequate answer to point to the fact that landscape as such (i.e., the actually perceived scene) was not appreciated or savored until a comparatively belated moment in Western history—a moment that is often identified as Petrach's scaling of Mount Ventoux in the middle of the fourteenth century.[5] The issue is not one of having the apposite *experience* but of creating the appropriate *representation* of experience.[6] Human beings have been experiencing landscape and doubtless enjoying it in certain respects from time immemorial. But any full-scale representation of such experiencing of the natural landscape was remarkably recent in the West—only several centuries ago. Why is this so? Clark himself supplies a hint in commenting that "the painting of landscape cannot be considered independently of the trend away from imitation as the *raison d'être* of art."[7] There is something about landscape that defies its representation in an imitative mode. To be true to it, forms of representation other than the purely mimetic have to be found. This is a plausible (though controversial) line of thought,[8] but it only pushes the question one stage further back by forcing us to ask: What is it about landscape that resists imitation? What kind of thing is landscape such that it discourages isomorphic representation?

Landscape is indeed an extraordinary sort of thing. Composed of particular objects—of animate and inanimate entities, of discrete shapes and colors, of distinctive configurations of many kinds—it exceeds any of them. Indeed, it even exceeds their totality. In this respect, landscape is an instance of what Sartre calls a "totality detotalized" and Jaspers "the encompassing": it is something that, while being experienced as a single whole, is nevertheless not reducible to the sum of its parts (a "totalization").[9] Thanks to its untotalizable status, it is tempting to consider landscape as (in Erwin Straus's apposite term) "invisible," that is, something not limited to literal visibility. But the ambiguity of a term such as "invisible" is considerable (does it mean altogether, or only provisionally, nonvisible?); and, in any case, we need to be more concrete.

As an encompassing detotalized totality, landscape proffers to us a maximized "circumambient array." I borrow this latter phrase from J. J. Gibson because of its admirable specificity in regard to the perceptual dimension of human experience.[10] Precisely as maximally circumambient, that is, as surrounding the human subject on all sides in an actively comprehensive way, landscape exceeds the scope of any given perceived object. Appealing as it does to all bodily senses and to their synesthetic unity, landscape is *panperceptual*.[11] No wonder landscape painting has always been tempted by the panorama—a theme to which we shall shortly return. As a panperceptual unity, landscape defies any simple imitation, that is to say, any effort to reduce it to the kind of definite object (or group of objects) that can be grasped in a single apprehension and, therefore, in a single image. That which appeals to the full bodily sensorium must itself be an encompassing whole: if it is an object at all, it is an object *of a higher order* (as Meinong might have put it). It follows that to such an "object" no simple imitative representation can adequately correspond, given that such representation obtains only for things of the same order of reality. Isomorphism between the representation and the represented is at best a regulative ideal in such a situation. And yet this is the very situation of landscape itself.

Viewed in this light (which is to say, as not *determinately* viewable in *any* light), the representation of landscape will be predictably problematic. Not only is any straightforward imitation put into question as a matter of principle, but *any* representation of it becomes a matter of difficulty. How is one to do representational justice to such a complex, puzzling thing as landscape? How is one to represent, in what medium and style, something that is at once elusive and omnipresent, a whole and yet not a totalization, perceived by no single sense but by all the senses in a com-position that is itself problematic?

At least we can begin to understand why the painterly representation of landscape could be so long deferred in Western art, which always had on hand a plethora of determinate objects to represent—whether these objects were the material things represented in portraits or still life or even the highly symbolic (yet also highly determinate) objects of religious quest. But if it is true that "from these individual elements the first landscapes are put together,"[12] it also remains true that such landscapes have to be considered as mysterious creations. The mystery resides in understanding how "individual elements," that is, discrete objects, are assembled in such a way as to constitute a representation that can justifiably be called "a landscape" (to use the very term that John Constable applied as a title to fully half of his exhibited paintings); for a landscape painting in any modern sense, as doubtless also in the ancient Chinese sense, exceeds the sum of its individual parts by virtue of encompassing them. But how this happens is by no means self-evident.

The problem of representing landscape can be seen as a quandary of *containment.* How is the artist to contain something as overflowing as landscape within the very particular confines of a painting? The issue here is not one of how objects of three dimensions are to be represented in a two-dimensional surface: that problem is common to *all* painting and is by no means limited to landscape. Nor is it a matter of figuring out how the picture plane of representation relates to the nonpictorial "real," that is, to the inhabitants of a given world-region: this, too, is common to all painting that purports to be "of" anything that is materially real. Instead, the issue is how something that, by its very nature, overflows ordinary perception can be represented by something else that, by *its* very nature, can only present itself to the viewer as a discrete object with definite dimensions and often within a delimited frame.[13] How can the decisively determinate contain the inestimably indeterminate? It is evident that the verb *contain* cannot be meant in anything like the Aristotelian sense of the term, for a painting could contain a landscape in Aristotle's sense of a strict "surrounder" *(periechon)* only if it were to wrap itself around it in its entirety and thus literally exceed, however infinitesimally, the totality of objects in it. Despite Christo's efforts to "wrap" enormous objects (e.g., the Pont Neuf in Paris, the Reichstag in Berlin, open fields in Marin County), he has not, and in principle cannot, wrap an entire landscape. Not only would it be hard to find sufficient material for such an omni-wrapping, one simply cannot contain an inherently indefinite dimension of the landscape such as its depth.[14]

THE PANORAMIC IN PAINTING

Given this impasse, it is hardly surprising that landscape artists of many different persuasions have been tempted to paint explicitly panoramic representations of landscape. In order to do this, they have attempted to piece together a set of representations of individual scenes so as to create the condensed representational analogue of the circumambient array that is essential to the experience of landscape overall. I have already noted one early effort toward panorama in the case of the late-fifteenth-century Dutch triptych that allowed landscape to seep in only from the wings, as it were. Many others could be cited: Breughel's sweeping vistas of landscapes of changing seasons, Bosch's equally panoramic depictions of nightmarish worlds of fantasy, John Trumbull's and Frederick Church's commanding paintings of Niagara Falls, and many Chinese scroll paintings, which, in their very unfolding, facilitate the presentation of panoramic views.

Perhaps the single most striking case of a panoramic painting is that created by Henry Lewis in 1849. His *Mammoth Panorama of the Mississippi River* was painted on forty-five thousand square feet of canvas and was slowly unrolled each evening during a sequence of several hours at the Louisville Theater. This gigantic work depicted the Mississippi from Saint Louis to the Falls of Saint Anthony; it was based on the painter's painstaking sketches of both sides of the river as he traveled downstream. It was the most successful work in a genre that had enormous popular appeal at the time.[15] Possibly the very first oil painting of Yosemite Valley (not yet Yosemite National Park) was Antoine Claveau's panorama commissioned by the Mann brothers and advertised as "Twelve

Different Views including the most remarkable features of the wonderful Yosemite, closing with a View of the Great Falls."[16] So widespread was the appeal of panoramas that when Albert Bierstadt's monumental painting *The Rocky Mountains* was first put on public view in 1863, it was assumed by many viewers to be merely the first scene of an unfolding panorama. James Jackson Jarves remarked about this curious circumstance: "The countryman that mistook 'The Rocky Mountains' for a panorama, and after waiting a while asked when the thing was going to move, was a more sagacious critic than he knew himself to be."[17] Indeed, American landscape painting exhibited a panoramic aspect ever since the epic cycle of Thomas Cole's, *The Course of Empire,* was first presented in 1836 in New York City: one had to circumambulate between the five scenes depicted in this epic, much as one has to walk around in nature to capture a full view of a single dramatic landscape, such as that found in Yosemite Valley.[18]

Circumambulation by walking is the bodily counterpart to the circumambience of landscape as perceived. Whether moving or standing still in a landscape, one is constantly surrounded by things that come from every which way—as also happens in a cityscape such as an arcade or a shopping mall.[19] Panoramic painting in the nineteenth century United States was an effort to re-create this very situation of being continually encompassed and exceeded by a given landscape. Alexander von Humboldt, whose book *Cosmos* was much admired by American landscape painters of the mid-nineteenth century, draws this conclusion as if it were self-evident: "Panoramas are more productive of effect than scenic decorations, since the spectator, inclosed, as it were, within a magic circle, and wholly removed from all the disturbing influences of reality, may the more easily fancy that he is actually surrounded by a foreign scene."[20] For "foreign scene" we need only substitute "natural scene" in order to make the application to landscape painting complete. Humboldt even advised that "besides museums, and thrown open, like them, to the public, a number of panoramic buildings, containing alternating pictures of landscapes of different geographical latitudes and from different zones of elevation, should be erected in our large cities."[21] Without having to espouse such extreme architectural literalism, we can readily agree that the representation of landscape is strongly tempted to extend itself to panoramic displays, either in paintings having the grandiose scale of Bierstadt and Cole (see Plates 5 and 9), or in panoramic spectacles such as those constructed so ingeniously by Claveau and Lewis. Either way, the representation, in its time-taking character and in its invitation to contemplative patience, embodies a significant analogue of being located in an actual landscape, which also takes time to perceive in its entirety.[22] No wonder a letter to the *Missouri Republican* could declare in 1849 that Lewis's panorama "had succeeded in imposing on the sense of the beholder and inducing him to believe that he is gazing, not on canvas, but on scenes of actual and sensible nature."[23] It follows that in viewing a panorama "the totality could not be apprehended instantaneously."[24] *So too with landscape itself*: its "actual and sensible nature" is such that it cannot be grasped *totum simul,* all at once as a single object. To its spatial unencompassibility must be added a temporal nonsimultaneity: just as a given landscape cannot be contained in a finite part of space, so its full perception cannot be confined to an instant of time. By its radical circumambience, landscape exceeds both kinds of unit, being the indecomposable and uncontainable excess of the natural world over chronometric and spatiometric delimitations. And if this is so, it is only to be expected that its representation in painting will be at once deferred historically and problematic philosophically and that various detours to its representation, such as that pursued by the panorama, will be very tempting to take.

LANDSCAPE AS TOPOGRAPHIC AND DECORATIVE

Two other detours call for consideration at this point. I refer to topographic painting and to the employment of landscape views in interior decoration. Each of these exemplifies a representation of landscape construed not just as background or scenery but as having a distinctive function of its own. In the United States, even more than in Europe, the advent of landscape painting as an autonomous genre was a belated phenomenon, yet all the more revealing for this very reason. It was delayed in large measure because of a continuing fascination with topographic and decorative uses of landscape.

Topographic painting. The use of painting for topographic purposes, that is, for conveying more or less accurate, exactly depicted, pictorial representations of a given landscape, is by no means restricted to early colonial times in the United States, for it continues intermittently until at least the arrival of photography in the middle of the nineteenth century; it is still very much in evidence in a painting such as Joseph H. Hidley's *View of Poestenkill, New York* (c. 1865–72; Luce Collection of American Art, Metropolitan Museum of Art), in which every street and every house of Poestenskill are represented in detail. Not that topographic painting was confined to the American scene: we have already noted how popular various "views" of Europe, or even of Bhutan, have been in the last several centuries. In the early eighteenth century in the United States, however, topographic art held a quite special interest. It reflected a widespread curiosity about the New World, especially on the part of those Europeans who were considering emigration to America: "As prospects of financial success continued to attract Europeans to the Colonies, the resulting urban expansion called for topographic views. It was these views that became the first authentic landscapes of America."[25]

It is striking that many of the first topographic artists were engravers by training, as is manifest in such collections of topographic views as William Burgis's *Scenographia Americana* (1768) or in J. F. W. Des Barres's *Atlantic Neptune* (1763–84). John Smibert's 1738 *Vew [sic] of Boston* (see Plate 2) attracted a great deal of attention precisely because of its detailed graphic qualities. The linearity of engraving, transposed to topographic painting, is ideally suited to the aim of topography: the discrete exactitude of etched lines provides a precise and easily rememberable representation of a given scene. An engraving fixes this scene in its literal traces. Like human memory itself, such a representation exists between perception and reflection; it both depicts and guides—and, again like memory, it does so by means of a network of traces.

In addition to its graphically descriptive character, topographic painting also tends to be panoramic. This is evident in Smibert's view of Boston—a vista that attempts to take in the entirety of the city from an elevated and distant vantage point. This vantage point is reinforced by the inclusion in the foreground of four figures who are themselves viewing Boston from afar: the viewer of the painting views the viewers viewing. Here the operative idea of any panorama—that is, the idea of looking around—is *enacted* in the representational space of the painting: the depicted viewers' looking around induces and facilitates our own survey of the scene. (This device continues to be employed in later panoramic painting: for example, in Jasper Francis Cropsey's *Starrucca Viaduct* [1865], in which two foreground figures and a dog gaze on the newly constructed viaduct in the distance.) Indeed, there is a mutually enhancing relationship between panorama and topography. As an anonymous contemporary of the panoramic painter John Vanderlyn put it:

> Panoramic exhibitions possess so much of the magic deceptions of the art, as irresistibly to captivate all classes of spectators, which gives them a decided advantage over every other description of pictures; for no study or cultivated taste is required fully to appreciate the merits of such representations. They have the further power of conveying much practical, and topographic, information, such as can in no other way be supplied, except by actually visiting the scenes which they represent, and if instruction and mental gratification be the aim and object of painting, no class of pictures have a fairer claim to the public estimation than panoramas.[26]

Despite the mention of "magic deceptions" (reminding us of Humboldt's description of a panoramic painting as a "magic circle"), the author of this statement stresses that panoramas provide "much practical, and topographic, information." It is as if the very limitlessness to which panoramas approximate in their striving to represent the circumambient array of a landscape brings, in such uniquely sweeping "vews," topographic information not otherwise available. This information is *"such as can in no other way be supplied, except by actually visiting the scenes which they represent."* Notice that the options allowed here are limited to two: visiting a scene in person (which only the artist-explorer has done in the case of exotic places) *or* (if this is not possible) representing that scene in such as way as to convey precise topographic information about it. The assumption is that such a representation must be *isomorphic* in character if it is to serve as an accurate purveyor of the scene and that its function must be *communicative* if its purpose is indeed to supply information. (*Isomorphic* signifies likeness of form, which is what is communicated in paintings that purport to be realistic representations.) It is precisely these two traits of panoramic-topographical representation that will be called into question with the rise of Romanticism in European and American art. In refusing to be topographic (though sometimes continuing to be panoramic), nineteenth-century landscape painters will engage in nonisomorphic representations that do not have as their primary purpose the communication of topographic (or any other kind of) information.

An exemplary case of such departure from the topographic norm of communicative isomorphism is that of Albert Bierstadt, who is notorious for romanticizing the Western wilderness, which he visited many times and which he professed to represent accurately. Not only did he combine several different views of one place—views that were not geographically compossible (as, for example, in his *Valley of the Yosemite* [1864], Plate 5, in which standpoints several miles apart are amalgamated)—but he imported major elements from European landscapes such as Alpine scenery into his American landscapes. Moreover, he managed to misidentify some of the very geographical places he purported to represent.[27] In his imaginative and often majestic works, the topographic impulse in American painting has already begun to exhaust itself, leaving us at the far end of the spectrum from Smibert's exact and informative view of Boston. That the desire for topographic accuracy does not die easily, however, is evident in Mark Twain's sardonic comment in an 1867 review of Bierstadt's *The Domes of the Yosemite*: "As a portrait I do not think it will answer. Portraits should be accurate . . . We do not want this glorified atmosphere smuggled into a portrait of Yosemite, where it surely does not belong . . . I believe that this atmosphere of Mr. Bierstadt's is altogether too gorgeous."[28] Indeed, even a present-day writer as sensitive as David Robertson in his book *West of Eden* can despair of Bierstadt's wholesale departures from topographic fidelity: "Finally, his liberties became outrageous. Yosemite became unrecognizable. Anything grossly exaggerated tends to evoke laughter, and some of Bierstadt's later Yosemite paintings do just that. Clearly the danger of trying to boost an already grand nature is that one may appear ludicrous."[29] That this could still be said in a book published in 1984 indicates an abiding interest in a form of landscape painting that is informed, at least implicitly, by topographic ideals. Even if the practical aspect of such painting (i.e., its geographical informativeness) has ceased to be pertinent after the advent of photography, questions as to the faithfulness of painterly representations of a given landscape remain alive.

Not only do such questions continue to be asked, but contemporary American landscape painting still manifests an important topographic strain. In certain instances, notably those of Neil Welliver and Eve Ingalls, this strain is even paramount and has reached new heights of attainment. Such exquisite landscape realism rivals the photograph in its fidelity to an original scene; indeed, it may make preliminary use of photographs, much as earlier artists relied on sketches made "in the field." In all such cases, isomorphism of the image is at work, as is communication (though a communication shorn of any ambition to inform the viewing public). So too, especially in Ingalls's later work, is a panoramic tendency.[30]

I allude to this abiding topographic idealism not merely to underline the historical continuity of landscape painting in the United States. Its existence also contests the claims of critics who assume that topography in landscape indicates either naïveté or degeneration. The thesis of naïveté presumes that individual artists (e.g., Turner in his early European studies) or whole traditions of painting (e.g., early American art) pass through a preliminary topographic phase—a kind of literalistic topophilia—which is soon, and fortunately, replaced by a more mature and creative phase. In Twain's terms, "portraits" are superseded by "pictures." If the work of Welliver and Ingalls as well as other highly realistic landscape painters is taken into account, however, this first thesis is seen to be quite dubious. Rembrandt's sketches of his environs in Holland are at once topographically accurate *and* art of great sophistication. The second thesis, according to which topography signifies artistic degeneration, is just as questionable. A critic of landscape painting as perceptive as Kenneth Clark, however, adopts this thesis in passing a negative judgment on late-seventeenth-century Dutch painting:

> By the end of the seventeenth century the painting of light had ceased to be an act of love and had become a trick . . . In this strange Indian summer of humanism man was so well satisfied with his own works that he did not wish to look beyond

them . . . No wonder that landscape painting became mere picture-making according
to certain formulas; and that fact was confined to topography.[31]

Clark singles out Canaletto as a case in point, an artist for whom, with several notable
exceptions, "the landscape of fact degenerated into topography."[32] But must topography
signify degeneration? Topographic landscape is doubtlessly degenerative when it is reduced to *(a)* the depiction of "site" alone (where this latter term implies a lack of sensitivity to the numinosity of place in landscape, including the dimension of depth); *(b)* the
merely "picturesque"—which is precisely what we witness in some of Canaletto's more
popular works: it is their pretentious prettiness, not their pictoriality as such, that undermines their status as serious art. But topography per se is not the villain. Far from being a
sign of degeneration or of naïveté, it may be a source of unsuspected strength in the
landscape painting of any period and in any country—and in the works of artists of
widely varying degrees of formal achievement.

   Landscape as interior decoration. In colonial times in the United States there were as
yet no full-time painters. Painting was not prized for its own sake in a situation where
bare settlement and sheer survival were at such a premium. Thus it is not surprising to
learn that "It was well along in the eighteenth century before any painter managed to live
solely by his earnings as an artist. Also, for most of the colonial period little if any distinction was made between the fine arts and the crafts."[33] Just as engravers became
topographers on their way to becoming painters, so aspiring painters were at first indistinguishable from craftsmen. The landscapes they painted often took the form of decorative panels that were called "landskips."[34] These paneled landscape scenes assumed various forms, depending on the decor of the room in which they were located. In the living
room at "Marmion" in King George County, Virginia—a room first constructed in the
years 1735–70 and said to be "the most ambitious decorative scheme to survive from
eighteenth-century America"[35]—we find three different examples of landskip. The first is
the "overmantel" panel, depicting a windmill on a pond; at first glance it looks like an ordinary painting hanging over a mantel, but in fact it is an integral part of the woodwork,
the oil paint having been applied directly to a panel that was already in place. The same
is true of the "overdoor" landskip, a sea scene boldly painted in a horizontal format to
match the length of the lintel over which it is situated. Also at Marmion is a pair of tall
vertical panels approximately five feet in height on which rather elaborate scenes of
houses, rivers, trees, and distant mountains have been painted. The dark tonalities of all
three landskips (especially the tall panels, which are replete with dark grays and greens,
subdued reds and yellows, with touches of white for detail) ensure their suitability in the
dimly illuminated living room. Here landscape qua landskip, a paneled representation of
nature in its external visibility, is an integral feature of the internal world of architecture.
At the same time, these painted panels open up glimpses of a world beyond the dark
interiority of the colonial room. Yet in the end these glimpses are as meager and partial
as those that are granted in Fra Filippo Lippi's *Annunciation.* In both cases, the outside
world is a literally liminal presence, clinging as it does to various threshold features.

   Landskips were not restricted to wall panels. They also formed part of pieces of furniture. Commodes in the American colonial period featured side and top panels decorat-

ed with miniature landscapes. The same was true of dressing tables and china closets. Even chairs possessed "crest panels" that took the form of landskips.[36] Painted directly onto the wall as murals, landskips took the place of the tapestries or wall hangings that are so prominent in other cultures. In these multiple modes of landskip representation, landscape was incorporated into virtually every corner of domestic architecture.

Such pervasiveness of landskip as interior decoration is accompanied by an almost complete lack of appreciation of it as an art form in its own right. Landscape is not just let in by the back door or by the side windows, it is allowed to flourish in the midst of daily living. But it prospers there precisely on the condition that it not be granted an autonomous status—that it not be considered "landscape" in the sense that we now unhesitatingly ascribe to the term in the wake of the revolution in painting effected by the Dutch painters of the mid-seventeenth century and pursued so vigorously in the nineteenth and twentieth centuries. (One concrete sign of this revolution is the linguistic fact that "landscape" and "landscape painting" have become interchangeable in ordinary English.)[37] Purely decorative landskip is not painterly in status, even though it is painted in fact. As in the case of a purely picturesque topographic work, it resists consideration as fully engaged, uncompromised landscape painting.

But the situation is in fact more complex than this. Just as topographic imagery shades into landscape painting in its fully accomplished forms (e.g., in Smibert's *Vew of Boston,* which, beyond its topographic intention, can be claimed to possess "a convincing representation of three-dimensional space"),[38] and just as later landscape paintings will continue to exhibit topographic traits (e.g., Frederick Edwin Church's *Cotopaxi* [Plate 7], which conveys accurate topographical information of this Ecuadorean mountain despite its predominantly painterly ambition), so the decorative and the painterly may converge in their respective trajectories. Not merely can one stand in for the other—or even be mistaken for the other, as when the contemporary visitor to the reconstructed Marmion Room easily misconstrues the overmantel panel as a freestanding painting—but the decorative itself has painterly dimensions (and conversely, as we see in paintings of Matisse's Moroccan period).

The tenuous line between decorative landskip and painterly landscape is exemplified in the work of Edward Hicks, the remarkable Pennsylvanian folk artist. Hicks was by profession a painter of decorative panels, but he was also a remarkable painter of landscapes. He was notable not merely because of his capacity for doing both in alternation, exercising two distinct sets of skills on two different kinds of occasion; he was also able to merge the two skills on the *same* occasion: for instance, in his painting *The Falls of Niagara* (see Plate 3). A brilliant landscape representation of Niagara, witnessed by animals and humans alike, is bordered by a Romantic poem of the period that addresses itself to the sublimity of the scene. The border itself is highly decorative—its corners cleverly contain the date of the painting—and yet it is perceived as continuous with the landscape whose perimeter it is. Here the decorative and the painterly merge, being as fully integrated as are topography and landscape in Cole's *The Oxbow* (see Plate 9). Landskip and landscape have become coeval.[39]

Hicks and Cole are exemplary in their capacity to mediate oppositions between the decorative and the painterly, the topographic and the landscapelike. Each puts together in one coherent work what had remained for the most part disparate and separate in earlier times. It is significant that they both flourished in the first half of the nineteenth century, by which point the marked discrepancies that existed earlier between decoration and topography on the one hand, and accomplished landscape painting on the other, had become less disruptive. The uneasy mixture of these elements in Smibert's 1738 painting or in the interior at Marmion of exactly the same period here gives way to a mutually enhancing embrace. The way is cleared for the representation of landscape in its own terms—terms no longer dictated by, or even having to be compatible with, the specific concerns of topography or interior decoration. Landscape painting, uninhibited by such concerns, can at last begin to come into its own, as it does so memorably in the Hudson River School and in the Luminist painters closely associated with this school.

THE PATHS OF INSTRUCTION AND PLEASURE

The two "detours" to landscape just examined—detours that have proved to be less devious than we suspected at first—appear to be very different, even if complementary, routes to the appreciation of nature on its own terms. The first, the topographic route, may be called "the path of instruction." Its original motive, if not always its final destiny, is one of instructing viewers as to the features of a particular landscape, so particular that the landscape itself is in effect reduced to a site or set of sites. By "site" I here mean a geographically determinate position, that is, a specific *location* that, by various means of projection and transformation, occupies a specific spot on a map.[40] The topographic instructs us as to cartographic fact; it tells us *just where,* at what precise point, something is located in the space of nature. It hardly needs to be added that it can be instructive in this manner only insofar as the representation of a given landscape is limited to a particular object or group of objects. One can instruct topographically only about that which is definite enough as an object to be situated, by transposition, on a cartographic grid. The space of nature, in being represented as a scene of sites, colludes with the literally objectifying demands of cartography.[41] Put otherwise, topographic painting in its pristine form is not just situated between the landscape of felt fact and the objective map but contributes directly to cartographic concerns. Site, its proper unit, sides with space and stands apart from place.

The alternative detour, that of decorative landscape, may be called "the path of pleasure." The wall panels and parts of furniture at Marmion do not pretend to instruct us as to the exact location of anything. Instead, they aim at pleasing us, at making domestic life more bearable because more enjoyable. The goal of education is replaced by the aimless aim of diversionary pleasure. But pleasure, no less than instruction, requires an object, often a quite determinate object. The overmantel or lintel panel has to fit the exact spatiality of the room in which it is situated, and it has to blend more or less fully with this room's local colors and shapes. Objecthood predominates, for the representation of nature has itself become an object—an ornamental object to be admired and enjoyed as an integral part of interior architecture.

It will be recalled that the anonymous critic cited earlier spoke of "instruction *and* mental gratification" as "the aim and object of painting"; he felt that panoramic painting in particular fulfills *both* goals.[42] If this latter claim is true, it indicates that the two *teloi* are by no means incompatible in painting. Indeed, the two may actively support each other. What else were viewers of the *Mammoth Panorama of the Mississippi River* paying all of fifty cents for if not for a combination of learning (about the exact configuration of the river) and of pleasure (at the perception of its unfurling)? Besides sharing these two goals, topographic painting and decorative panels, each a form of "landskip," both refuse to represent landscape as a detotalized totality—as something of maximal circumambience. Their common stress on objecthood stands in the way of such a representation. In this, they part company with the panorama, which at least attempts to convey the diffusely encompassing presence of nature construed as something more than a totality of discrete objects.

The commonality between the decorative and the topographic may be stated more positively. Both exhibit a comparable concern for representing what is, quite literally, *superficial* in landscape: its actual *superficies* or outer surface. Consider only the fact that topographic painting in its purest guise focuses on strictly visible and traversible features of the terrain—where *terrain* signifies the manifested outer surface of the earth in a given region. Such painting attempts to replicate this surface as faithfully as possible. The isomorphism that is sought is one existing between two surfaces, that of the earth and that of the topographic painting itself. Similarly, decorative landskips are attached to the physical surface of a wall or an item of furniture so completely as to become part of the wall or piece of furniture. A landscape scene painted on the surface of a panel belongs in turn to the surface of something *else* that acts as its architectural host. In their obsession with the surface structure of the (exterior) earth or of (interior) domestic space, both topographic and decorative painting show themselves to be doubly superficial: their own surface is dependent on, and often literally contiguous with, another pre-given surface.

To be such a bound variable contrasts with the situation in which no precedent surface (e.g., that of the earth or of a living room) dictates in advance the terms of representation and in which no host–parasite relationship prevails. This is the situation of landscape painting that comes into its own and determines its own destiny, free of requirements of resemblance and not bound to tasks of instructing or pleasing. It is the risk that Albert Bierstadt, for example, took in painting *Valley of the Yosemite* (Plate 5), in which the autonomy of the depicted image expresses the heteronomy of natural place.

We need to notice a final aspect of the common concern with surface that is exhibited by the two modes of landscape painting under examination here. This is their conspicuous neglect of *depth*. Such neglect is most conspicuous in topographic painting done in a "primitive" style. For example, *The Plantation* (see Plate 4), an anonymous American painting of circa 1825, is noteworthy for being at once topographically precise—it conveys the exact configuration of a plantation with its outbuildings and in relation to a river at its base—and yet bereft of any representation of depth. The image clings to the picture plane at every point, with the only hint of depth being the differential size of houses and trees. No schema of recession is presented, and we are left with a sheer depthless tableau on which are situated the terrain and buildings of the plantation.

This is admittedly an extreme instance, but it is symptomatic of much topographic painting: even Smibert's *Vew of Boston* (Plate 2), a more sophisticated painting, is curiously depthless, despite the claim cited earlier. The same is true of the wooden panels from Marmion; any suggestion of depth in their representational content is effectively suspended by their resolutely planar character and by their enforced continuity with the walls of the room they serve to decorate. In all such cases, we find an elision of the dimension of depth—as if depth were the very last thing to consider in decorative or topographic painting.

Yet depth is the very first consideration in an uninhibited landscape of natural place, part of its being and essential to its representation. Nowhere is this more evident than in Bierstadt's *Valley of the Yosemite* and other like paintings of his that were so much regretted by Twain and Robertson. Bierstadt's extraordinary handling of light—"what distinguishes Bierstadt from all other painters of Yosemite," asserts Robertson, "is his use of light for dramatic effects"[43]—brings with it a nuanced appreciation of the uniquely configured depths of Yosemite Valley. Even Twain ruefully acknowledges that "some of Mr. Bierstadt's mountains swim in a lustrous, pearly mist, which is so enchantingly beautiful that I am sorry the Creator hadn't made it instead of him, so that it would always remain there."[44] This mist creates an atmosphere that transmits the special sense of depth enlivening the landscape of Yosemite Valley as a place of nature. Such a sense, bearing on such a place, is indispensable to any representation that claims to do justice to the heightened circumambience that is felt by artist and traveler alike, especially when this extraordinary valley is experienced from its floor; for depth is an uneliminable dimension of natural landscape taken as something encompassing and incommensurable. To stay on the surface of such a landscape, even if it is to gain topographic fidelity or decorative diversion, is to lose its depth. It is to gain as site what is lost as place.

The task of landscape painting that has become something more or other than topography or decoration is to represent the place of natural landscape as a place of depth. Which is not to abandon surface but to recognize (as Wittgenstein said) that "the depths are on the surface."[45]

ART AS REPRESENTATION

The two forms of "landskip" just explored are certainly not the only ways in which landscape comes to be represented by purely pictorial means. Its representation also occurs as the setting of many paintings with specifically historical content (e.g., *The Oath of the Horatii* or *Washington Crossing the Delaware*) or as the background of religious art; a background that, in the case of Memling's *Virgin and Child with Saints Catherine of Alexandria and Barbara,* begins to pervade the foreground as well. In all four cases of decorative, topographic, historical, and religious painting, the representation of landscape is at stake. But what is it to represent landscape? Indeed, what is representation in the visual arts? These are basic questions that we must begin to address, if only in a preliminary way.

In the most general sense, "representation" signifies a circumstance in which one thing is a surrogate for something else. The first item substitutes for the second and is therefore its proxy, its "representative." Moreover, the representative, to the extent that it is successful, stands in for what is represented in the relevant regard, much in the manner in which an elected representative is charged with standing in for the pertinent economic, social, and political interests of an electorate that does not need to appear in person in the legislative process itself.[46] In contrast with political representation, however, artistic representation does not stem from an election by those who are being represented but from a choice by an artist vis-à-vis the subject matter that is to be represented (where the subject matter can be a landscape, a living person, a historical scene, a religious passion, or, for that matter, an abstract idea or a formal concern of painting itself). The representation is therefore mainly responsible to the subject matter and not to other human beings as in the political arena. The artist may be sensitive to the interests and needs of viewers, collectors, museum directors, and so on, and even possess a personal bonding to them. But the *work* as a work of representation is *for* and *about* the subject matter of what is represented: created for its sake and intimating certain things about it. Thus the primary bonding is to the "topic" of the painting and not to those who would savor the finished work or profit from it, including the artist himself or herself.

Being a work for and about something, being an implication (a "folding in") as well as a complication of its topic, an artistic representation is never a transparent window onto its own subject matter. Standing before even the most realistic painting— for example, an obsessively accurate topographic painting such as a view of Venice by Canaletto—we are still not led to believe that by viewing it we are simply present to the scene it represents. Instead, the painting stands before us *in place of* this scene as the self-announced surrogate of it. Any other experience of it would have to be classed as perceptual illusion or, indeed, as delusion or hallucination. When Edward Harnett attempted to deceive his viewers into believing that they were perceiving not a representation of a wall but an actual wall to which odd bits of paper were stuck (so convincingly that it is reported that those unpacking his paintings attempted to tear off these presumptive pieces of paper), he only managed in the end to test the limits of representation, not to trespass them; illusionism, like topographic painting, fails at its own game. Or, more exactly, its very success demonstrates its ultimate failure, because what we marvel at in Harnett's work or Canaletto's is how close the artist comes to the extremity of convincing representation. Just as we do not mistake the elected representative for the people he or she represents, however closely that representative may reflect or resemble those people in important respects (e.g., in regard to political party, ethnicity, or gender) and however effectively he or she promotes their interests, so we do not mistake even the most accomplished illusionist painting for what it represents.[47]

A painting that represents something at once *stands for* and *stands in for* that which it represents. As standing *for,* it serves as a sign of its represented content or subject matter: in this regard, it is a semiological entity.[48] As standing *in for,* it is a perceived object, a material object in its own right that has taken the place of something else material (even, at the limit, a landscape as an embracing detotalized totality). As a sign, a painting participates in a nonphysical relationship of signification; as a perceived object, it is undeniably physical and is prized as such.

From this double conception of representation in art, three consequences follow forthwith. First, not only are there many possible kinds of subject matter to represent in art—"landscape" is to be ranged alongside such other genres as "still life," "portrait," "allegory," "historical action," and "conceptual art"—but within a given genre there are also many possible *species* and *modes* by which that subject matter may be represented. By *species* I mean the various types and subtypes of the genre in question. In the case of landscape painting, for example, there are such species as (in Kenneth Clark's nomenclature) "the landscape of symbols," "the landscape of fact," "the landscape of fantasy," "the ideal landscape," and so on.[49] In each case, distinctively different kinds of landscape are addressed and treated; some of these bear on actually experienced landscapes (e.g., "the landscape of fact"); others fabricate the landscape itself from conceptual and notional factors; and all, doubtless, involve some degree of interplay between fact and figment, experience and concept, self and society.[50] The "modes" of treatment are the specific ways in which a given species is brought to aesthetic completion; they are tantamount to what we usually designate as artistic "styles." Even within a particular species of landscape, say, the landscape of fact, there are decisive differences between El Greco and the Breughels; differences that we would not hesitate to call stylistic in character. What is most striking about this first consequence of painting's basic representational nature is therefore the sheer multiplicity of realizations to which it may give rise. Where an open window on a given scene would provide us with a relatively fixed view of this scene, representations of the same scene can proliferate endlessly.

A second sequela of the representational status of art is that no criterion of isomorphism need obtain in a given genre, species, or mode. Only if "representation" means a literal *re*presentation, that is, a second presentation of the same thing, would such a criterion hold across the board. In this case, a painting would be an exemplar—some would call it an icon—of an original that resembles the original as much as possible, thus that can *take its place,* even to the point of being indistinguishable from it.[51] Such is, admittedly, the ideal of topographic painting. But the impossibility of fully realizing this ideal, as signified in the constant admixture of decidedly nonisomorphic elements, should alert us to its unattainability. For this ideal to be accomplished in fact, a painting would have to instantiate a landscape literally; and this is not possible in principle, not only because every landscape is an undecomposable totality, but also because any given landscape uniquely embodies its own diverse elements and cannot be fully replicated in any other medium.

Indeed, the failure of panoramic painting to represent a landscape completely, to take its place altogether—along with its inherent tendency to associate itself with entertainment and with profit[52]—has already signaled to us reasons for believing that no such strict second presentation of a landscape is possible. And if this is so, the criterion of isomorphism, that is, of the precise part-to-part matching of original and copy, of (perceptual) presentation and (artistic) *re*presentation, must also fall under suspicion. Even if exact resemblance may serve as an important motive for a particular painter, and even if it may be more or less successfully achieved in certain great paintings (notably in Vermeer's *View of Delft*),[53] it cannot be invoked as a criterion by which all artistic representations are to be judged.

The third consequence is that painterly representations are genuine *presentations.* They seek to present, not to *re*present; they strive to show, not to replicate. They are forms, not of *eikastikē,* but of *phantastikē*—exhibiting the *phan-* root of this last word in its connotation of outright appearing or sheer showing.[54] This is not to say that artistic presentations are inherently fantastic in the specific manner of "landscapes of fantasy."

They may be perfectly naturalistic in the form of their representation, and yet still function as showings, as is luminously evident in the landscapes of Jakob van Ruisdael or various members of the Hudson River School. The crucial element in presentation is that in it the topic or motif is allowed, indeed encouraged, to come forward: this in contrast with the implicitly backward movement of *re*presentation (*"re-"* signifies "back" as well as "again"). The topic is *ushered in,* as it were—*made manifest* in its own terms, whether these be fantastic or naturalistic, real or surreal, ideal or symbolic, secular or spiritual. It was just such manifesting of the Yosemite landscape that Albert Bierstadt attempted in his characteristic mode of grandiosity, which contrasts with the modest modes of Thomas Hill or of William Keith, who painted the same landscape in much more subtle ways.[55] Each of these three early painters of Yosemite *presented* this landscape as their common motif, displaying it by means of painted images and bringing it forth into what Heidegger might call the "clearing" or "open" of the painting: the "earth" of the intensely experienced landscape was presented in the "world" of their works.[56] In landscape painting that is more than purely topographic or decorative, and other than historical or religious, there is a literal demonstration of the topic, a showing forth in its very presentation.

If this presentation of nature can also be seen as its self-presentation (because it is the natural landscape itself that comes forth in the painting), it is in and through such a representation that, without taking its place, *stands in its place.* The topos of the topic, the place of the landscape, is taken up in the representation that both stands for this place and stands in for it. In this way, place is at once signified and reinstated, reinstated-as-signified, *assigned* in a painting that represents it. Place is not replicated but transmuted in the work.

It is also re-placed, given a place of its own—a place to the second power that is none other than its own place *in* the very place of another. If this is still a matter of representation, it does not occur as *re*presentation but *re-implacement,* that is, by means of finding another place in the painting that represents it.

Finding Place for the Elemental

I saw the seal of evening on the river. There was a quiet beauty in the landscape at that
hour in which my senses were prepared to appreciate . . . The greater stillness, the serenity
of the air, its coolness and transparency, the mistiness being condensed, are favorable to
thought.

—Henry David Thoreau, journal entry of 1851

PHOTOGRAPHING THE VALLEY

The assignment and re-implacement of landscape, its transmutation, is nowhere more
dramatically evident than in its photographic representation. Given our usual beliefs
about photography, this might seem to be a most unlikely place to look for such transformative actions. Is not a photograph a paradigm case of isomorphic representation? So
it would seem to be, not only at first glance but also after looking at thousands of ama-
teurish snapshots of landscapes of all kinds. But it need not be so. Landscape photographs can be genuine representations in all three senses discussed at the end of chapter 1.

That this is so is powerfully evident in the work of Carleton Watkins, who has been
called "the first important photographer of the Far West."[1] Watkins was first known, and
still is best known, for his photographic views of Yosemite Valley. These views had a decisive impact on Bierstadt and inspired him to visit the valley in the first place; they also
deeply influenced Hill and Keith, as well as many other painters and photographers
down to the present day. This inspiration does *not* derive from their topographic precision or panoramic comprehensiveness. Watkins's most influential photographs were shot
on his very first trip to Yosemite in 1861—not later on in 1865–66, when he had become a
member of the California State Geological Survey team. On his inaugural visit, Watkins
took no sweeping views, tempting as these might have been; he confined himself to particular places on the valley floor, perhaps the most unlikely vantage point of all. From
this ground level, hemmed in by the narrowness of the valley, it is hard to convey the
monumentality of the rock formations overhead because there is little with which to
contrast them except other equally outsize formations. The problem for Watkins was
how to achieve a fitting sense of scale in a representation of an incredibly precipitous
landscape. Basic to such scale is depth, also very difficult to capture when at the bottom
of Yosemite Valley: ordinary depth cues, along with whole grids of depth, are eliminated
by the gigantic cliffs and by the twisting turns of the valley floor itself. And the one-eyed

character of a camera, which flattens out the represented image, only makes matters worse from the standpoint of an isomorphic theory of truth in art.

None of these problems of representation daunted Watkins. In his 1861 photographs, he "took a frontal approach to individual landmarks. He set up his tripod directly in front of the object, aimed the camera straight ahead, and shot."[2] The results are altogether remarkable, especially in view of the state of the art at that time. Struggling with clumsy "mammoth plates" because enlarging was not yet practiced (i.e., the size of the negative had to be the same size as the eventual print, which Watkins preferred to be as large as 17 × 21 inches), he stationed himself strategically in face of massive formations and managed to convey a sense of their enormous size by the way in which these formations filled up, indeed overwhelmed, the frame of the photograph. There was nothing automatic about Watkins's procedure except for the operation of light and chemicals. The search for the right frame called for shrewd judgment on Watkins's part.[3] It often led him, for example, to round off the upper corners of his photographs when he finally printed them.

Watkins was led to this last step because the wide-angle lens he used in these photographs distorted the image in its marginal areas. There was no question of exact replication

Figure 2.1. Carleton Watkins, *Cathedral Rock,* 1861. Mammoth-plate. Reproduced by permission of The Huntington Library, San Marino, California.

here! Further, Watkins turned a defect into a virtue by allowing shots of the heights of rock formations to be overexposed (his emulsion was much more sensitive to the sky, and thus to these heights, than to dark foliage), thereby obscuring the image in the upper parts and so contributing to the impression of a given formation's formidable height.[4] In this instance, we observe how a feature of Watkins's apparatus that is undeniably non-isomorphic in its effects (in fact, it is manifestly distorting) contributed to a more vivid and convincing sense of the landscape. Similarly, Watkins succeeded in conveying the depth of a given natural scene by such ingenious devices as showing the water of Nevada Falls in Yosemite at two distinct points between which it disappears. "The device is almost stereoscopic," remarks David Robertson, "in that the sharp juxtaposition of far and near forces us to imagine the distance the river has travelled after it falls and before it reappears."[5] Being led to imagine this distance, we project a corresponding depth of which the distance is itself an immanent feature. More generally, Watkins conveys depth in his photographs by adroitly positioning foreground objects in such a way that they introduce the spectator to a more distant vista from *their* point of view—which the spectator assumes by active projection.

Even if it is true that Watkins himself "thought mainly of providing his customers back in San Francisco with adequate likenesses of Yosemite's domed landscape,"[6] he can-

Figure 2.2. Carleton Watkins, *Nevada Falls,* 1861. Mammoth-plate. Reproduced by permission of The Huntington Library, San Marino, California.

not be considered a mere topographer of this landscape, much less its exact cartographer. He photographed vistas rather than objects as such or their sheer spatial layout. In this way, he represented nature as a detotalized totality that, as we have seen, is no object at all, not even a total object or totalized set of objects. In accomplishing this, he showed that photography is no more wedded to strict replication than is painting. Like painting, its subject matter has a number of genres, which in turn branch out into many species and modes. Proof of this latter trait is given by the tradition of Yosemite photography to which Watkins's seminal efforts gave rise in the subsequent work of Eadweard Muybridge and Ansel Adams, both of whom treat the same subject matter in perceptibly different ways.

Most important, Carleton Watkins's photographs make it evident that photography, far from being a mere technological aide-mémoire, is capable of producing genuine presentations that are not beholden to their subject matter in terms of pictographic accuracy. In spite of their highly mechanical character, photographs are not bound to be literal *re*presentations of independently situated objects: they can express the manifold ways in which nature is an encompassing presence of maximum circumambience for its willing witness. Moreover, thanks to framing and scaling, a photograph can become the effective re-implacing of a landscape in the new world of the work.

If this is true of photography—if it is able (especially in the hands of someone as talented as Watkins) to make the fateful transition from landskip to landscape—then a fortiori the same transition can be found in painting, which is not dependent on an apparatus that tempts its user to be content with sheer pictographic accuracy. In painting as in photography, a representation of a given landscape *stands in the place of that landscape,* standing in for it in a new material form (being itself a material representation) and standing for it (as a sign of it). In both instances as well, landscape may be said to advance its position, to be *promoted,* by taking up a new residence in the artwork. In being moved into this work, it moves to a place of more nuanced, more fully differentiated, power. Instead of being diminished in this act of relocation, landscape comes into its own.

REPRESENTING THE COVE

Asher B. Durand, one of the leading landscape painters of the first half of the nineteenth century in the United States, had this to say in his "Letters on Landscape Painting": "Paint and repaint until you are *sure* the work *represents* the model—not that it merely represents it."[7] In this section I want to explore a concrete case of painting and repainting that has everything to do with *representing* something in the strong sense intended by Durand. Merely to "represent" in its unitalicized form is for him tantamount to giving an isomorphically accurate analogue of the subject matter or topic and thus to an act of *re*presentation that merely duplicates the same attributes in another medium or format: it is, in short, to create a copy of the original, an image of diminished status. To *represent* is something else again, as Durand implies by his emphatic, honorific use of the term. It is to create not merely a likeness of a discrete initial object but something that, in contrast with a copy, would have to be called a *phantasma* or, better, a "phantom thing" in a nonpejorative and nonfantastic sense.[8] It is also, in my own preferred nomenclature, a matter of re-implacing something, that is, finding another place for it, a place of a different order, within the frame of the canvas or photograph. This enframed place-of-representation stands in/for the place-of-origin; it stands in its place. As Derrida might put it, it "supplements" the original scene, being both additional to it (insofar as the work of art is itself a thing or thinglike) and necessary to its aesthetic well-being (as its transmuted presentation).

It is one thing to claim that all this belongs to representation in Durand's positive sense of the term. It is something else to witness it at work in a given landscape painting. Yet such an experience is crucial to understanding what Durand had in mind. In order to gain this understanding, I shall pursue a close analysis of three closely related works by Fitz Hugh Lane, a considerable but still not fully appreciated mid-nineteenth-century landscape painter. All three works have as their identifiable subject matter Norman's Woe, a perilous rock jutting out of the sea near Gloucester, Massachusetts. The first of these works is a sketch made in 1861, the same year in which Watkins first set foot in Yosemite Valley. But where Watkins made mammoth-plate photographs of the dramatic features of the Far Western landscape, Lane attempted to convey the quiescence of a comparatively undramatic Northeastern cove in this diminutive drawing and in two paintings realized subsequently in 1862.[9]

Lane, who helped to inaugurate what has come to be called "Luminism," began his career as "an uninspired topographer of town and harbor views and a painter of ships' portraits."[10] The heritage of his topographic origins is still evident in the sequence of works with which we are concerned. The initial drawing in particular is painstakingly pictorial: not only the placement of Norman's Woe vis-à-vis the coastline but the position of each visible rock on the beach and the grouping of various flowers in the foreground are recorded with exactitude and finesse.[11] To the finished drawing are added grid lines so that the transfer of the image to a canvas could be done with maximum fidelity. Nevertheless, already present at the stage of the sketch are powerful factors of design, of overall composition. As Lisa Andrus has analyzed it,

> The selection of the scope of the panorama demonstrates that the process of design was already at work. Lane chose a point of sight with an eye for the balance of con-

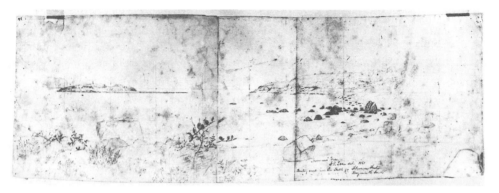

Figure 2.3 Fitz Hugh Lane, *Norman's Woe*, 1861. Pencil on paper. 8½ × 25½ inches. Cape Ann Historical Association, Gloucester, Massachusetts. Reprinted with permission.

trasts: the solid curve of the shore against the void of sea and sky, the reach of land anchored by the island, one rock in the foreground silhouetted against the water and another enclosed within the line of the shore.[12]

As with Watkins, the choice of a "point of sight" is critical, for it determines the entire outlook of the drawing and of the paintings that stem from it. However, in contrast with Watkins's proclivity for constricting the point of view, the outlook in Lane's work is panoramic, albeit on a small scale. This is partly in keeping with his topographic training—we have witnessed the natural alliance between panorama and topography more than once—and partly in keeping with the subject matter. As in early Dutch seascapes (which probably influenced Lane),[13] a comparatively low horizon, along with a resultant wide sky, encourages an open view of maximum visibility. Where mountains discourage panoramic vistas (at least when viewed from a deeply ensconced position such as that of the Yosemite Valley floor), coastal regions lend themselves to open, sweeping compositions. In Lane's work and that of the Luminists generally, "the coast and marshlands replaced the mountains as the primary theme."[14]

Replaced: this word is appositely chosen by the author of this last statement, for it is not a merely trivial difference that is at stake when the subject matter of a represented place is changed from mountains to coastline. Not only will the likelihood of painting panoramas be directly affected, but other basic parameters of landscape painting will be altered. Depth, for example, becomes a much more conspicuous dimension in Lane's Luminist horizontalism. Instead of having to rely on sheer suggestiveness, as did Watkins in his photograph of Nevada Falls, Lane is able to represent depth in a consistent and convincing manner in his sketch of Norman's Woe: "the forms are situated on the paper with a precision that marks off the recession of space and locks the coast and island into the plane of water. The smaller rocks act as units of measure charting the distance across the inlet."[15] In other words, the presence of several horizontal planes facilitates setting up units of implicit measurement *between* such planes—units that, taken together, constitute a grid of depth cues not unlike that which J. J. Gibson imputes to our untutored perception independently of representation.[16]

In his mere sketch, then, Lane has already given us a portrait of a place: a "portrait" in the sense of a topographically reliable depiction. He has created a landskip. Instead of resting content with this, however, Lane took decisive steps to transmute landskip into landscape, accurate portrait into imaginative painting. In his subsequent paintings of Norman's Woe, he effected a "transformation from topography to poetry and from fact to truth."[17] How did he do this? To begin with, he limited himself to subtle variations on the landscape schema already at stake in the drawing—so subtle that some of these variations are not at all apparent on first viewing. By this act of schematic carryover, he ensured that the painted variations had the sanction of the place, that is, that they were *its* variations and not arbitrary ones superadded by the artist. As Andrus puts it, "Lane discovered abstract relationships *only [as] intimated by the site*."[18] In fact, sameness of site (Andrus's capacious sense of "site" is equivalent to my own invocation of "place") not only preserves self-identity of location but actively suggests to the attentive artist and viewer creative variations on its own structure. In contrast with invariant topographic representations, these variations respect both sameness *and* difference: the difference plays within the sameness and is legitimated by it.[19]

The paradox, of course, is that it is precisely by variation that the true invariancy of a given place is realized. Just as Husserl claims that the abiding *eidos* or essence of a phenomenon is attained by a close scrutiny of the "free variations" performed by the phenomenologist upon its factual exemplars—variations in imagination as much as in perceived reality—so a painter such as Lane freely varies the representation of a place in getting at its material essence. In this way he "elevate[s] the factual to the poetic."[20] He moves, we might say, from the topographic to the topopoetic: from drawing or "graphing" *(graphēsis)* a place in its strict self-identity (thereby rendering its topography) to the painting or "creation" *(poēsis)* of a place in its open selfsameness (thus attaining its topopoetry, the poetry of the place). Becoming other than itself as mere position or site (in the constricted sense of this latter term), the place as painted becomes, finally, the same as itself: it is realized as same in becoming different. Which is to say: it is re-presented, presented again as fresh. Or, more exactly, it is re-implaced.

What are the differences that bring about such re-presenting and re-implacing? In the case of Lane's renderings of Norman's Woe, they are several. Each successive version incorporates significant differences and explores them in their qualitative and placial properties. For example, the first painted version, titled *Norman's Woe, Gloucester,* introduces reflections and shadows that are not included in the sketch. These areas of darkness anchor the land masses and provide the distinct impression that the painted view is of the landscape at a particular time of day (daytime shadows being a function of the sun's trajectory through the arc of the sky). Reinforcing this anchoring in space and tethering in time are ripples in the water, that is to say, moving forces that enliven the paint-

Figure 2.4. Fitz Hugh Lane, *Norman's Woe, Gloucester,* 1862. Oil on canvas, 28 × 50 inches. Private collection.

ing in comparison with the drawing. Onto this water two ships are now added; these connote a specifically human agency, further underlined by the depiction of two figures in the nearer vessel. These signs of human life (and a group of flowers in the extreme foreground) are juxtaposed with the ruins of a shipwrecked hull, lending the scene a pathos altogether absent from the drawing. The pathos is accentuated by the hushed atmosphere induced by the restrained but intense colors of Lane's discriminating palette. The dark values of the landmasses embody a dramatic difference from the luminous atmosphere of the sky and its lustrous reflection in the sea.

In this first re-implacement, therefore, we witness the initially represented place, that is, the drawn place, transmogrified. To be sure, the painting represents *the same place* as the drawing, and yet it has become importantly different. It has taken on the life of color, the animation of human presence, the clarity of daytime, the deepening of space. In the process, "this scene reaches beyond the mere description of topography to become a vehicle of contemplation."[21] In topography, one measures the terrain (recall the grid lines in the drawing), whereas in topopoetry one contemplates the form of this same terrain. The locus of fact has become the place of form.

A second painted version (known as *The Western Shore with Norman's Woe*) is something else again. A reduction of size (to 21½ × 35½ inches) puts this version in between the diminutive drawing (8½ × 25½ inches) and the first painted version (28 × 50 inches). But the scale, the spatial scope of the depicted scene, is now greater than that of the drawing or the larger painting: space has been opened out by bringing the horizon still

Figure 2.5. Fitz Hugh Lane, *The Western Shore with Norman's Woe*, 1862. Oil on canvas, 21½ × 35½ inches. Cape Ann Historical Association, Gloucester, Massachusetts. Reprinted with permission.

lower than in the first painting, thus augmenting "the push into space and spaciousness."[22] The sky is correspondingly larger, and the central cove has been emptied out so as to reflect the expanded sky more completely. In keeping with this evacuation of the center of the painting, the main ship of the first version has been pushed to the far left margin. Just as crucially, its passengers have vanished, leaving no direct human presence in the painting—not even the trace of such presence implicit in an abandoned hull on the beach. A distant ship on the far horizon only emphasizes the emptiness of this scene of becalmed existence. Concomitant with the evacuation of the painting's physical and spiritual center is a more forceful framing of this same center. The middle-distance ship frames the cove from the left, its vertical masts reinstating the verticality of the left edge of the painting itself; the rocks in the water of the cove are now continuous with the dark land mass on the right; and the distant ship acts to close the circle at its topmost point. With the central body of water at once opened up and enframed, it takes on an increased presence of its own, one of transcendental quietude: even the ripples, evident in the first version, have been removed and placed instead in the tide lines of the bare stony beach.[23] We are invited to contemplate an exquisitely tranquil central sea in its sheer glassy essence.

With these subtle but significant changes, we observe in the space of two paintings created in the same year a development that holds true for Lane's overall evolution as a painter. As one critic puts it, this is a development "from an early narrative style, with generally cluttered and active compositions, to more serene and spacious designs in the later work."[24] Thus the two painted versions of *Norman's Woe* reflect a shift from a busier, denser composition to something considerably more placid and receptive. A place that tells an implicit story—a tale of pathos, of life and death, of danger in sailing—has been re-implaced in a scene that resists narrativity. Just as its center has been emptied of objects and other complications, so it has been emptied of story. The diachronic has been superseded by a timeless moment that cannot be tied to a particular time of day: it is a moment of contemplative sublimity, of the presence of nature to itself and to us as its absorbed onlookers.[25] From such sublimity mediating figures are properly removed, leaving a mirrorlike, opalescent medium in which to contemplate the transparent form of the land- and seascape. *This form is the transformation of the place.* It transforms the factual to the poetic, the graphic statics of drawing to the color dynamics of painting, and (within the painting itself) place to its own re-implacement.

Otherwise put: the transcription of topography has given way to the transmuted *representation* of topopoetic creation. Fitz Hugh Lane, in moving from drawing to painting and then to repainting, has heeded Asher B. Durand's advice. He has painted and repainted until he has become "*sure* [that] the work *represents* the model—not that it merely represents it."

THE CONTEMPLATIVE SUBLIME

I just spoke of "a moment of contemplative sublimity, of the presence of nature to itself and to us as its absorbed onlookers." Grandiose as this might sound, I had something quite specific in mind. This is the way in which, in Lane's capable hands, the low-lying landscape of the coastline is represented in such a way that it *includes us.* It includes us not just as projectively situated at the "point of sight" chosen by the artist (as we so often experience in paintings of the Italian Renaissance) but as *ourselves implicitly implaced* in the full scene of the painting, as an integral part of that scene even if not expressly represented in it. This is especially true of *The Western Shore with Norman's Woe*. Thanks to the diminished presence of the foreground flowers, the foreground itself seems to stretch *toward* us and even to go *under* us.[26] The effect is to make us not just indifferent viewers of the scene but its privileged spectators, so privileged that we begin to feel ourselves active participants as well.

Any such participation as we may feel, however, has two important limitations built into it. First, its terms have been dictated by the painter in advance: we take up the position that he allows, indeed requires, us to assume. If it is true that Lane "scarcely composed at all except to choose, like a photographer, his place in the landscape and the extent of his view,"[27] still the place in the landscape he chose is one that sweeps us in, much as we are also swept in by Watkins's sumptuous photographs of Yosemite. In both cases, we experience an ushering in that is of the essence of many painterly presentations of landscape or seascape. Just as Henry Lewis's *Mammoth Panorama of the Mississippi River* entranced spectators in Louisville in 1849, so Lane's diminutive panorama of 1862 entices us to enter it by making us feel part of a place that is situated in a seascape in which the horizon dominates. But (and this is the second limitation) we can fully enter Lane's work only by an act of empathic mentation, that is, by sensing ourselves to be cerebrally, or, more exactly, spiritually, at one with the represented scene, immanent to its sublimity.[28] What we contemplate in such a scene (especially in its second painted version) is not just a physical body of water surrounded by a coastline, and it is not even, strictly speaking, the painterly representation of the scene. We contemplate instead the representation as sublated into a common place that is shared by the scene as represented and by ourselves as its appreciative spectators. At this precarious but precise point, we can speak of *our* place in the painting and of *our* point of view; for the artist has invited us into his place of perception and put us in that place—put us there at his own expense, because *we* now occupy *his* place. We have replaced him in the scene of his own creation. Re-implacement here occurs as *replacement*: not the re-placement of place but the re-placement of one figure (the artist) by another (oneself).

The contemplative sublime—in comparison with its predecessor, the romantic-Gothick sublime, and its contemporary competitor, the apocalyptic sublime—is modest in content and scope. Instead of expressing itself via dramatic (or, more likely, melodramatic) landscape vistas such as those of Bierstadt or Church, it seeks the unpretentious and undramatic as the proper vehicle of sublimity. More than this, the contemplative sublime is a matter of plumbing one's own soul instead of plundering external nature: "Luminist vision came from within."[29] But if so, how can we speak of a "common place" in describing such vision? Is this not oxymoronic?

The oxymoron is only apparent. For one thing, inward vision is not necessarily an individual affair; such vision can be collective, as we know from interpersonally experienced trance states and other shared states of mind. The "within" of Luminist vision can just as well be located in the community of like-minded appreciators of Luminist art as in the isolated interiority of the artist or a single viewer. For another thing, the extreme realism of Luminist painting facilitates rather than obstructs the creation of a truly visionary representation that is accessible to all who are able to share in the vision.

To be sure, the contemplative sublime is anchored in the fact of place. Yet fact is not here brute fact but the very locus of truth, above all, poetic truth; the subtly structured representation of place creates a domain in which truth can appear as in a clearing. Such a clearing is what we experience in viewing *The Western Shore with Norman's Woe,* a painting whose serene center clears a place for contemplating truth. Nevertheless, this place is an abstract place: "abstract" in comparison with the detailed particularity of the actual place sketched by Lane in his preliminary drawing. "In the painting," as Novak avers, "all the natural actualities of the scene have been manipulated into a more abstract planar order."[30]

It is just such an order that is a fit topic of the contemplative sublime. It is the right place for experiencing a sublimity that is at once interior and shared, experiential and representational, hyperreal and ideal.[31]

THREE WAYS TO RE-IMPLACE

To understand how painting attains such sublimity we must distinguish between three senses of re-implacement in art: "place at," "place of," and "place for." In the case of place *at,* a painter is concerned with the precise location of a place—thus with its exact topography. When a painter proffers to us the place *of* landscape, in contrast, he or she reinstalls this place in a representational transformation that modifies some of its aspects while keeping the place itself recognizable to the viewer. This second sense of re-implacement is, as it were, the upper bound of the process of painting as it moves from eikonic to phantasmic production, from topography to re-presentational truth. But it is at the same time the lower bound of the movement from that goes from re-presentational to poetic truth: that is, to topopoetry.

To attain poetic truth, a third sense of re-implacement is required: creating a place *for* the vision that such truth calls for. Just as the "of" at stake in the second kind of re-implacement looks back toward a topographic origin that is given detailed depiction in the phenomenon of place "at," so the "for" of place-for is forward-looking: looking forward, for example, toward the sublime of contemplation.

Norman's Woe, in its ultimate rendition as *The Western Shore with Norman's Woe* (Figure 2.5), achieves an equipoise between the three modes of re-implacement just distinguished. Insofar as it looks back to its actual topographic origin in Lane's detailed sketch of 1861 (Figure 2.3), this is a painting of a place *at* which the sketch was made. But it also refers to the first of the two paintings, *Norman's Woe, Gloucester* (Figure 2.4), manifesting characteristics that belong to a painting *of* a place—a painting that transforms the topography of an actual place while retaining its overall recognizability. At the same time, this second painting in the series looks forward to its appreciation by contemporary and future viewers, for it truly *represents* a place, transfiguring it into a place *for* the contemplation of its sublimity.

The Western Shore with Norman's Woe not only proffers each of the three kinds of re-implacement but it gives them to us *all at once,* in the clearing constituted by its own composition. It incorporates into its midst these different ways of attaining place-of-representation—and yet stays the same. It is simultaneously place-at, place-of, and place-for, just as it is at the same time at once topographic and transformative and topopoetic: in painting of this fully realized sort, *les extrêmes se touchent*.

In such triple re-implacement, an actual sea-*cum*-landscape like that found at Norman's Woe comes into its own: no longer just a place-of-origin, it *becomes its own place,* not so much a new place as a renewed place—a place to the second power. This happens, however, only *in the place of another*: in the place-of-representation that is the manifest content of the painting.

But that is not to say *in place of the painting.* On the contrary: such a place as Lane beheld at Norman's Woe comes to such realization in the very place of the painting that represents it, the place of the accomplished work.

EXPLORING THE ELEMENTAL

A landscape representation—whether a drawing, a painting, a photograph, or any other form of *representamen* (in Peirce's term for any kind of sign that serves as a representation)—can present to us varying commixtures of place-at, place-of, and place-for. These latter relate to each other as if they were members of a "complemental series": the more of one, the less of the other, yet without being incompatible with each other. Thus, Vermeer's *View of Delft* is richly freighted with its place of origin, the town of Delft, and in particular the view of it that Vermeer chose to paint: this was the very place in which Vermeer himself lived. In this intensely concrete, almost photographic representation—bristling with place-at and place-of virtues—there might seem to be little room for sublimity, that is, for the etherealizing place-for factor that (as we have just seen) carries with it a certain abstractness and ideality. In contrast, Bierstadt's *Valley of the Yosemite* is an embodiment of apocalyptic sublimity (which is doubtless why Mark Twain could not bear to look at it). It presents to us a place for beholding such sublimity, meanwhile downplaying questions of topographic (i.e., place-at) accuracy or re-presentational (i.e., place-of) loyalty. In a painting such as this, we witness *the transcendence of place by place itself.*

Nevertheless, every significant landscape *representamen* must, in keeping with the model of the complemental series, contain at least an echo of the other, less salient senses of re-implacement. Without this echo, it would fall into a deadly literalism on the one hand, or a merely tenuous sublimity on the other. In fact, Vermeer's masterpiece of topographic realism does not lack sublimity—and precisely the contemplative sublimity that will be so indispensable to American Luminist painters two centuries later—while Bierstadt's effusive masterwork, on the other hand, retains recognizable traces of the actual Yosemite Valley he visited. A comparable complexity is evident in even more unlikely instances, for example, in Carleton Watkins's photographs of the same valley (these photographs, for all of their direct frontality and identifiable detail, are also imbued with the contemplative sublime, most notably at their misty edges) or in Willem de Kooning's abstract expressionist landscapes of East Hampton (these latter, for all of their dazzling sublimity, contain concrete emblems of features endemic to the East Hampton area, e.g., its highly luminous atmosphere, its high-keyed coloration, the muted light of beaches, and the sparkling presence of the ocean beyond).

Identifiability and contemplatability, concreteness and sublimity, go hand in hand:

extremes rejoin once again. Not only are they compatible, they call for each other and are mutually enhancing in the complementary and coeval formats of place-at, place-of, and place-for. How is this possible? What binds together three such disparate senses of the re-implacement effected by painting or photography? In this impasse, it would be falsely reassuring, as well as conceptually opaque, to posit an overarching metaplace (e.g., "New England coast"), which is nothing but an ad hoc entity, posited to paper over the disparity here in question. It forms no part of our spontaneous experience of landscape or its representation. But what, then, is the source of the binding that holds the three modes of re-implacement together?

One such source is to be found in a set of factors that are "elemental" in status: elemental in a sense that is most clearly articulated in a phenomenology of wilderness that describes natural places in their comparatively pristine state.[32] When I am actually present in a given wildscape, say, a cirque high in the Rocky Mountains (wherein various features cohere to form a scooped-out basin in a semicircle of precipitous mountain walls), I find the elemental to be deeply binding. Each phenomenal element,[33] whether it be air or water or rock, is intimately linked to every other and in this capacity helps to tie together the landscape I witness, making it into that *one* scene I am experiencing (or remembering: as in this very case). If it is true that, as Karl Jaspers has claimed, "the image and model of all world orientation is orientation in space,"[34] the most pervasively qualitative dimension of such orientation is given by the elemental factors that furnish the "image and model" of orientation in landscape of a wild cast.

And of the *representation* of such landscape! Let us see how this is so by examining the role of three elemental factors—light, water, and earth—as these figure into American Luminist paintings of the mid-nineteenth century, especially those of Fitz Hugh Lane, whose paintings can be considered exemplary in their treatment of these basic ingredients of the natural world.

Light. In landscape representation of any kind, it can be truly said that light is "the great organizer."[35] It figures not just as the illuminator of physical objects (in which instance it is often reductively identified with the sun and its rays) but as the basis of the perceptibility of such nontangible items as air and atmosphere, which gain visibility only by grace of light. (Because air and atmosphere are themselves binding within any given landscape, light, insofar as it makes this binding possible, may be said to *bind their binding* or, to put it otherwise, to *pervade their pervading*.) Light is also more and other than a medium: it is not restricted to conveying phenomenal properties in an unchanged state to our vision. It alters what it transmits. Such alteration affects not just the secondary qualities singled out by Locke and Berkeley (e.g., color, texture, shape). It affects the things themselves: "more than any other component, [light is] the alchemistic medium by which the landscape artist turns matter into spirit."[36]

Whether as sheer illumination or as source of alteration, light is essential to the representation of the *place of things,* their secure situatedness in a finite region of space. As such, it subtends the factors of place-at and place-of: it allows us to say that the things represented in a painting are located *just here* in represented space, that is, just where the phenomenal light of the representation reaches them. But light as "alchemistic medium," as transformative of what it illuminates, is essential to things in their place-for-us as objects of our contemplation. In this second guise, light not only helps to distinguish discrete objects, it *clears the place for the meditative contemplation of the scene they coinhabit*: it manifests such place by its luminous action.

Clearly, such action is basic to Luminist painting in its special concern to draw our attention to places that we can contemplate as sublime. But more yet is involved in the Luminist use of light. A painter such as Lane not only employs light for the sake of clarity and manifestation. He also strives to represent the activity of light itself. His work is "enhanced by a portrayal of light."[37] This is more extraordinary than it might appear at first blush. It is one thing to paint *by means of* light or to paint objects *in the light*: this occurs in all painterly representation, whatever be the particular genre, species, or mode (including the night scenes of El Greco, de la Tour, and Ryder). It is quite another thing, however, to paint how light acts. In this latter instance, the painter must manage to represent light in its very activity of *lighting*: not just objects that are illuminated but the process of illuminating itself. This is a difficult demand, one that is first met fully successfully in Western art in Vermeer's interiors, in which instreaming light from an open or partly open window is displayed at the moment of lighting up and transmuting ordinary objects (see Plate 11A). Here the action of light, its lighting, not only makes visible *things in a place,* but also creates *places for things.*

Lane's contemporary Frederic Edwin Church was also intimately attuned to light and attempted to portray its placially specific effects (see Plates 6–8). The result, however, is a very different one from what we observe in Lane or Vermeer, in whose works light is evenly suffused over an entire scene, in keeping with a predominant mood of quietism. The light in Church's paintings of the apocalyptic sublime is an actively dissolving light. Instead of illuminating by suffusion—a constructive and supportive role—it deconstructs the objects on which it falls, divesting them of their objecthood. The result is a literal *bewilderment,* that is, the transfiguration of a *natural* into a *wild* place (where "natural" connotes the merely uncultivated and "wild" the radically untrammeled).

The same is true of the light in Bierstadt's equally apocalyptic paintings. As Novak notes, "in the large paintings by Church and Bierstadt light moves, consumes, agitates, and drowns. Its ecstasy approaches transcendence"[38]—a transcendence over the particularity of things as well as a transcendence of the merely natural by the utterly wild. In Church's South American paintings, large portions of the landscape are dissipated by the intensity of a light that has a "diffusive, vaporous quality."[39] In his celebrated *Heart of the Andes* (Plate 6), wild place is powerfully present in its very lack of discrete detail: present in its very in-discretion. In keeping with the principle of complemental series, however, Church also furnishes just such detail in certain delimited parts of the same paintings, for example, the exquisitely portrayed flora in the foreground of the work.

This monumental, sweeping kind of painting stands in stark contrast with Lane's intense and decidedly unapocalyptic coastline paintings, in which we are treated to things seen through the light and in the light as well as to the spectacle of light itself. Rather than being a dissolving and disseminating influence, light for Lane is a clarifying reflective power: "light itself partakes of the hard shiny substance of glass."[40] But it has not thereby become inanimate; it has a quiet power of its own, a "silent, unstirring energy"[41] that is forcefully at work in Lane's entire *Norman's Woe* series; for it is light, more than any other single element, that animates the Gloucester oceanside as represented by Lane. As the veritable *anima mundi* of this serene seascape, it not only clarifies it, but brings it alive and enables it to become a place for meditation: for what Baudelaire would have called *recueillement,* a reflective withdrawal from the world. By this alchemical action, Lane has indeed "turn[ed] matter into spirit." Painting light, he has fashioned one place out of another, re-implacing one *in* another, and thereby attained the contemplative sublime.

Water. The literal as well as the sublime focus of Lane's sketch and paintings of Norman's Woe is the central body of water that fills the cove as defined by the beach, the surrounding landmass, the island bearing the series's name, and two ships. It is *in* this body of water, emptied of any object other than itself, that the luminousness of the painting resides. The water not only reflects this light; it *exhibits* it, creating in this way a clearing, an opening that is the primary place for contemplation of the sublime. Such an alliance between light and water is not an accidental or even an unusual one: "In American art, especially, light has often been used in conjunction with water to assist spiritual transmutation, either dissolving form, as in some of Church's large South American pieces, or rendering it crystalline, as in the works of Lane."[42] In European painting, by contrast, the role of water is often that of defining a stable plane in the middle distance, as in the classical schema of Claude Lorraine and those who followed in his footsteps.[43] With the notable exception of Turner, water lacks the alchemical status that it came to assume in American art—starting with the Hudson River School and reaching an apogee in Luminism. What is it about water, seemingly so heavy and listless, so "sullen, untamed, and intractable"—precisely in comparison with light—that gave rise to such sublime use?

In fact, water is no more a monovalent element than is light. Like light, water possesses plasticity of form, as is evident when one considers that water takes on the shape of whatever contains it. Moreover, water changes character continually in keeping with seasons, local weather, air, and light itself. Is it surprising that Proteus, god of constant change, is depicted by the Greeks as a sea deity? In fact, the place of water in American landscape and seascape painting of the first half of the nineteenth century is a complex and changing one. It varies from the dramatic torrents of Church's famous renditions of Niagara Falls to the quiet waters of Sanford Gifford's exquisite *Lake George* and of George Caleb Bingham's studies of life on the Mississippi, culminating in the preternaturally still waters of Lane's coastline paintings. The range is indeed protean, but everywhere the spiritual significance of water asserts itself:

> Even in the large, dramatic compositions [e.g., of Church or Bierstadt], which maintain contact with the older sublime, water often inserts a quota of stillness, symbolizing a spirit untroubled in its depths and unifying both surface and depth in its reflection of the world above. The artist and spectator, after scaling the picture's height and descending to its valleys in an empathy that was encouraged, could here find rest. Thus when we see pockets of still water in nineteenth-century American landscape we may speak of a contemplative idea, a refuge bathing and restoring the spirit.[44]

Novak here suggests that water is always, even in its more agitated avatars, a redoubt of tranquillity. If so, this surely stems from the prominence of its depth dimension, a depth that underlies even the most troubled surface. It ensues that when both surface and depth are presented as placid, as in both of Lane's painted renderings of Norman's Woe, we are doubly reassured.

This reassurance is not only psychological; it is cosmic. We sense that water is somehow the ultimate medium, where *medium* is now taken in the literal sense of standing *between* things. Water is, after all, the go-between of earth and sky, their middle term. This is why Thoreau can speak oxymoronically of a lake both as "earth's eye" and as "sky water."[45] Where air and light permeate the other elements, water exists in their midst. It is the ultimate elemental mediatrix.

No wonder, then, that water has such a central presence in Lane's paintings, acting as the nearly exclusive material-*cum*-spiritual cynosure of a work such as *The Western Shore with Norman's Woe* (Figure 2.5). Being placed as it is between the foreground of the near shore and the middle ground of the encircling stretch of that same shore, water is also continuous with the sky through the opening left between the arm of the coast on the right and the rock mass of the island just to the left of center. Everywhere we look in this painting, water is found *in between* other elements and other things.

To be "in-between" is not to fail to have a place, as might seem to be the case were we to construe the idea of the "between" in a strictly site-bound manner. To be between two mathematical points, or two edges in an indifferent geometric space, may well be to lack a place—to be a-topical, exiled and having no location. But to be in-between in a landscape or seascape is not only to *have a place* in the literal sense of occupying a pre-designated position (e.g., on a cartographically precise map). It is to *be a place* in a formative way. This is what the "in" of the in-between indicates. To be in-between in a painted representation of land or sea is to be *in* a place that is at the same time *between* other places. It is to constitute an "interplace." Here the "in" is the actively formed "in" of *in-habitation*, not the passively produced "in" of merely *occupying* pre-given positions in an indifferent space. The action of this "in" is such that the "between" with which it is associated is likewise an active affair, effecting as it does the copresence of places.[46]

Water is the in-between of the elements of landscape. As Lane's Luminist masterpiece shows so vividly, its representation creates a new kind of place, an interplace, in the work of art. The result is a subtle but effective reinstatement of landscape within the work, this time by an action of constituting a uniquely mediatory place existing between its various co-elemental constituents.

Earth. It is a striking fact that earth, which might seem to be the most elemental of elements, is rarely discussed by art critics and historians who write on American landscape painting in the nineteenth century. The reason cannot be that it fails to be represented in such painting. It is as continuously present there as it is in Claude Lorraine or Jakob van Ruisdael or John Constable. Indeed, it is often massively present, as we witness in paintings such as Cole's *The Oxbow* (Plate 9) or Church's *Twilight in the Wilderness* (Plate 8), which are among the most admired landscapes of the period. Nor does the American representation of the earth lack drama, as Bierstadt's paintings of Yosemite (and Watkins's photographs of the same place) so amply demonstrate. Why, then, is explicit recognition of this primordial element so conspicuously lacking on the part of art critics and theorists?

One reason may be that earth is less affected by light than is, say, water. Where water takes on almost any conceivable hue and texture by virtue of its interplay with light—think only of how the character of a given body of water not only changes from day to day but often from minute to minute in accordance with the quality of light that suffuses it—earth is more recalcitrant to the influence of light. Often on its very surface, earth *resists* penetration by light and is thus impervious to many of the effects of luminescence that occur as a matter of course in the case of water because of the latter's transparency or translucency. Earth exhibits no equivalent of the instantaneous "sea change" that is so characteristic of aqueous phenomena; it is closed in on itself in relation to light and thus to evanescent changes in its phenomenal texture.

This marks an important elemental difference in the eyes of many nineteenth-century American landscape painters. As we have seen, their assumption is often that light is the truly transformative element—that it, and it alone, bears the burden of the radical transfiguration of matter into spirit at which they were aiming. Whatever resists such transfiguration would be not so much ignored (how could earth be ignored in *land*scape painting?) as not fully thematized, left to take care of itself.

Emerson's seminal text *Nature* was published in 1836 (the same year in which Cole painted *The Oxbow*) and was widely read by American painters of the period. In this essay, Emerson writes that, in experiencing nature, "We come to our own, and make friends with matter, which the ambitious chatter of the [idealist] schools would persuade us to despise. We never can part with it; the mind loves its old home: as water to our thirst, so is the rock, the ground, to our eyes and hands and feet. It is firm water."[47] This seemingly innocuous observation is quite revealing. The simple phrase "firm water" combines both sides of the ambiguity of earth. As *firm,* earth is that solid element on which we count for support, for orientation, ultimately for our secure being-in-the-world. When Husserl speaks of the unmoving character of the earth beneath our feet—of earth as the "ark" *(Arche)* of our existence—he is echoing Emerson's reference to the firmness that earth provides.[48] As "the rock, the ground" that pushes up solidly from below, earth is both densely unyielding and reassuringly supportive. On the other hand, as "firm *water,*" earth shows itself to be inferior to water in regard to such protean qualities as lambency, transitoriness, motility, and the like. Its very firmness, a decided virtue when we are seeking support, becomes a liability when we are interested—as most painters after Constable and Turner were passionately interested—in rendering the continually changing effects of light upon the natural world. No wonder, then, that earth can be a major factor in the representation of landscape while at the same time failing to possess the luminous/numinous presence of water in interaction with light. The ambivalence of critics and historians of art, in their comparative neglect of earth,[49] only reflect its ambiguous status in nineteenth-century sensibility and thought.

One begins to suspect that the root of such ambivalence is, paradoxically, none other than the extreme importance of earth itself as an element in landscape painting. If the representation of landscape has no choice but to be *about* earth, earth cannot be just one element or region among others in a landscape painting. Concerning any such massive and unavoidable presence, one cannot help but be ambivalent—much as on Melanie Klein's model of the equally massive caretaking parent toward whom the young child has such intensely ambivalent feelings (e.g., emotions directed both at the Good and the Bad Breast).

Not only is the monumental presence of the earth indispensable in landscape painting, but it can even be said to form for it a *topic of topics.* By "topic of topics" I do not mean a general or supertopic that would exceed the particular topicality of a given landscape and, in surpassing this topicality, betray it. Instead, I have in mind something closer to what I have elsewhere called a "matrix of matrices."[50] Such a matrix is at once materially encompassing (in terms of depth) and formally ordering (by way of region). The depth is that furnished by the earth itself, that is, a distinctively downward-tending depth, while the region is that scene of world-orientation within which all particular regions come to be ordered. Yet it is precisely this crucial twofold importance that goes unrecognized—in keeping with the all-too-human tendency to look away precisely from what matters most.

The earth is, then, a topic of topics in this double sense: it *gives depth* as that which underlies water and sky, air and light, always being beneath these elements as their ground; and it *regionalizes* in the manner of a topological matrix—that is, by determining and delineating the boundaries of everything that fits within its compass. In this double respect, the earth is the ultimate topic of landscape painting, its first and final subject matter. It is the formal-material matrix of the "wherein" of what is painted.[51]

That this is so is evident in *The Western Shore with Norman's Woe.* The earth, here present as a shore, serves as a wherein of represented space. Its action is two-fold. First, by continuing under our own presumed viewing point, it gives depth and orientation to ourselves as the included onlookers of the scene. Second, by sheltering and surrounding the central cove of water, the earth regionalizes this cove, situating it as well as the particular objects (rocks, ships, and, in this painting's earlier version, the rotting hull) that are located within it. Where the first action, that of depth-giving in the context of orientation, belongs to the earth as a material matrix, the second locatory action accrues to the earth in its role as formal matrix. In these two ways—as the source of depth and of regionalization—earth serves as the underlying resource of landscape as it is represented in painting. In this doubly basic role, it literally *holds* our aesthetic attention, stabilizes it, and allows it to come to rest in the painted landscape. If aesthetic attention focuses on various particular places within this landscape, it is the earth that is the subtending vehicle, the sensed *chōra,* of these places, holding them together, thereby embracing them (as well as the things they proffer). And if the earth is the source (and resource) of the determinate places of a represented landscape, from its firm depth stems the same landscape's capacity to be a privileged occasion for contemplating the sublime in all of its indeterminacy.

The elemental factors we have just considered are resolutely plenary in character. They fill up or fill out any given place in a landscape or seascape. Light pervades, water reaches to the outermost limit of its container, and earth is a plenum of depth and region. Everywhere we encounter the fullness of the elemental. Such elemental plenitude is very much in evidence in American painting of the nineteenth century. This is hardly surprising: when viewing landscape painting of any kind, we enjoy being presented with a full-bodied exemplar, a plenary re-implacement, of a landscape that we could in principle experience bodily and in first person. But, in the very midst of such plenitude, there lurks something else quite different—a factor of the void, a matter of radical nonpresence.

The void at stake in Luminism is of two sorts. First, it is a void of material presence, or, more exactly, a void *in* such presence. In *The Western Shore with Norman's Woe,* such a void is dramatically displayed in the form of the central cove. Emptied of all distinguishable content—whether rocks or ships, even the ripples that had cluttered its presentation in the first version—this cove may be likened to an enormous Cyclopean eye that stares vacantly at the sky overhead, not altogether unlike Emerson's celebrated "transparent eyeball" that looks intently at the surrounding cosmos. Second, an evacuation of a very different sort occurs insofar as any sense of separate self, of solid personal identity, is dissolved within this same emptied-out space—an aqueous vacuum that reflects an equally vacuous atmosphere above it: "The clarity of this luminist atmosphere is applicable both to air and crystal, to hard and soft, to mirror and void. These reversible dematerializations serve to abolish two egos—first that of the artist, then the spectator's."[52] This is a most remarkable moment in the history of landscape painting. Even in contemporary works by members of the Hudson River School (including its most preeminent figures, Cole and Church) there remain many traces of the artist's ego, for example, in the form of actual strokes visibly displayed on the canvas.[53] In contrast, the virtual elimination of visible strokes that characterizes Luminist painting—foreshadowed by Vermeer and picked up again by American photo-realism a century later—is tantamount to the elimination of the insistent self of the artist. Reinforcing this egoless state is the presence of visual voids—open water, open sky, open field—that augment the depersonification of the work and contribute to the calm vacuity that suffuses so much of Luminist landscape painting.

The voidness of object and ego invites contemplation rather than action, tranquillity rather than awe or fear, reassurance rather than dispossession. Earlier versions of the sublime—to wit, apocalyptic and romantic-Gothick—here give way to a contemplative sublime of which Lane is the American master. The void at the heart of this sublimity, far from being an ineffective and merely nugatory absence, is constitutive in a landscape art in which the ponderous elementarities of a directly perceived landscape, its felt thickness and thereness, cede place to an unballasted scene of alleviation, of sheer nonaction. Even such explicitly represented elements of landscape as light and water and earth are sublated and sublimed in this radical transfiguration of nature.

Or let us say that these elements are *transplaced* in the circumstance. The void, in its nonentitative power, effects a transplacement of elements of the landscape that amounts to the virtual annihilation of place in its strictly topographic sense as place-at, that is, as location or position. But this annihilation does not end in sheer nothingness. It enhances already existing propensities of landscape as place-of and as place-for—place of re-representation and place for becoming a scene of contemplation. The ethereal brings the material into the transplaced realm of what Emerson, in the wake of Kant, called "the transcendental."[54]

In attaining such concerted and complete transplacement, *The Western Shore with Norman's Woe* is a painting that, in the judgment of Earl A. Powell, "illustrates the ultimate influence of . . . transcendental philosophy."[55] If so, this is because the painting creates—

indeed, finally *is*—a transfigured arena for the contemplative sublime, its own transcendental domain.

Fitz Hugh Lane, starting as a lithographer who depicted historical scenes and dramatic events, ended by persistently painting and repainting a single place, bringing forth an impersonal, transcendental *representation* of that place in the void of its very manifestation. But still more than representation is at stake here. The topography of the place, so painstakingly sketched in a preliminary drawing, is transformed into painted images that are topopoetic in their contemplative sublimity. By these transfigurative actions, the decoration of landskip becomes the decorum of landscape: landscape that speaks with the silence of its own voice from the void of its own place.

CHAPTER 3

Apocalyptic and Contemplative Sublimity

All the different degrees of Goodness in Painting may be reduc'd to these three General Classes. The Mediocre, or Indifferently Good, the Excellent, and the Sublime. The first is of a large Extent; the second much Narrower; and the Last still more so . . . the Sublime therefore must be Marvellous, and Surprising. It must strike vehemently upon the Mind, and Fill, and Captivate it irresistibly.

—Jonathan Richardson, *The Connoisseur,* 1719

ASTONISHMENT AND TERROR BEFORE THE SUBLIME

I have been speaking, casually and almost in passing, of "the sublime" and of two of its primary modes, the "apocalyptic" and (especially) the "contemplative." Although only employed obliquely thus far, the sublime is far from peripheral to the representation of place in landscape painting in the West. The very painters whom I have taken as emblematic of the transformation of landskip into landscape in nineteenth-century American art thought of their paintings as inroads into the sublime. In their various writings, they employed the language of sublimity that they had learned mainly from Emerson and Thoreau. But the use of *sublime* as pertinent to the experience of landscape occurs even before the publication of Emerson's celebrated *Nature* in 1836. A year earlier, Thomas Cole's "Essay on American Scenery" invoked sublimity in the following dramatic terms:

> Shut in by stupendous mountains which rest on crags that tower more than a thousand feet above the water, whose rugged brows and shadowy breaks are clothed by dark and tangled woods, they have such an aspect of deep seclusion, of utter and unbroken solitude that, when standing on their brink a lonely traveller, I was overwhelmed by an emotion of the sublime, such as I have rarely felt.[1]

Cole here describes a sense of sublimity that could have come straight out of Edmund Burke's treatise of 1757, *A Philosophical Enquiry into the Origin of our Ideas of the Sublime and the Beautiful.* In this epoch-making essay, Burke speaks of "greatness of dimension" or "vastness" as "a powerful cause of the sublime."[2] He singles out height and depth as the two dimensions most powerfully linked to sublimity. Although he is inclined to regard depth as the more effective dimension of the sublime—"we are more struck at looking down from a precipice, than at looking up at an object of equal height"[3]—Burke, like Cole, is quite sensitive to the sublimity of height: "the view of a bare wall, if it be of a great height and length, is undoubtedly grand . . . it is therefore great, not so much upon the principle of *infinity,* as upon that of *vastness.*"[4] Burke also devotes two sections of his essay to "darkness" as productive of sublimity: these are titled "Darkness Terrible in Its Own Nature" and "Why Darkness Is Terrible." The reason why darkness is "more productive of sublime ideas than [is] light"[5] is that it contributes greatly to obscurity, which in turn fosters terror: "to make anything very terrible, obscurity seems in general to be necessary."[6]

It is terror that is the prime motor of the sublime, in Burke's view. "Terror," he writes, "is in all cases whatsoever either more openly or latently the ruling principle of the sublime."[7] But what does terror have to do with sublimity?

In feeling terror, we are so overwhelmed that we cannot resort to our usual strategies of ratiocination, or even of self-preservation.[8] When we confront something terrible, we are stymied—rendered incapable of a discerning response. Burke's statement of this matter constitutes the opening lines of the celebrated part I, section 6, of his *Enquiry,* a section titled "Of the Sublime":

> Whatever is fitted in any sort to excite the ideas of pain, and danger, that is to say, whatever is in any sort terrible, or is conversant about terrible objects, or operates in a manner analogous to terror, is a source of the *sublime*; that is, it is productive of the strongest emotion which the mind is capable of feeling.[9]

"The strongest emotion which the mind is capable of feeling" is important not because of any sheerly quantitative consideration but because of its immediate effect. This effect is named "astonishment" by Burke and is described thus in Part II of his work: "astonishment is that state of the soul, in which all its motions are suspended, with some degree of horror. *In this case the mind is so entirely filled with its object, that it cannot entertain any other,* nor by consequence reason on that object which employs it."[10] In being astonished, the mind is so overwhelmed by the sublime as to become wholly saturated by it, allowing Burke to conclude that "hence arises the great power of the sublime, that far from being produced by them, it anticipates our reasonings, and hurries us on by an irresistible force."[11]

Even though Burke's description of the sublime expressly privileges the literary representation of the sublime (as when Burke contends that, in contrast with a mere sketch of a landscape, a verbal description of this same landscape "raises a very obscure and imperfect idea"[12] to greater articulation), his description nevertheless applies admirably to painted images of the sublime, especially of that sublime which was called "Gothick" in the first half of the nineteenth century. Take, for example, the following passage from the section titled "Colour": "An immense mountain covered with a shining green turf, is nothing in this respect, to one dark and gloomy; the cloudy sky is more grand than the blue; and night more sublime and solemn than day."[13] This reads like a set of practical instructions to aspiring painters—instructions that seem to have been closely heeded in the case of a committed night-scene painter such as Albert Pinkham Ryder! But night scenes, however favorable they may be as vehicles of the sublime, do not form its exclusive setting. Light, and the representation of light, is also "a cause capable of producing the sublime."[14] Although ordinary illumination is unlikely to have a sublime aspect, "such a light as that of the sun, immediately exerted on the eye, as it overpowers the sense, is a very great idea."[15] To be a very great idea is to be the very basis of the sublime, for, as Burke adds, *"without a strong impression nothing can be sublime."*[16]

The words I have just emphasized amount to a declaration of principle, to be followed by all painters of what Novak calls "the older romantic-Gothick sublime,"[17] whether William Blake in England, Caspar David Friedrich in Germany, or Washington Allston in the United States. Each of these artists, foreshadowed by such precursors as Salvator Rosa and Pieter Breughel (the elder), sought to create a maximally strong impression that captivates the viewer: an impression that corresponds to (and is reinforced by) a sublime landscape's dramatic dimensions of depth and height.[18]

Frederic Edwin Church's *Cotopaxi* (see Plate 7), a painting of an active volcano in Ecuador, is so dramatically presented that, in viewing it, "the senses are blurred in a paroxysm of activity."[19] Not only is the impression strong but the fuming volcano as the central subject of the painting is such as to preoccupy the mind, bringing it about that the mind is "entirely filled with its object." In this case the object itself is not only of great height (it is almost twenty thousand feet high and rises in splendid isolation from the Ecuadorean tropics) and located at considerable depth (i.e., when seen from the implicit viewing point assumed by Church's painting), but is manifestly threatening in nature: dark reddish-brown clouds emanate from the volcano, forming an ominous mantle over the surrounding landscape. The sulfuric cast of these clouds obscures the sun's illumination, suggesting something infernal or Satanic. Thanks to a pervasive glow of Mars red, the entire landscape of *Cotopaxi* appears molten and on the verge of dissolving into red-hot magma. Even the lake, normally a source of tranquillity, has taken on a sinister air: it could well be a pool of blood, over which the anxious sun is perched precariously. Or it could be an extinct crater, an ancient relic of previous volcanic activity, which has now become a stagnant pool. And the canyon of the foreground seems to have no bottom: a desolate abyss over which the spectator hovers hopelessly at an indeterminate distance.

In a painting such as this, the two prerequisites of Burkean sublimity are undeniably present: the *impression* is powerful, and the *object* that conveys it demands aesthetic attention. The eighteenth-century vocabulary of "impression" and "object" needs only to be replaced by different terms in order to be more fully convincing today: say, by "perception" and "scene"; for it is our perception of the entire scene that grasps it as sublime—sublime in a truly apocalyptic manner whereby we seem to be witnessing an event situated at the beginning or the end of the known geological world.[20] On the other hand, Burke's nomenclature for the affective dimensions of the sublime remains apt and intelligible. The seething scene still occasions astonishment in us: it is capable of moving us to emotional depths that answer to the perceptual depths of the scene itself. For the contemporary viewer as well, "strength, violence, pain and terror . . . rush in upon the mind together."[21]

These last words provide an apt description of nature in its wilder reaches. More than any other kind of landscape, wildscapes are capable of captivating us to the point where we are unable to "entertain any other" impressions or "by consequence [to] reason on that object" which causes them. Thus it is not surprising that an entire genre of nineteenth-century landscape painting, focusing on wilderness, attempted to recapture and to reinduce this kind of experience. This is precisely what I have been calling the "apocalyptic sublime," and its origins lie in the Gothic sensibility of Burke's successors among the painters (most notably, John Martin and William Blake, but also Benjamin West, Philippe Jacques de Loutherbourg, and the later work of Turner).[22] Its full flowering occurred just after the middle of the century in the United States, where it was extended further by a Christianized New World fervor for which North (and sometimes South) America were regarded as the Promised Land.[23]

Seen in this perspective, Church's panoramic-melodramatic scenes from the Ecuadorean Andes or Bierstadt's equally grandiose "cosmoscapes" of the American Far West bring to an appropriately grand conclusion the conception of the sublime so forcefully articulated by Edmund Burke a century earlier. Church and Bierstadt would doubtless have agreed with Burke's pronouncement that "I know of nothing sublime which is not some modification of power."[24] The sublime they painted in such "greatness of dimension" was indeed a matter of a power instilled in strong impressions that were ineluctably linked to terror: "power," adds Burke, "derives all its sublimity from the terror with which it is generally accompanied."[25]

TWO KINDS OF SUBLIMITY

In view of the apparent continuity of this development from Burke to Bierstadt, it would seem plausible to rest the case here and not seek further complications. But the circumstance dictates otherwise. Just as landscape painting has many species and modes, so it supports various forms of sublimity and not one only. It would be as mistaken to impute a single kind of sublimity to the painterly representation of landscape as to limit all forms of landscape to its wilderness form alone. Tempting as such attribution may be in both cases, it would result in a partial and reductive view of the natural world and of the paintings that represent it. Landscape painters themselves teach us to resist this temptation.

Thomas Cole, for his part, furnishes a Burkean description of his experience of two lakes in Franconia Notch, only to suggest a quite different way of construing this same scene: "It was not that the jagged precipices were lofty, that the encircling woods were of the dimmest shade, or that the waters were profoundly deep [these words directly recall Burke's descriptions of the sublime]; but that over all, rocks, woods, and water, brooded the spirit of repose, and the silent energy of nature stirred the soul to its inmost depths."[26] At stake here is a very different, nonapocalyptic conception of the sublimity of landscape, one that, as indicated in the last chapter, is expressly taken up in American Luminist painting. But we need not revert to Luminists such as Gifford or Lane to witness the emergence of a quieter sublime, one based not on power or terror but on the "tranquillity"[27] that Cole found in his Franconia Notch experience. Cole, true to this experience, painted a number of deeply quietistic lakescapes, for example, his *American Lake Scene* of 1844 in which the surface of the water is almost as mirrorlike as that in Lane's 1862 *The Western Shore with Norman's Woe* and from which there is a conspicuous absence of daunting objects. Even the redoubtable Bierstadt produced calm and diminutive studies, for instance, his *Cloud Study: Sunset* (c. 1870–90), in which light, instead of being something that "overpowers the sense" (in Burke's phrase), is explored in a markedly muted manner.

In fact, every major painter of the apocalyptic sublime in the nineteenth-century United States was at the same time drawn to that very different sublime which I have been calling the "contemplative sublime." This is even true of Frederic Edwin Church, who is in so many respects the archapocalyptic painter of the century. Still more significantly, Church sometimes sets forth a convincing presentation of *both* sorts of sublimity in a single painting, for example, his *Twilight in the Wilderness* (see Plate 8). This painting, often regarded as an apogee of American landscape painting in the nineteenth century, presents a brilliantly apocalyptic sunset sky, filled with intense cadmium reds, oranges, and yellows (as well as bluish-green moments in the upper sky), all in such an astonishing array that Burke would doubtless have approved: here, indeed, "the cloudy sky is more grand than the blue; and night more sublime and solemn than day." Yet underneath this sky of exceedingly strong impression is a lake scene that rivals Lane's quietism in its cool complacency: not bloody or infernal, it reflects the sky—yet in decidedly soft tones and textures. The light that in the upper part of the picture is precisely (again in Burke's words) "a light which by its very excess is converted into a species of darkness"[28] is transmuted in the picture's lower half into a light of sheer reflection and contemplation: a light of "silent, *unstirring* energy."[29] The powerful stirringness of one region of the painting (a region where all is "strength, violence, pain and terror") exists in equipoise with the calmness of the region below it. The vastness of the sky, which has greatness of dimension in both height and depth, cohabits remarkably well with the detail of the comparatively diminutive objects in the near and middle distance. As a result, each part of the painting is effectively yet differently sublime. How can this be? Do we have to do here with two distinct *kinds* of sublimity, or do we have to do with two different aspects of the *same* sublime?[30]

A clue is provided by a neglected section of Burke's *Enquiry* titled simply "Privation." Here he asserts that "all *general* privations are great, because they are all terrible; *Vacuity, Darkness, Solitude* and *Silence.*"[31] We might add to this list the desolation experienced in certain forms of wilderness.[32] Just as desolation is by no means the negation or undoing of wilderness (it is often its very epitome), so the four general privations cited by Burke contribute to the sublime instead of detracting from it—a detraction we might presume to be the case were we to rely too exclusively on Burke's own rhetoric of terror and power. Indeed, the privations he lists are strikingly close to the leading values of Luminism and thus of their distinctive sublime of contemplation. This is especially true of the last two privations, "solitude" and "silence," each of which contributes directly to such a sublimity, as is so tellingly illustrated in the work of Fitz Hugh Lane. But the other two privations are also crucial to the contemplative sublime: "vacuity" is but another name for what Luminists were wont to call "space" (i.e., the empty space of nature as signified by an infinitely receding horizon), while "darkness," as Burke himself argues, answers to light in its most sublime power: "extreme light [i.e., the only light capable of sublimity], by overcoming the organs of sight, obliterates all objects, so as in its effect exactly to resemble darkness."[33] No wonder Church could choose as an archetypal wildscape what can be seen only at the moment of "twilight," that is to say, the very moment in which we most poignantly observe what Burke calls the "transition from light to darkness"![34]

If it is true that Luminism represents "the intuited unification of light, space, and silence,"[35] then Edmund Burke has anticipated this holy trinity of American Luminist landscape painting by his list of privations pertaining to the sublime. He also anticipates Luminism by espousing the basic notion, mentioned earlier, of the way in which the sublime can so overwhelm consciousness as to deprive it of any other object. Now this particular state of being engorged with an object (or more properly with a *scene*) has the remarkable property of leading to an experience of emptiness. To become fully preoccupied with just *one* thing to the exclusion of all others is to lose track of that variegated set of identifying and locating references by which any scene is set. With the withdrawal of the context established by other items, the single item of attention itself founders, leaving vacuity, darkness/light, solitude, and silence. To contemplate *one thing alone* is in effect to contemplate no-thing. To be a thing at all is to be a thing among things; not to be so situated is to be no thing at all. That this is so has been known from times immemorial in cultures in which meditation on a single item gives rise to what Buddhists call "formless concentration" *(arupa dhyāna)* on endless space.[36]

A painting attains "light, space, and silence" in two major ways: either by presenting a scene that lacks a grippingly dramatic central thing (the body of water in Lane's *The Western Shore with Norman's Woe,* though central, is not dramatic but ethereal) or a scene that blatantly exhibits such a drama (as in Church's *Cotopaxi*). Whereas painters of the apocalyptic sublime demand that we concentrate on a super-thing that engages our consciousness to start with—only to witness its eventual dissolution—their Luminist colleagues eliminate any such thing from the initial presentation and invite us to enter immediately into a formless consciousness of silence and the void.

In this respect, the transition from apocalyptic to contemplative sublime is less disruptive than we might think. The super-thing of apocalyptic painting melts down in its lonely superiority just as surely as the sunset in *Twilight in the Wilderness* gives way to the becalmed water beneath it. Whether they occur in the same painting or not, the two sorts of sublime can be considered two phases of one process or transition: a first phase of object assertion and emphasis giving rise to a second phase of object dissolution and disappearance. Both testify eloquently to the fact that the experience of landscape (and all the more so wildscape) is an experience of something that cannot be reduced to a finite set of discrete objects, much less to a single total object. They bear witness to the fact that landscape is more scene than thing, more event than object, more place than site.

Far from there being two incompatible kinds or types of sublimity in the experience of landscape and its representation, one of which is characterizable as exclusively "apocalyptic" and the other as equally exclusively "contemplative,"[37] the apocalyptic and the contemplative can be considered two aspects of the massive sublimity of nature itself. Or, more exactly: they are two ways of *representing* nature's sublimity—a sublimity that ranges from the rugged to the reposeful, from the agitated to the tranquil, from the immense to the diminutive. This is why an exemplary painting such as *Twilight in the Wilderness* can so successfully combine the apocalyptic with the contemplative sublime. As "twi-light" brings darkness and light together at the most poignant moment, so this painting shows the two modes of sublimity to be not only compatible with one another but capable of enhancing each other's presence.[38]

Moreover, the apocalyptic and the contemplative sublime are mutually sustaining *in their conjoint representation,* that is to say, in the very place that a landscape painting uniquely provides. Wild places have been transformed in an apocalyptic-luminist vision; this has happened in a revealing representation of a twilight space that is empty and silent, yet also intensely meaningful to the point of melodrama.

In this way sublimity—a redoubled and thus overdetermined sublimity—has been attained. The re-implacement of nature in such painting, the supersession of landskip by landscape in two modes of its sublimity, is a dual sublimation of place itself.

KANT ON THE SUBLIME

We have glimpsed something of the historical and conceptual complexity of the sublime in painting, especially in the revealing case of nineteenth-century American landscape painting. But do we yet know what the sublime itself is? We need to find a viable working model if we are to speak meaningfully of the representation of the sublimity of nature in landscape painting and, indeed, in representations of any kind. So let us begin again, this time basing our reflections on the celebrated treatment of the sublime to be found in Immanuel Kant's "Analytic of the Sublime," which first appeared in 1790 as an integral part of his *Critique of Judgment.*

Kant's analysis of the sublime builds on Burke's *Enquiry* by agreeing that the primary distinction in matters of aesthetic judgment is that between the beautiful and the sublime. But whereas Burke had defined beauty as "that quality or those qualities in bodies by which they cause love, or some passion similar to it,"[39] Kant places the emphasis on the factor of *form* in beauty: "the beautiful in nature is a question of the form of the object."[40] With form comes limitation, and it is precisely with regard to the issue of limits that the sublime is to be distinguished from the beautiful: "the sublime is to be found in an Object even devoid of form, so far as it immediately involves, or else by its presence provokes, a representation of *limitlessness,* yet with a super-added thought of its totality."[41] Burke had spoken of "greatness of dimension" and of "vastness," but Kant subsumes the Burkean intuition under the overall heading of the "mathematically sublime." Such a sublime is to be thought as "what is *absolutely great.*"[42] To be absolutely great is to possess a magnitude with which nothing else can compare, with the result that "that is sublime in comparison with which all else is small."[43]

Just here we reach a crucial parting of the ways between the Burkean and Kantian conceptions of the sublime. So long as the sublime is a matter of "vastness of extent, or quantity,"[44] as Burke claims, it can be securely located within physical nature, that is, within the landscape we experience. On this conception, nature itself is sublime, whatever be the exact form in which it presents itself, and we apprehend its intrinsic sublimity with that "delightful horror" which Burke considers both the effect and the proof of the sublime.[45] Kant concedes the paradoxical pleasure we take in the sublime—"a pleasure that is only possible through the mediation of a displeasure"[46]—but he is much more interested in the issue of absolute magnitude, which he carefully distinguishes from greatness of quantity. Although it is a fact that we discover things of enormous quantity in nature such as oceans or mountains, something of absolute magnitude cannot reside, strictly speaking, within the natural realm. Were it to do so *per impossible,* it would so dwarf every other thing in that realm as to *put it out of place*—in effect, to situate it be-yond nature altogether; for such a thing would be not just comparatively larger than any natural thing (as the sun, say, is larger than the moon by a determinate factor of size) but so large as to transcend *any* determination of quantity derived from the domain of sensible experience.[47]

Where, then, are we to locate the mathematical sublime if not in nature? *In the mind.* Only in the mind, in its ideas, is there room for something of absolute magnitude:

> If . . . we call anything not alone great, but, without qualification, absolutely and in every respect (beyond all comparison) great, that is to say, sublime, we soon perceive that for this it is not permissible to seek an appropriate standard outside itself [i.e., in nature], but merely in itself. It is a greatness comparable to itself alone. Hence it comes [about] that the sublime is not to be looked for in the things of nature, but only in our own ideas.[48]

The ideas here at stake are ideas of reason rather than concepts of the understanding, which are inapplicable to that which cannot present itself as a natural phenomenon. Such ideas are able to deal with the "limitless" *(grenzenlos)* as a totality for which there is no exact quantitative determination or even an accurate estimate. The idea of the mathematical sublime, therefore, is an idea of an undelimited totality: so undelimited in fact that not even our own imagination, which on its own terms knows "no bounds to its progress,"[49] can catch up with it! (In this way we again reach, by a very different route, the notion of a detotalized totality: detotalized this time thanks to the unquantifiability of the limitless.)

By conceiving of the sublime as absolutely great, Kant is thus led to seek its basis within the human mind, wherein an appropriate conceptual space is cleared by the possession of an idea of reason that exceeds any possible sensuous presentation.[50] In this fertile place of the mind, a dialectical struggle ensues between imagination in its striving to live up to the specific idea of the limitless and the idea per se, which always eludes what imagination can achieve. (By way of contrast, in the case of beauty, imagination and understanding accomplish an accord or harmony with each other.)[51] Although this situation is bound to be frustrating and a matter of "displeasure"—given that imagination can never fully embody what reason proposes—it also contains a positivity that Burke had overlooked. This is that the mind "feels itself elevated in its own estimate of itself on finding all the might of [its] imagination still unequal to its [own] ideas."[52] Even as it notices its shortcomings qua imagination, the human mind appreciates its powers of rational ideation. What it cannot attain in imagination, it can esteem in itself as reason. "True sublimity," in short, "must be sought only in the mind of the judging Subject, and not in the Object of nature that occasions this attitude by the estimate formed of it."[53]

The sublime is to be conceived either mathematically or dynamically: it is a matter of "a double mode of representing an object as sublime."[54] Much as we have seen the painterly sublime to occur in at least two forms of representation (i.e., the apocalyptic and the contemplative), here we have to do with two "affections of the imagination"[55] that are not contradictory but complementary. In the case of the dynamical mode, we confront something not of "absolute greatness" but of sheer "might" *(Macht).* As Kant remarks, "Nature considered in an aesthetic judgment as might that has no dominion over us is dynamically sublime."[56]

The might of nature has no dominion over us precisely because of our effective "resistance" *(Widerstand)* to it. Such resistance, once again, is internal and is found in the very same ideational capacity of reason we have just examined. Thanks to this capacity, the human subject realizes "a dominion which reason exercises over sensibility with a view to extending it [i.e., sensibility] to the requirements of its own realm."[57] By successfully resisting the putative dominion of nature in its sheer physical force, we demonstrate that we have our *own* considerable cognitive might. In making this point, Kant evokes dramatic images of nature that might have been painted by Church or Bierstadt, or by his fellow countryman Caspar David Friedrich:

> Bold, overhanging, and, as it were, threatening rocks, thunderclouds piled up [in] the vault of heaven, borne along with flashes and peals, volcanoes in all their violence of destruction, hurricanes leaving desolation in their track, the boundless ocean rising with rebellious force, the high waterfall of some mighty river, and the alike, make our power of [physical] resistance of trifling moment in comparison with their might. But, provided our own position is secure, their aspect is all the more attractive for its fearfulness [i.e., it occasions "delightful horror" in us]; and we readily call these objects sublime, because they raise the forces of the soul above the height of vulgar commonplace, and *discover within us a power of resistance of quite another kind,* which gives us courage to be able to measure ourselves against the seeming omnipotence of nature.[58]

The power we discover "within us" is indeed of "quite another kind," that is, another kind from that which the unvarnished power of nature manifests when we are overwhelmed in the presence of its most majestic spectacles. But even if these spectacles are genuinely threatening, we need not actually *fear* them—provided that "our own position is secure" and given that we can muster a considerable power of our own in the circumstance.[59] This power is that of reason in its idea-generating, supersensible status and thus belongs to ourselves as free rational subjects. Indeed, our rational freedom, which is ultimately moral in character, is such as to assure our "pre-eminence above nature."[60] Such preeminence, which Kant also characterizes as "a superiority over nature within and thus also over nature without,"[61] is the true source of sublimity and in particular of the might that characterizes the dynamically sublime.

But if this is so—if our free and rational self is, in Merleau-Ponty's expression, "the absolute source"[62]—why do we continue to attribute sublimity to external physical nature, that is, to that which is "improperly called sublime"?[63] The answer to this critical question is twofold. On the one hand, nature itself encourages such attribution, especially when it presents itself as "rude nature merely . . . involving magnitude,"[64] that is to say, as possessing the precipitous heights and depths that Burke had singled out as emblematic of "greatness of size." Instead of taking such heights and depths as the literal bearers of the sublime (as they were by Burke and his followers), however, Kant sees them as merely supporting the sublimity that stems from within: "the [natural] object lends itself to the presentation *[Darstellung]* of a sublimity discoverable *in the mind*."[65] On the other hand, human beings play a transcendental trick on themselves: they project the sublimity that properly stems from rational ideation onto natural things and then forget that they have done so. Kant calls this act of self-deception "subreption," and describes it as follows: "the feeling of the sublime in nature is [implicitly] respect for our own vocation [i.e., as rational beings], which we attribute to an Object of nature by a certain subreption (substitution of a respect for the Object in place of [respect] for the idea of humanity in our own self—the Subject)."[66] The result of subreption is there for all to witness: we take the wild ocean to be "sublime" in its frenzied movement, but it is properly speaking only "horrible" (Burke would have said "terrible"). It certainly has dynamic force and power, just as it has determinate magnitude, but it does not *possess* sublimity as a definite property. Instead, it merely *exhibits* a sublimity that is, in the end (or rather, in the very beginning), an expression of our own mind as it responds to the natural spectacle, grafting onto this spectacle a sublimity of our own making. Strictly speaking, then, the broad ocean agitated by storms cannot be called sublime: "Its aspect is [indeed] horrible, [but] one must have stored one's mind in advance with a rich stock of ideas, if such an intuition is to raise it to the pitch of a feeling which is itself sublime—sublime because the mind has been incited to abandon sensibility, and employ itself upon ideas involving higher finality."[67] Not only the roiling ocean, but *any* natural thing, no matter how impressive or even astounding it may be in its perceptual presentation to us, lacks sublimity as an inherent property: "we express ourselves on the whole inaccurately if we term any Object of nature sublime."[68]

The sublime is in us and not in things. Such is Kant's challenge: "to locate the absolutely great only in the proper estate of the Subject."[69] The same holds for the might at stake in the dynamically sublime. It is as if Kant has made artists of us all by insisting that only by our own creative (albeit unself-knowing) action does the sublime come into being, much as works of art arise only through the free agency of the artist. Just as the artist exercises the countermight of art in the face of the daunting power of nature, so we, ordinary subjects of aesthetic judgment, exert the counterdominion of our own (ultimately rational) freedom in confronting the ostensible dominion of nature over our finite and fragile lives. Whether as artists or as mere spectators of art, we draw upon the unlimited resources of our own ideas of reason (whose capacious scope outreaches even our "wildest imagination," as we say revealingly) by establishing the sublime as absolutely great and as dynamically mighty. The sublime is one of the seemingly most extra-human and transcendent features of our experience, a feature that is both "contra-final" and "an outrage on the imagination,"[70] yet in dealing with it, we are in fact dealing with what, by an act of subreption, we have brought about ourselves. What we do within ourselves shows up outside ourselves, as if it were an indissociable and objective feature of the natural world.

DRAWING LESSONS FROM KANT

I have here followed the lead of *The Critique of Judgment* for three principal reasons. To begin with, Kant's treatment of the sublime in its double aspect of the mathematical-*cum*-dynamic parallels our earlier suspicion that the apocalyptic and contemplative sublime in nineteenth-century American landscape painting are not two entirely distinct phenomena. Not only this, but it turns out that the Kantian "double mode of representing an Object as sublime" bears an uncanny resemblance to apocalyptic and contemplative forms of sublimity. The dynamically sublime, which Kant finds exemplified in precipitous rocks, violent volcanoes, and boundless oceans, infuses paintings by such apostles of the apocalyptic vision as John Martin and J. M. W. Turner in England, and Church and Bierstadt in the United States. The mathematically sublime, on the other hand, is evident in Lane's and Heade's concern with the specifically "mensurational" properties of their work (as revealed by their careful, gridded drawings)—a concern that in the end, however, succumbs to the realization that their sublime subjects exceed any determinate measure, much in the manner of Kant's idea of the "absolutely great."[71] The fact that certain painters, such as Turner and Church, expressly combine both forms of sublimity in their work only confirms the deep complicity of the mathematical with the dynamical aspects of the sublime. The same is true of certain subtler combinations of the apocalyptic with the contemplative—an echo of one being present even in the emphatic employment of the other—that we find in many other artists of the era, including Richard Wilson, Thomas Moran, Jasper Cropsey, and Carleton Watkins.

We have followed Kant for a second reason as well: in order to oppose the naive view that the sublime is "simply located" (as Whitehead would say) in external nature where it can in effect threaten us. Such is the position of Burke, for whom the sublime is said to bear importantly on "self-preservation," given that the danger and terror on which the sublime is based are located outside us in life-threatening circumstances.[72] The consternation and near paralysis that result from these circumstances are said to constitute "the effect of the sublime in its highest degree," while admiration, reverence, and respect (among the very emotions singled out by Kant) are said to be merely "inferior affects."[73] Here Kant would ask Burke: how can anything aesthetic be discerned, much less enjoyed, if we are really threatened by it, if it is in fact dangerous to life and limb? Even in regard to that "rude nature" that lends itself most readily to artistic representation as sublime, Kant is led to add the cautionary note that such representation can arise "only insofar as [such nature] does not convey any charm or *any emotion arising from actual danger*."[74] This is not only an act of deliteralization; it is also a matter of relocation from the strictly physical site of *rerum natura* to a psychical place, a *locus mentis* that is "the proper estate of the Subject."

This suggests a final reason for following the "Analytic of the Sublime" as closely as we have done. "Relocation" is in effect the converse of subreption. Whereas in subrepting, we unwittingly bestow upon things of nature a trait properly belonging to us, in relocating we take back the results of subreption itself. We take these results back upon ourselves, assuming responsibility for what we have done in an initial act of displacement. Displacement outward—the very basis for Burke's simple location of the sublime squarely *in* nature—gives way in Kant to inward reinstatement. Such reinstatement, let us note, is a further form of what I have been calling "re-implacement." In this case, the re-implacing is not from the (actually beheld) scene to the (representing) work but from this scene to the mental space that subtends any such work by making artistic representation itself possible.

It is a matter, in sum, of *the sublimation of place itself.* In contradistinction to Burkean sublimity (and its legacy in the apocalyptic sublime of British and American painting), we have to do here with an act of sub-liming in the literal sense of going *beneath the threshold* of the physical in order to enter the domain of the mental. Thus it is a question of going into a space that is neither strictly physical (e.g., the space of external nature or, for that matter, of internal neurological activity) nor strictly imaginary (Kant has warned us that imagination is not up to the task of adequately apprehending the sublime). It is a space that occurs *as place,* that is, in particular topoi of inward location: "psychotopics," as I am inclined to call it. What Bergson says of "extensity" in *Matter and Memory* holds true for such psychical re-implacement: it exists on "the nearer side of homogeneous space," while being itself "concrete and indivisible."[75]

To be concrete in matters of space is to occur as place, that is, as something spatially heterogeneous—hence as something that refuses to be abstracted and unified as a single universal space. Mental space is an indivisible (and invisible) totality whose first and last unit is that of place. In contrast with the homogeneous space of physical nature, which is in principle infinitely divisible,[76] mental space is indivisible *in its very capacity of generating the sublime*: this is a direct implication of Kant's conception of the sublime as at once limitless and totalized. To lack a limit, whether of the great or of the small, is to lack the basis for divisibility: to be divisible at all is to be divisible *in accordance with a limit*. To be a sublime totality is to resist any act of division—any act that dissolves or diminishes its state of amassment. (A perceived landscape, as a detotalized totality, exhibits the same resistance to subtraction and delimitation.)

To speak of re-implacement via sublimation is to speak of an activity of internalization that is comparable in many respects to Freud's idea of character as based on identification. Just as character is built from "a precipitate of abandoned object-choices"—a precipitate that "contains the history of these choices"[77]—so re-implacement that is at once subliminal and sublimated brings experiences of the external spaces of nature, including those of "natural scenery," into internally held histories of places in the mind. These latter places, the loci of identification in memory, are the concrete locations of the sublime, its relocalizations within the subject.

In these places, *in* them and certainly *with* them, the artist paints the sublimity of nature. Without them, there would not be enough room, room of the right psychical extensity, for the representation of the sublime in art. It is by means of such sublimatory space that the resplendent cove of Norman's Woe or the deep valley of the Yosemite or the vast plain of Cotopaxi become for certain painters places of apocalyptic/contemplative sublimity. Such painters have so thoroughly experienced and taken in a given place that it has become one with his psychic space, a psychotopia; it has become a place in that space. Only *from* this place of intense interior identification can the painted place come to have the character it exhibits and be the place it is, a place-of-representation: a place-by-proxy that stands surety both for the initially experienced and subsequently internalized place.

In landscape painting, three kinds of place are at stake: the place-of-origin, the place-of-identification, the place-of-representation. The latter stands between the two former insofar as it is a middle term between the externality of the one and the internality of the other. But there is no simple sequence here: all three forms of implacement are simultaneously in play.[78]

Therefore, even if Constable can claim truly that "these scenes [i.e., scenes of his youth and early manhood in East Anglia] made me a painter,"[79] it is also the case that the same painter *made these scenes*: he remade them by representing them in paint. But he could do so only because of the internal extensity into which he had sublimated their inaugural and ongoing perception and through which he re-implaced them in the compositions of the landscape paintings he created.

WHERE IS THE SUBLIME?

This is not to say that Kant's treatment of the sublime, suggestive as it undeniably is, is unproblematic in its application to landscape painting. Far from it! A direct consequence of his analysis of sublimity strictly in terms of internal processes of reason and imagination is that he has difficulty resituating the sublime in the realm of nature. Indeed, if it is true that the sublime "is not to be looked for in the things of nature, but only in our own ideas,"[80] then we must even wonder how it is located in a *painting* of landscape. More basically yet: why do we so resolutely continue to take the sublime to be located in the natural world—and to *represent* it as being there in paintings of it—when its provenance and proper province are supposedly found only in the human subject? Has not Kant, by his own ingenious hypothesis of subreption, too drastically detached the sublime from its moorings in actually experienced landscapes by displacing it from nature to mind? Has he not thereby skewed its representation in landscape paintings that allude, however indirectly, to a previously beheld scene, or at least to a natural world that we presume to be external to the mind? However important mind can be in the representation of place—and we have seen just how crucial it is in the internalization of place and identification with it: that is, in psychotopics—must we not finally give priority of place to the sublime *in nature*?

Kant admits priority to natural place, that is, nature in its wildness, only insofar as it *occasions* or *inspires* ideas of the sublime—not as itself embodying the sublime, much less as providing its ultimate locus. In an important passage of the "Analytic of the Sublime," Kant claims that whereas we can legitimately impute beauty to nature,[81] we cannot ascribe sublimity to the natural order itself. His example is precisely that of wilderness:

> In what we are wont to call sublime in nature there is such an absence of anything leading to particular objective principles and corresponding forms of nature [as can occur via the imputation of *beauty* to nature] that it is rather in its chaos, or in its wildest and most irregular disorder and desolation, provided it gives signs of magnitude and might, that nature chiefly excites the ideas of the sublime.[82]

Nature, then, does not house the sublime but instead "excites" it as an idea. Kant carefully chooses the word "excites" (*erregt*: also translatable as "arouses" or "agitates") given that excitation, being a matter of force, operates in terms of the very "magnitude and might" that Kant recognizes to be the language itself of the sublime. But just as such—as a force that can get out of hand (i.e., in wilderness), in contrast with the discrete "form" that beauty entails—it *resists representation*. And Kant does not fail to draw this very conclusion: the sublime "does not yield a representation of any particular form in nature, but involves no more than the development of a final employment by the imagination of its own representation."[83] Moreover, if it is true that sublimity is a matter of excitation *by* nature but not of the representation *of* nature, it follows that it is not a fit subject for landscape painting. For Kant, such painting, inasmuch as it seeks to render certain formal properties of nature, aims at representing "the beautiful in nature." It also follows, still more generally, that "the concept of the sublime in nature is far less important and rich in consequences [i.e., for its representation in painting] than that of its beauty."[84]

But how can this be? Has not the entire Romantic movement in painting (and still earlier in literature) taught us otherwise? Is not one major lesson of nineteenth-century American landscape painting—created in the immediate aftermath of the Romantic movement in England and Europe—that the sublime has every right to be regarded as belonging to nature, and thus to the paintings that purport to represent it? Has not Kant here gone too far, not merely relocating the sublime within ourselves but displacing it to such an extent that, were Kant to be right, it could no longer play any effective role in the representation of nature in landscape paintings?

Confronted with Kant's radical internalization of the sublime, we may well be tempted to return to Burke's merely "empirical"[85] imputation of sublimity to external nature as the literal source of terror and as the rightful origin of the astonishment we feel before its awesome spectacles. Instead of seeking sublimity within the aesthetic subject, should we not, after all, attempt to find it without, in that raw wildness which Kant ascribes to the natural world? Does not such unmitigatedly "rude nature," especially in its prototypical form as wilderness, do more than merely "excite" the feeling of sublimity in us, that is, merely precipitate the process of imagination and ideation? Does it not house sublimity itself as an intrinsic quality that is capable of representation in painting and in other media such as photography?

Instead of seeking the basis for the representation of sublimity in the natural world as its apparently obvious source—obvious to Burke and perhaps to all who first ponder the subject—Kant leads us to an impasse wherein the sublime is seen as a merely subjective matter, depending on us alone in all our emotional vulnerability and cognitive frailty. It ensues that any putative representation of the sublime that purports to stand in/for its presence in nature, as being its painted proxy, is rendered suspect.

Nevertheless, it is only in losing ourselves in this labyrinthine maze of subjectivity that we shall find a way out of this impasse. For it would be a mistake to return to a model of nature as obdurately and simply *there*—there to manifest, indeed to possess, all the supposedly objective characteristics and qualities (including both beauty *and* sublimity) that we as perceivers or artists need only register and document. That way lies the seduction of topography, of landskip as the accurate depiction of an already given and predetermined world. It is the way of "the prejudice of the world," as Merleau-Ponty calls it: the belief that the perceived world is a determinate whole that it is our task to replicate in the form of an isomorphically precise rendering.[86] The effect of this prejudice is to presume that all meaning and value lie before us transparently displayed in our surroundings (of which the world of nature is a conspicuous part) and that we, as the faithful observers of such instantly accessible meaning and value, make no constructive contribution to its constitution—thus that its representation in painting is merely a matter of reinstating the identical in a literal re-presentation. As if the painter and the viewer do not contribute to the experience of the sublime from within the realm of their own subjectivity—a subjectivity that is not only psychical but social and cultural in its formation![87]

Any return to hard-edged nature by means of a doctrine of direct or naive realism cannot be an effective basis of genuine landscape art. If Kant moves too far in the opposite direction—so far that he can only claim that the sublime, having no proper place in nature, incites us to "abandon sensibility"[88]—he is surely right to insist that in the representation of nature "one must have stored one's mind in advance with a rich stock of ideas."[89] We do not have to concede that sublimity is exclusively "discoverable in the mind"[90] in order to affirm, rightly, that such ideas (which include socially and culturally informed ideas) can enrich more than the mind itself. In particular, they can enrich the representation of nature by affecting the course and character of that representation. Hence they can enrich our conception of nature itself, given that "representation" in Asher B. Durand's strong sense of *representation* means to be a worthy stand-in, a fit surrogate, the equal of nature, as it were, in the realm of the painted image.

Moreover, nature itself, far from being indifferent to our representational activities, is deeply influenced by the very "mental movement" that Kant attributes to the sublime.[91] Such movement takes place in the form not just of ideas, but also of phantasms that cannot be reduced to merely reproductive icons: a phantasm, as Aristotle first argued explicitly, has a perceptible form *common to* sensuous appearances and to the mind that apprehends them and is not based on likeness in the manner of strictly iconic images—a likeness that is established only across differences of form rather than within the bounds of a common form.[92]

The movement here at issue is also more than "a final employment by the imagination of its own representation." Rather than aiming at a determinate human end *(Zweck),* it is a movement of the mind *into* nature; by this movement of mind, nature is endowed with more than it can present on its own pristine perceptual terms. As the natural world exceeds what reason and imagination construct independently of it, so the mind exceeds what this world conveys in pure sensation: each eclipses the other, albeit in different respects.[93]

Among the mind's "rich stock of ideas" is the idea of the sublime, an idea that is not only *imputed* to nature (i.e., superimposed on it) but actively *bestowed* on it. Such bestowing is only possible if it is the case that, as Kant says expressly in a passage I have already cited, the natural world "*lends itself* to the presentation of a sublimity discoverable in the mind" (my italics). Yet such sublimity, grafted onto a world lending itself to its presence, cannot be discoverable in the mind *alone.* Like the image or phantasm that conveys it, it must exist somewhere between mind and nature—somewhere and somehow in their mutual interaction, their intense interplay, their intimate interplace. Viewed from this middle position, both the "seeming omnipotence of nature" and our own supposed "pre-eminence above nature" are seen to be misconstruals of a deeper accord wherein the sublime is rooted. "True sublimity" exists neither in the mind taken by itself alone (as Kant holds) nor in the empirical world at large (as Burke presumes). It exists in their coeval commixture—a commixture that is located and manifested in the places of landscape, in whose circumambience mind and nature meet.

The challenge of landscape painting lies in the representation of such circumambience, a representation that at once discerns and honors the sublime even as it constructs and deconstructs it. Landscape painting is painting that has moved beyond the delimited and utilitarian purposes of topographical rendition (which seeks to depict site rather than to convey place) in order to concretize the "purposiveness without purpose," the topopoetry, that is at stake in all artistic representation.[94] Such painting has moved from the promotion of landskip to the production of landscape: landscape construed in its own right and comprehended in its own place.

CHAPTER 4

Pursuing the Natural Sublime

Thomas Cole's *The Oxbow*

Anything which elevates the mind is sublime, and elevation of mind is produced by the contemplation of greatness of any kind; but chiefly, of course, by the greatness of the noblest things. Sublimity is, therefore, only another word for the effect of greatness upon the feelings—greatness, whether of matter, space, power, virtue, or beauty: and there is perhaps no desirable quality of a work of art, which, in its perfection, is not, in some way or degree, sublime.

—John Ruskin, *Modern Painters*, book 1

A NEW SUBLIMITY

Thomas Cole, though trained as an engraver in England, became a landscape painter precipitously, soon after his arrival in the United States in 1818. Unlike Smibert or Lane, Durand or Kensett—each of whom was an earnest topographer, a painter of exact detail, in early (and sometimes continuing) phases of their careers—Cole did not pass through the rigors of a purely pictorial phase. His early landscape paintings of the Catskill mountain region are remarkable for their freshness of conception and vigor of execution. They precede Cole's stay in England and Italy in 1829–32 and thus do not show any influence of earlier European painters such as Nicolas Poussin or Claude Lorraine, whom Cole came to admire greatly. Exhibited in a New York shopwindow in 1825, they caught the eye of John Trumbull, the most celebrated painter of his day. As Trumbull realized right away (he is said to have remarked that "this youth has done what I have all my life attempted in vain"), the paintings he saw in the window are fashioned in an original idiom that has no real predecessor and that transcribes the American landscape in an especially compelling way.

Cole is important not just because of his eminence as the founder of the Hudson River School but also because his work offers a most instructive instance of what I shall call the "natural sublime" in landscape painting. At stake here is a new mode of sublimity that calls into question any claim that the apocalyptic and contemplative exhaust the field of the sublime. However convenient these latter two appellations are as descriptions of nineteenth-century American landscape painting—and of certain precedents in British art as well—they tend to dichotomize the phenomenon of the sublime into presumptive opposites. Despite the emended formulation I have suggested (whereby they are two ways of representing the sublime and not two disparate kinds of sublimity), it remains the case that any analysis of landscape painting that has only these two alternative modes of sublimity at its disposal will still find itself strung out between them, forced to choose on occasions when choice is difficult or even impossible, or else driven to compromise (a compromise, however, that can be quite creative, e.g., as in Church's scintillating employment of both modes in his *Twilight in the Wilderness*).

Moreover, the place of the beautiful in relation to the sublime is unduly complicated by the binary pair apocalyptic/contemplative. What, for example, are we to make of Kant's claim that "the beautiful presupposes that the mind is in restful contemplation"?[1] It seems evident that the idea of the contemplative sublime that characterizes Luminist painting—an art form that undeniably attempts to induce "restful contemplation" in the mind of the viewer—is also a candidate for the *beautiful,* even though virtually every aesthetician who has been concerned with these matters has insisted on a basic, irrecusable difference between the sublime and the beautiful. A consideration of Cole's work will help to clarify such questions, as well as deepen our understanding of the sublime itself.[2]

The natural sublime is the sublime as it inheres in the *land* of landscape, that is to say, the primary region of the earth in which sublime phenomena are experienced (and represented) as appearing.[3] When the land as the place of sublimity is also wilderness, the natural sublime takes on elements of what Kant called the "rugged" and a recent author "sublime rawness."[4] But it is crucial to notice that the painting of wild land in no way precludes human, and more specifically moral and religious, values from appearing in such painting. Cole himself, a master of depicting what he called the "Savage State" (the title of the first of his celebrated series *The Course of Empire*), declared that the wilderness is "a fitting place to speak of God."[5] Indeed, as Barbara Novak remarks, "by the time Emerson wrote 'Nature' in 1836, the terms 'God' and 'nature' were often the same thing, and could be used interchangeably."[6] Thus paintings of nature from this period were regarded as partaking of a religious character. "The paintings of Cole," said William Cullen Bryant in his funeral oration on the occasion of Cole's early death in 1848, are "acts of religion."[7] Not surprisingly, Cole's aesthetic pantheism also had a decidedly moral cast. A mid-nineteenth-century art critic noted with reference to Cole's influence that "numerous modern artists are distinguished by a feeling for nature which has made landscape, instead of mere imitation, a vehicle of great moral impressions."[8] This observation would not have surprised John Ruskin (who links the sublime to "the noblest things") or Immanuel Kant, who had already anticipated the deep link between nature and morality in his claim that beauty (including natural beauty) is ultimately a "symbol of morality."[9]

At the same time, the natural sublime of Cole's wilderness paintings, suffused as they are with moral and religious dimensions, is quite compatible with an allegorical or narrative structure of presentation. Even if it is true, as Cole wrote, that "all nature is new to art, no Tivolis, Ternis, Mont Blancs, Plinlimmons, hackneyed and worn by the daily pencils of hundreds; but primeval forests, virgin lakes and waterfalls,"[10] such untarnished natural newness may still call for a narrative format, as Cole demonstrated most notably in thematically linked series of paintings such as *The Course of Empire* (1836) and *The Voyages of Life* (1840), and also in explicitly religious paintings such as *Expulsion from the Garden of Eden* (1827–28), which was realized at the same time as the early Catskill paintings that so impressed Trumbull. The title of another painting of this early period expresses Cole's untroubled admixture of narrativity with spirituality: *Landscape, Composition, St. John in the Wilderness* (1827).

But more than a mere merging of the narrative and the religious is at play in a

Figure 4.1. Thomas Cole, *Landscape, Composition, St. John in the Wilderness,* 1827. Oil on canvas, 36 × 28¹⁵⁄₁₆ inches. Wadsworth Atheneum, Hartford. Bequest of Daniel Wadsworth. Reprinted with permission.

painting such as this: the drama of Cole's self-discovery as a painter of the natural sublime. This drama represents a movement away from the temptations of topography—the spatial analogue of narrative and the predominant concern of previous generations in American painting—toward a conception of landscape painting that is at once naturalistic and psychologically astute. The compelling character of the resulting psychological naturalism is doubtless what drew so much admiring attention to Cole's paintings in the first place. As Bryan Jay Wolf has written:

> Cole's sublime landscapes, in fact, are profoundly non-realistic [that is, for all of their overt naturalism]. They draw their energy from the drama of the psyche in the struggle of self-definition, and they reach into the uniqueness of an American topography only as that topography reinforces their own aesthetic or psychological needs . . . The triumph of the sublime depends ultimately on its ability to appropriate the energy of antecedent art forms for its own ends . . . Through the Romantic sublime, Cole invents his own story, filling the silence of nonnarrative vistas with the clamor of self-discovery.[11]

Indeed, one of the identifying aspects of what Wolf here calls the "Romantic sublime" (which I take to be equivalent to what I am calling the natural sublime) is its irreducibly *personal* character, its capacity to express not only Nature as an impersonal force that is sublime in its terrifying transcendence of the human but the natural world as a concrete scene of the artist's struggle to clarify his own consciousness at the same time that he forges his own style. In this light, it is not surprising that Kant's conception of the dynamically sublime, by emphasizing the interplay of imagination and reason, foreshadows (and in fact directly influences) specifically Romantic versions of the sublime. Cole presumes in effect that Kant's battle to transcendentalize the sublime—to make it into an interior condition of artistic creation—has been won, thus giving the painter freedom to psychologize it as well. Beyond the formal demands of composition and the blandishments of moral and religious allegory, there is room for a drama of oppression (e.g., of being crushed beneath a juggernaut of natural forces) and of release from this oppression (through the creation of a psychological sublimity that is Romantic in its core). In Wolf's words, "behind the terror and exhilaration of Cole's paintings the viewer encounters a moment of psychological reversal when an oppressive burden is left and the soul receives an influx of power, which it experiences in an ecstasy of liberation and release."[12] In other words, beyond the Burkean conception of an empirical (i.e., physicalized and externalized) sublime, and beyond the topographic tradition that ignores the role of the artist's psyche in its desire to depict a landscape in isomorphic verisimilitude, Cole's work embodies a self-assertion of the artist in which the naturalistic and the psychological join forces in creating a new form of sublimity.

In this equiprimordially physical and psychical domain, the natural sublime arises and flourishes. It is the same domain of which I spoke at the end of the preceding chapter as a landscape place that exists between mind and nature. In Cole's capable hands, this is a place in which mind and nature, spirit and matter, are one: his natural sublime is both fully naturalized and wholly psychologized. But what kind of a place is this? How is it represented? What does it tell us about the role of place in the painterly representation of landscape?

The only way to answer these difficult questions is to attend to the spatiality, and more particularly the placiality, of Cole's actual compositions. Despite the undeniable importance of the allegorical and narrative aspects of his paintings, we must now take a more concerted look at their compositional structure. As Wolf remarks, "narrative functions for Cole as a form of thematic self-accounting, explicating in temporal fashion *an experience otherwise accessible only in the spatial structure of the work.*"[13] The issue, then, is how Cole, as a paradigmatic painter of the natural sublime, captures the place of landscape in the representational space of landscape painting.

THE PSYCHOPHYSICS OF PLACE IN *THE OXBOW*

We could not do better than to consider Cole's magnum opus, *View from Mount Holyoke, Northampton, Massachusetts, after a Thunderstorm (The Oxbow)* (see Plate 9), usually referred to simply as *The Oxbow*. Painted in 1836, the same year as he completed *The Course of Empire*—indeed, partly as a diversion from the labor of finishing that monumental series—it is perhaps Cole's most celebrated single painting and is regarded as "one of the earliest statements of an identifiably American landscape painting tradition."[14] In fact, it revives themes that Cole had developed in his inaugural Catskill paintings of 1825–27, carrying them forward in new ways that in part reflect his experiences in Europe in the intervening years, and in still larger part his increasingly generous sense of the full scope of landscape painting itself.

The major topoi of *The Oxbow* are a rockbound foreground (to the left of center) on which graphically presented stones, trees, and other vegetation are depicted in a tormented and stormbound state; a middle ground that includes a concealed abyss and a green hill in the middle distance; and a scene in the far distance (above and to the right of center) with a meandering oxbow and tranquil, pastoral land viewed at a halcyon and sunlit moment. Taken as narrative—as we are certainly encouraged to do by the artist[15]—such a schema tells an implicit story that moves from a situation of distress and danger (evident in the storm bursting overhead, the effects of previous such storms being manifest in the blasted and crippled trees in the foreground), through a circumstance of uncertain and unknown resolution (i.e., the abyssal mid-distance), and into a time of safety, redemption, and peace (the distant scene of sunny tranquillity: the promised land). Such a story has a beginning, middle, and end; and it moves from left to right in the painting much as we read printed lines on a page. Accentuating the implicit narrativity is the piquant placement of the tiny solitary figure of Cole himself painting at the edge of the abyss—reminiscent of the single Indian in Cole's *The Clove, Catskills* (1827) and in his *Falls of Kaaterskill* (1826). Such a figure is too diminutive to humanize the massive scene in which it is situated; it serves, rather, to emphasize the scale of the scene by its very subordination to the larger landscape. Not contributing centrally to the narrative itself, it is poised precariously on the cusp between the temporality of the story and the spatiality of the surrounding landscape.

But we need not construe such a painting as a story.[16] The narrative element, even if unquestionably suggested, is subordinated to a presentation of the sublime that is accomplished by mainly spatial means: "The essential drama of the sublime is achieved in Cole through the painting's composition: the arrangement of space, the massing of forms, the play of light, and the pattern of movement directing the viewer through the landscape—all create an experience of sublime epiphany independent of the narrative subject that Cole tends on occasion to introduce."[17] In contrast with a story, whose narrative structure requires a diachronic, event-by-event time of enactment, an epiphany occurs in a single, synchronic moment. It possesses an all-at-once character that a landscape painting, even a comparatively complex one, is well suited to convey. Not only does such a painting, as an intact perceptual object, present itself as a spatial whole that can be taken in by one single glance, but, as a "com-position," it co-posits the items it represents as simultaneously present to each other in a literal compresence. Instead of a reassuring development from the fearful to the peaceful, in which each successive moment puts its predecessor behind it as what has *already* occurred, we are presented with a much more challenging situation in which each item coexists in an uneasy equilibrium with every other. Thus, rather than reading *The Oxbow* as a story of salvation from a storm that has elapsed by the time we reach a place of serene security, as spectators of this complex composition we are invited to take in the moment of storm and danger together with the moment of tranquillity—and both with the moment of the unknown abyss. All three moments compenetrate in a single flash of sublime epiphany. Viewed as aspects of this epiphany, an epiphany that is coextensive with the entirety of the presented scene, they are "moments" not just in a temporal but also in a spatial sense: contiguous parts of the complete landscape they conjointly represent.

The spatiality of these parts is not, of course, the same throughout. Their very difference as three sorts of space contributes to their complementarity with one another.

1. The foreground space is *agonized*: it is a violent space of contorted directions and twisted vectors whose outlines are indicated by the abrupt angles of the tree trunks and limbs that gesture with a distinctly pathetic power. These distressed members gesture both downward toward the rugged earth on which they stand (as if to say: here, at least, is a patch of solid soil) and outward toward the middle and far distance (as if to say: there, over there, is both the unknown and the redemptive). This turbulent space, heightened by the turmoil of the storm overhead, is not only structured but literally overstructured, suggesting the special frenzy that being at risk in space can bring with it. The risk is exactly what being exposed on the top of a mountain entails; and it is significant that in *The Oxbow* the artist has placed himself on the margin of the exposed space in which the danger of the oncoming storm is located: he is ensconced between two rocky protuberances that mark the outer limit of the summit even as they protect him from its tribulations.

2. The near-space of turbulence stands in stark contrast with the middle-range *abyssal* space of emptiness. Where the foreground is agonizingly overwrought, the middle ground is largely unstructured. In fact, part of it is altogether invisible: that is, the descending moutainside as it leads downward to the Connecticut River. We sense that the drop is virtually perpendicular and thus dangerous in its own right, and this reading is reinforced by the presence of the intensely green wooded hill (to the left of center) whose exposed side leads rapidly downward toward the river. The compressed and denuded character of the initiatory space

up front thus gives way to a verdant but risky space of transition. At the same time, the contorted lines of the foreground dissolve into an imputed plunge downward: the danger of exposure is replaced by the danger of falling.

3. The verticality of the fall stands in stark contrast with the resolute horizontality of the river and plain below. These latter bring with them the decidedly *serene* space of the flatlands. Danger has been eliminated from these areas, which are carefully cultivated and filled with signs of prosperous farming. Their placidity is underscored by the presence of a slow-moving ferry on the river at the far right and, more dramatically, by the oxbow curve of the river itself, a relaxed shape that underlines its leisurely and slow movement. Also contributing to the serenity of the scene is the presence of mountains in the far distance: their lavender color and the pink sky just above them sublimate the becalmed space of the river into a luminous haze of repose.

The three kinds of spatial representation just distinguished (agonized, abyssal, and serene) serve to situate three sorts of place. They *define* such places—give them a local habitation, if not a full-fledged name. Let us call the first of these the *place of exposure.* This is the open place of the foreground that, being depicted as close *to us,* draws us into its menace. In certain of the earlier Catskill paintings the exposed place is even more threatened than in *The Oxbow,* which cushions the menace by folding it into the protective embrace of the undamaged wooded hill in the middle distance and both in turn into the calming presence of the far landscape. On closer inspection, however, it becomes apparent that a distinct spatial break occurs between the foreground and the nearby hill, thereby leaving the foreground to be its own distressingly unsheltered spot. As such, it is not only uninhabitable, but actively hostile to habitation, as is symbolized by the violently twisted tree stumps. Even the intrepid artist has to take up quarters on its periphery, leaving his knapsack and an umbrella on one of the outer rocks. In this fore-region of the painting we witness nature on its own terms, submitting to its own storm and absorbing it without human mediation. This is, as Hegel might put it, "nature outside itself," that is, sheer spatial exteriority; it is all surface, and juts out toward us from the picture plane. The agony of space thus opens onto the exposure of place.

This place of overt menace is dramatically juxtaposed with the *place of the abyss.* It is tempting to regard the latter as a nonplace. On many notions of place—such as that of Aristotle, for whom place depends on a determinate content—an abyss would indeed be a nonplace, a sheer void of nonbeing.[18] But we have already seen that the abyss in Cole's painting has certain definite properties: it exists on the *other side* of the foreground and it falls into a decidedly *vertical axis.* It also serves to connect various regions within the painting's overall pictorial space. As inherently dangerous, it is a place *into which to fall.* For all of its literal invisibility, it possesses a quite positive presence. If the foreground, for example, solicits an act of standing up on it defiantly (e.g., to get a better view or to mock the weather), the abyss beyond it calls us just as forcefully to fall head over heels into its injurious depths. The two places support two types of detrimentality: exposure and fall. Both are obstacles to the enjoyment of the nondetrimental life of unthreatened security that is exhibited in the world of the river and plains below.

For this world the appropriate designation is the *place of prosperity.* By "prosperity" I do not mean economic wealth as such but "the good life" that brings with it features of comfort and warmth. In contrast with the first two kinds of place, movement in this latter place is unobstructed and safe, as is signified by the ferry on the river and the open roads on the flat plain. Fear of exposure and fear of falling here give way to confidence in productive action. This action is essentially social, requiring the cooperation of many persons, in contrast with the lonely activity of the artist on the promontory. The place way down below is a pastoral place, being tied to the seasons and to cycles of plant growth on the land, whose well-cultivated surface betrays the life of germination just underneath it in the subsoil. But the vista is not altogether pastoral, because Cole (who is known to have been intensely concerned with the inroads of industrialization) also depicts tiny funnels of smoke rising from what are probably homes but what could be fledgling factories. Overall, we are presented with a downward depth in the river as well as a depth outward in the open movement toward the far-off mountains.[19] Not to mention an expansion skywards—thanks to the upward tilt of the distant plains, giving to this third kind of place a distinctive aura of promise that is the positive counterpart to the closed and fear-inducing character of the first two kinds of place. If there is any menace here, it is the menace of a future that has not yet arrived, that of satanic mills and overpopulated towns that will come to dominate the still unmarred landscape—a menace perhaps foreshadowed in the storm that dominates the left third of the painting.

The three basic kinds of place represented in *The Oxbow* are complementary with regard to what each actively solicits. In the place of exposure, a *saving action* is called for: either that of sheltering oneself from the storm or redeeming oneself as an artist. In the place of abyss, *no action* is recommended, because *any* action (save that of backing away from the edge) could lead to a disastrous fall. In the place of prosperity, *cultivating actions* and social *interactions* are called for in order to assure the continuation of the settled life that is adumbrated there—and to protect against the encroachment of industrialization. In each case, the link between place and action is so intimate that we could say that each place is a place for a certain sort of action (or nonaction). This is so despite the fact that the painting as a compositional whole exhibits spatial stasis in the stillness of unmoving and interlocked parts; its natural sublimity, condensed in an epiphanic but comprehensive moment, is at once kinetic and static. Reinforcing this double character is the fact that the prospect is both panoramic—even "panoptic"—and telescoped in its attention to minute detail.[20]

The Oxbow thus makes it clear that represented places introduce possibilities of movement into what would otherwise be static spatial totalities. In mobilizing these totalities, they condense and convert them into scenes of action. It is within these active scenes that the distinctive drama of the natural sublime unfolds. This drama need not be narrative in format to be forcefully presented; it can be built up just as well from the nonnarrative moments of pictorial space and, in particular, from the places into which this space resolves itself in a given painting.

It has been a continuing temptation since Descartes first identified nature with space (and space with congealed and obdurate *res extensa*) to assume that anything having to do with nature must exclude mind construed as *res cogitans,* that is, mental substance.

The drama of the natural sublime in its late-Romantic and specifically American forms, however, calls for the inclusion of mentation—if not for mental substance as such. The agency of mind in the genesis of the sublime is something for which we have seen Kant arguing vigorously, and it is reinforced by Ruskin's claim in the epigraph to this chapter. It is also implied in my previous ruminations as to the need to sublimate place as externally perceived if it is to play a meaningful role in landscape painting. A significant aspect of what I have been calling "re-implacement" in the representation of landscape has to do with the sublating of exterior location, via an interior mental space, into the kind of place in which matter and mind freely meet and exchange influences.

Once a painting has been created—once it has reexteriorized what began as "natural scenery" that is perceived, remembered, and reconsidered by the painter: once it has become a genuine psychotopia—it presents to its viewer a series of possible movements by laying down virtual tracks through the represented particulars of the painting. In the case of *The Oxbow* these tracks would include the downward motion of the storm on the upper left, the irregular path through the thicket of branches and stumps in the foreground, the threatened fall into the abyss, the meander of the river along its oxbow, and the subtle movement outward toward the distant mountains. In all these cases, the movement is made by our viewing eye, which guides our phantom body along the routes it discerns: the more we adhere to the painting, the more we feel the subtle kinesthesias of this virtual body as it follows out the implied or represented paths of the visual scene.

Beyond these quasi-physical motions, motions of the mind are also elicited by a painting such as *The Oxbow.* These are actions not of Cartesian thinking substance but of the actively imagining psyche, and they are as intrinsic to the natural sublime as are the (virtually) corporeal motions just described. To demonstrate this, we need not revert to reductive psychologistic interpretations.[21] We need only notice that a full understanding of a given landscape painting must include what Kant calls "mental movements" in *The Critique of Judgment.* But such movements are not limited to explicit acts of cognition *(Erkenntnisse)* of the sort Kant had in mind when he described the activity of imagination and reason in bringing forth the mathematical or dynamical sublime. They also include actions of appreciation and evaluation, recognition and understanding. All such actions, far from being enclosed within *res cogitans,* bear directly on what Wordsworth called the "presences of nature."[22] It is in and on, indeed *through,* such presences that the psychical actions crucial for the experience of sublimity are most fully realized—rather than being confined to the interiority and isolation of cogitation as conceived by Descartes and Kant.

In landscape paintings of the natural sublime, the actions of the psyche become conterminous with the places of the landscape in which the presences of nature are situated. Movement through these places—the drama of the sublime—is as much psychical as it is physical: as much a matter of imagination and memory, appreciation and evaluation, recognition and understanding, as it is of the phantom body whose virtual actions carry out the topographic surveys of the actual eye.

STAGING *THE OXBOW*

A striking illustration of this profoundly psychophysical status of the natural sublime is found in the preparatory work for *The Oxbow.* Cole made two pencil studies and one oil sketch of the scene he beheld from the top of Mount Holyoke. Each of these studies represents a stage in the evolution of the completed painting. Each also represents a step away from landskip qua topography to landscape qua sublime. The first pencil sketch (c. 1829) is in fact a careful tracing of a plate in Basil Hall's *Forty Etchings, From Sketches Made with the Camera Lucida, in North America in 1827 and 1828.*[23] Because Hall's etching is already a conscientious topographic view of the Connecticut River as seen from Mount Holyoke—a view aided by the exact image cast by the camera lucida—Cole's slavish copy of it is topographic twice over: an icon of an icon. The next step is a pencil sketch, which Cole made on Mount Holyoke while on a trip to Boston in 1833 (Figure 4.2).

In this sketch Cole positions himself in approximately the same spot as had Hall, but now he depicts a hill in the middle distance (which Hall had altogether neglected). Already evident here are the three spatial regions that will structure the painting of several years later. There is a frozen quality to the sketch, which embodies a determined effort to capture the exact features of the total vista—a vista that is more of a panorama than a landscape. (The panoramic quality of the sketch is enhanced by Cole's decision to extend the image across two pages of his sketchbook, thereby doubling the angle of vision thanks to a 7:22 ratio for both pages taken together.)[24] Still, Cole saw promise in this second view—he wrote to his patron Luman Reed that it was "about the finest scene I have in my sketchbook"[25]—and he turned back to it for inspiration when he was seeking a new subject matter as relief from his arduous efforts on *The Course of Empire.* Just after turning back to the second sketch, he made a free-form oil painting that dramatizes the scene considerably, breathing life into it much in the manner of a Constable oil study. Clouds and mountains are added in the distance, providing depth; and an ominous shape, presumably a tree, is now stationed in the foreground at the far left.

The painter himself is represented under his umbrella on a rock to the far right— not yet at work on his canvas but staring the spectator in the eye (in the completed work, the painter glances coyly over his shoulder at the viewer of the painting). Here the drama is very much under way: already present is the combination of fear and benediction that will characterize the full-scale painting. Even if less encompassing, the view is still quite sweeping.

Figure 4.2. Thomas Cole, *Panorama of the Oxbow on the Connecticut River as Seen from Mount Holyoke,* c. 1833. Pencil on paper, 8⅞ × 13¾ inches. Founders Society Purchase, William H. Murphy Fund. Photograph copyright 1990, The Detroit Institute of Arts.

In the final painting (Plate 9), Cole lays out a theater of the sublime by effecting several further changes. First, in contrast with the pencil drawing, he makes the view as a whole more depthful; in particular, the land below is at once vertiginous in its downward thrust and auspicious in its outward and upward directionality. Second, the green hill in the middle distance is given a more pronounced place by the elimination of another hill behind it, thereby increasing the abyssal difference between the green hill and the foreground promontory, now rendered in much more telling detail such as that which depicts the blasted trees. Third, and importantly modifying both the drawing and the oil sketch, the storm is made more prominent and acts, in concert with the green hill and the foreground, to divide the canvas into two major visible regions: a leftward region that represents imminent danger and a rightward region that manifests pastoral quiescence. (I say two "visible" regions, for an invisible abyss lurks between these regions.) Instead of giving us a continuous panorama, Cole has in effect juxtaposed two disparate views, thus increasing the angle of vision but at the cost of forcing together two perspectives in manifest tension with each other.[26]

In this manner, the drama of the natural sublime, which infuses the entire canvas, is given the very particular form of a "drama of oppression and release."[27] The left-hand part of the pictorial space is oppressive because of the oncoming storm and the exposed rocks; the right-hand part represents release thanks to its open fields and generous light. But the movement from one to the other is not merely a visual transition between two differently painted parts of the same canvas. It is at the same time a mental movement from a condition of felt oppression to a state of beneficient, even ecstatic, epiphany. This movement of the psyche, which is not ultimately separable from the purely pictorial dynamics of the painting, enacts (in Wolf's words) "the drama of the psyche in the struggle of self-definition." This drama is the drama of the natural sublime itself.

In presenting this drama, Cole transcends topography even more radically than does Lane. Although both begin with detailed sketches of their subject matter (Cole even starts by tracing another artist's drawing), Lane directly transposes his sketch via grid lines onto the canvas and thereby keeps everything in its original position, whereas Cole feels at liberty to make important changes in all three major regions of his pencil sketch and to add a brewing storm. Lane, we might say, adopts a procedure of *transposition*, in which each object's specific location is retained and carried over from one representational space to another by means of a quasi-cartographic grid. Cole, on the other hand, proceeds by a *transformation* of place itself whereby the precise form of a place—that is, that which would demand strict preservation on a transpositional model—is altered in the course of creation.

Where Lane achieves re-implacement by transposition, Cole accomplishes it by actively *relocating* and *reconfiguring* objects in pictorial space rather than by their point-by-point projection onto that space. Fidelity to position is no longer a primary virtue; indeed, for Cole it gets in the way of the drama he wishes to present, for the preservation of the identity of positions brings with it an assumption of the sameness of the space within which these positions inhere. And it is just this assumption that Cole cannot allow, as we see in his effort to alter not just the particular locations of objects (e.g., his own self-represented position in the foreground) but the overall spatiality, the scenography, of the definitive version of *The Oxbow*. In the end, he changes both the positions of given objects *and* the spatial layout (by reconfiguring the left and the right regions of the pictorial space through the introduction of the storm in the leftward portion) while retaining sameness of basic zoning (the tripartite division into foreground, middle distance, and far distance). Instead of keeping the depicted scene the same or even simplifying it—as happens in the preliminary drawings, and as is Lane's characteristic tendency throughout—Cole complicates it by his eagerness to relocate and to regroup, a complication that first appears in the oil sketch and is then fully realized in the definitive painting.

This is not to claim that Cole's differential and highly dynamic handling of the representation of place is somehow superior to Lane's, any more than it is, say, to that of Bierstadt and Church, who carry relocation and reconfiguration to still further extremes. It is only to suggest that the ways of accomplishing re-implacement in landscape painting are manifold, ranging from the most rigorous transposition to the most radical transformation—ways that become increasingly difficult to map or predict in the wake of Monet and Cézanne (who pick up where Bierstadt and Church leave off) or of Kandinsky and de Kooning (whose transformation of features of the natural landscape reaches the point of outright unrecognizability). And, by the same token, it is to suggest that the forms of the sublime in modern Western landscape painting are legion: not only apocalyptic and contemplative, romantic-Gothick and natural, but also those still unnamed modes of sublimity to be engendered by Postimpressionist and abstract expressionist painters such as Van Gogh and Kandinsky and Ernst, Gorky and de Kooning and Rothko.

"THE SUBLIME MELTING INTO THE BEAUTIFUL"

Indissociable from landscape painting in the West is "the beautiful," a category that began to be contrasted with the sublime in the early eighteenth century—first of all in Joseph Addison's celebrated text of 1712, "Essay on the Pleasures of the Imagination," and more massively and notably a half century later in Edmund Burke's *Enquiry*.[28] The category of the beautiful was called for by the work of a painter such as Claude Lorraine, who exerted a profound influence on almost every subsequent landscape painter of importance in Europe and the United States, including Constable, Turner, Bierstadt, Church, and Cole. A prototypical Claudian landscape has no trace of the melodramatic tenor that we have seen to be essential to the natural sublime. Instead, it presents a calm and well-ordered scene of Arcadian serenity, set out according to a schema of trees acting as *repoussoirs* (i.e., graceful leads into the pictorial space), shepherds and other idealized figures in the foreground, a placid lake in the middle distance, all of this encircled by a horizon of low-lying mountains enclosed in a haze of subdued pastel colors.

This compositional schema is to be contrasted with that of Cole, who deidealizes each of these features. Cole's trees, as we have seen, are twisted and desolate, his foreground is a place of storm and risk, his middle distance collapses into a spatial abyss, and he refuses the hazy fade-out in the far distance that became a hallmark of a landscape painted by Lorraine or those inspired by him.[29] Nevertheless, Cole's work retains one crucial echo of Lorraine that serves as a reminder of his predecessor's embrace of the beautiful.[30] This is the way in which the most distant part of a landscape as rendered by both artists often portrays a place of redemption and promise in contrast with the mundanity—or menace—of other places in the same painting. This far-place is a place of beauty, of pleasing formal perfection. Such an invocation of the beautiful is present not only in *The Oxbow* but already in Cole's Catskill paintings of the mid-1820s on which the spatial schema of *The Oxbow* is modeled; indeed, in the earlier works, the beautiful far-place beckons tantalizingly as a salvific region that we glimpse at the end of a tortuous pathway through an intervening abyss.[31]

At the beginning of this chapter I remarked that the place of beauty in relation to the sublime is complicated in certain forms of landscape painting. Kant's notion of the beautiful as involving "restful contemplation," I suggested, suggests the contemplative sublime as this latter figures in the paintings of Lane and other Luminists. These paintings are no longer Claudian in format (in fact, they are much more influenced by seventeenth-century Dutch landscape) and yet their clarity and repose are akin to Lorraine's bucolic tranquillity in terms of feeling and mood. Are they beautiful or are they sublime? It appears that, being on the borderline between the conventional senses of these two terms, they challenge the very distinction between them. Another challenge arises from the fact that certain painters expressly aim to combine the sublime *with* the beautiful, among them Turner, Church, Bierstadt, and Cole himself. Yet Edmund Burke would be chagrined by such a combination. Although he had allowed for occasional dalliances between the beautiful and the sublime in works of art, he argued eloquently for their fundamental incompatibility: "the ideas of the sublime and the beautiful stand on foundations so different, that it is hard, I had almost said impossible, to think of reconciling them in the same subject, without considerably lessening the effect of the one or the other upon the passions."[32]

Just as we have been led to concede that there may be more than one mode of sublimity in a given painting—notably, in Church's *Twilight in the Wilderness*, which successfully combines the apocalyptic and contemplative wonder of the sublime—so I believe that, against Burke's view, we should allow for the real possibility of a creative conjunction of the sublime (in whatever mode) and the beautiful (in whatever form) in landscape painting. And we can do so by retaining Kant's conception of beauty in terms of formal perfection and restfulness, adding to this Burke's own much more particular list of seven traits:

> On the whole, the qualities of beauty, [insofar] as they are merely sensible qualities, are the following. First, to be comparatively small [i.e., in contrast with the sublime's "greatness of dimension," its "absolute magnitude"]. Secondly, to be smooth. Thirdly, to have a variety in the direction of the parts [in contrast with monolithic vastness]; but fourthly, to have those parts not angular, but melted as it were into each other. Fifthly, to be of a delicate frame, without any remarkable appearance of strength [so that such nondelicate things as mountains will be played down, or at least put into the distance]. Sixthly, to have its colors clear and bright; but not very strong and glaring [as is paradigmatically the case in Lorraine's roseate landscapes]. Seventhly, if it should have any glaring color, to have it diversified with others. These are, I believe, the properties on which beauty depends.[33]

What remains abstractly stated in Kant is here given sensuous specificity by Burke, in keeping with the latter's view that beauty is "some quality in bodies, acting mechanically upon the human mind by the intervention of the senses."[34]

Despite Kant's effort to locate the sublime securely in the human subject and the beautiful in nature ("for the beautiful in nature we must seek a ground external to ourselves"),[35] painters such as Church and Bierstadt and Cole—all three anticipated by Turner—believe *both* to be located in nature. In his "Essay on American Scenery," delivered the year before he painted *The Oxbow,* Cole is explicit about the compossibility, indeed the actual colocation, of the beautiful and the sublime within nature. Not only does he speak admiringly of "the atmosphere that softens the most rugged forms of the landscape,"[36] but he singles out Niagara Falls as an instance of the convincing combination of beauty and sublimity:

> And Niagara! that wonder of the world!—where the sublime and beautiful are bound together in an indissoluble chain. In gazing on it we feel as though a great [void] had been filled in our minds—our conceptions expand—we conceive immensity; in its course, everlasting duration; in its impetuosity, uncontrollable power. These are the elements of its sublimity. Its beauty is garlanded around in the varied hues of the water, in the spring that ascends the sky, and in that unrivalled bow which forms a complete cincture round the unresting floods.[37]

Cole here seizes upon what was regarded at the time as the archetypal manifestation of the natural sublime: every major American landscape painter was impelled to paint Niagara Falls, from Trumbull to Bierstadt and Church (though not, ironically, Cole himself).[38] He discerns in it not only classical features of the sublime such as immensity and power but specific traits of the beautiful in the domains of form and color. Most important, he avers that these two sets of attributes are not merely compatible but enhance each other's presence in the natural phenomenon.

As in the natural phenomenon, so in its representation as landscape. Let us return for a last time to *The Oxbow* in order to recognize in this painting a quintessential blending of the sublime and the beautiful in a single work of art. The blending is subtle and is part of this painting's numinous power. Within its capacious pictorial space, there is a progression from the sublime to the beautiful—but the beautiful in a form that is at one with the sublime itself.

To appreciate this progression, we need to reassess the three major places of this painting:

1. *Place of exposure.* The arena of the foreground, utterly ex-posed, presents to us one basic aspect of the Burkean sublime in an almost naked form. The danger and terror that Burke had built into his conception of the sublime are here graphically depicted in the form of tortured trees and a precarious precipice. The tortuous devastation of the trees in particular conveys the extreme exposure that comes with existing at the mercy of the elements, that is to say, nature in its unbridled "power" (in one of Burke's key terms) or "might" (in Kant's nomenclature).

2. *Place of abyss.* From this abrupt initiation into the sublime through violence—a violence reinforced by the storm breaking overhead—*The Oxbow* leads us to a second

aspect of sublimity: "obscurity," "vacuity," "darkness," and "silence" (in Burke's words). To do justice to this "privative" side of the sublime, Burke is led to cite Virgil's lines: "Give me your mighty secrets to display / From those black realms of darkness to the day." And he speaks approvingly of Milton's description of hell, where "all is dark, uncertain, confused, terrible, and sublime to the last degree."[39] In moving to the far and dark side of the promontory, we move to what Kant himself, normally little given to extreme descriptions, calls "the abyss":

> The mind feels itself set in motion in the representation of the sublime in nature . . . This movement, especially in its inception, may be compared with a vibration, i.e. with a rapidly alternating repulsion and attraction produced by one and the same Object. The point of excess for imagination (towards which it is driven in the apprehension of the intuition) *is like an abyss in which it fears to lose itself.*[40]

Concrete instances of such an abyss are also given by Kant: "deep ravines and torrents raging there, deep-shadowed solitudes that invite to brooding melancholy."[41] Cole's painting of 1836 includes such ravines and solitudes, located just over the edge of the precipice on which the painter is perched. They configure the very scenario of the dynamical sublime that Kant (here very much in agreement with Burke) regards as prototypical of its dark infinity.[42]

3. *Place of prosperity.* A place of this kind is constituted by the distant plains and mountains that lie becalmed beneath an open sky. This region of the painting seems to be a painterly transcription of the words from Cole's lecture of 1835 cited in my earlier discussion of the contemplative sublime in Lane's *The Western Shore with Norman's Woe*: "over all, rocks, woods, and water, brooded the spirit of repose, and the silent energy of nature stirred the soul to its inmost depths."[43] Cole amplifies on this reposeful spirit insofar as it rests on the natural presence of water (in which we perceive "the expression of tranquillity and peace")[44] and especially of mountains, about which he says that in apprehending them one "sees the sublime melting into the beautiful, the savage tempered by the magnificent."[45] Cole's own words explicitly envisage that very merging of the beautiful and the sublime allowed only begrudgingly by Burke and at least tacitly admitted by Kant when he writes that confronting the dynamical sublime is tantamount to "combining the movement of the mind thereby aroused *with its serenity*":[46] mental movement, as we know, is at stake in the generation of the sublime, whereas serenity signifies the "restful contemplation" that Kant links expressly to the beautiful.

The crucial phrase, however, is Cole's own: "the sublime melting into the beautiful"; for this is precisely what *The Oxbow* accomplishes in its sweeping movement from exposed promontory and abyssal depth—their compounded verticality signifying the sublime in its characteristic up/down dimension—into the settled horizontality of cultivated land and distant mountains, into a scene of beauty in short. The melting is at once meteorological (the storm ceases to threaten over the place of prosperity), painterly (the dark tonalities of precipice and abyss give way to the brighter hues and lighter tonalities of the plain and mountains), and spatial (the raw edge of the descending green hill yields to the smoothly sinuous oxbow below).

Indeed, we feel *ourselves* melting down in the warm prospect of the river world and its agrarian life, both situated in the same sunlit scene. At the same time, we feel transported from a setting in which clinging motions of the hands and feet—scrambling to survive in high places—are superseded by an open-eyed affirmation as it extends over the broad vista before it. As spectators of the place of prosperity, we take a decided step toward the visual and away from the tactile, while the artist himself, poised on the inter-place between the two worlds, participates in both sensory modalities as someone who at once looks at the landscape and touches his brushes to the canvas. Beyond both the visual and the tactile is the visionary, in which we engage once we bypass the immediate dangers of precipice and abyss. The vision is of a future beauty brought about by the realization of an archetypal American dream of peace and prosperity: a realization that combines the resources of natural place with the fruits of human labor.

In his celebrated lecture, given before the American Lyceum in New York, Cole spoke of Eden as a still attainable state, or, more exactly, as what we actually possess in American scenery if only we would open our eyes to it: "we are still in Eden; the wall that shuts us out of the garden is our own ignorance and folly."[47] To mention Eden not only imports a patently religious reference but also invokes a poetic dimension of landscape: where *poetic* still resonates with its root sense of *productive*. In *The Oxbow* the representation of life on the plains tells a productive tale, a projected narrative of *longue durée* that reaches out beyond the merely episodic incidents that take place on promontories and precipices. By moving us into this poetic/productive world, Cole is leading us not only from sublimity to beauty, and from hand to eye, but from rugged fact (i.e., the literalized danger of the foreground) to felicitous utopia—to a scene of idyllic prosperity. We are given a highly poeticized view of the promised land, a species of topopoetry. This is in keeping with an observation of Kant's in the "Analytic of the Beautiful":

> Beautiful objects have to be distinguished from beautiful views of objects . . . In the latter case taste appears to fasten, not so much on what the imagination grasps in this field, as on the incentive it receives to indulge in poetic fiction, i.e. in the peculiar fancies with which the mind entertains itself as it is being continually stirred by the variety that strikes the eye.[48]

To move to the topopoetic is to move to a deeply psychical dimension of the painterly field; it is to move to a place from which the world is viewed not only with one's physical eyes but with "the eye of the soul." The claims of Burke and Cole himself notwithstanding, the beautiful, no less than the sublime, is found both in the natural world *and* in us: in our poetic soul, in our own viewing. What comes under the heading of the mental and the transcendental in Kant is psychic and poetic and pictorial in Cole. We can say of the open vista of *The Oxbow* what Margaret Fuller wrote of Washington Allston's highly poetic landscapes: "The soul of the painter is in these landscapes, but not his character. Is not that the highest art? Nature and the soul [are] combined; the former [is] freed from slight crudities or blemishes, the latter from its merely human aspect."[49]

BEAUTY AS THE SUBLIMATION OF THE SUBLIME

A final question arises. Let us grant that Cole has achieved in one major part of his masterwork of 1836 a convincing representation of the beautiful in landscape—a beauty describable in Burkean or Kantian terms and supplemented by allusions to vision, poetry, and the soul. Does this mean that Cole's painting simply offers us a movement from sublimity to beauty, as my own narrative has implied, where "beauty" signifies what is somehow beyond the sublime, distinct from it in some far-place, and (as both Burke and Kant would aver) barely able to coexist with it? It is certainly tempting to hold that *The Oxbow* is a painting in tension with itself—not just formally or spatially but as riven by its own successful representation of the sublime, to which the beautiful is added as a distinctly post-sublime moment. I borrow the term *post-sublime* from Bryan Wolf, who says this concerning Cole's landscape painting in general:

> [In Cole's painting] the beautiful substitutes coherence for power: that is its essential move. It provides a continuous narrative about a discontinuous state, creating a fiction of timeless repose in a world marching inexorably to death and ravishment. As a post-sublime experience, it presents a nature divested of its original awe-fulness, substituting a myth of integration and domesticity for the repressed and demonic energies of the sublime. It domesticates the sublime by reconceiving its struggles in pastoral terms, providing us—unlike the sublime—with a world we can live in.[50]

Is it, then, the case, as Wolf claims, that the beautiful only "arises in Cole as a reaction to the sublime," a reaction framed in a "detraumatized narrative"?[51] Tempting as this interpretation may be, I think it is profoundly mistaken. Cole's genius as a painter is to refuse to submit to any rigid dichotomy of the sublime and the beautiful, just as he also refuses to choose between the apocalyptic and the contemplative sublime. Not only does Cole achieve a harmony between the sublime and the beautiful in *The Oxbow,* he blends them so deeply and so subtly that the beautiful can here be said to be a *subliming of the sublime itself.* Rather than being antithetically *post*-sublime or a mere reaction to an initial (but traumatic) sublimity, the beautiful in this painting is the very sublimation of its own vertiginous sublimity.

Such sublimation—whereby the sublime disappears into the beautiful, by which it is at the same time preserved—finds representation in the painting in two concrete ways. First, the storm at the upper left dissipates gradually as it moves toward the center and right; at the very center, it becomes a misty natural presence that is "sublime" in the original physical sense of the word: a sheer vaporization.[52] Second, the artist's self-representation is the human counterpart of this transitional phase of the storm: not only does the figure of the artist invite us into the drama of the sublime by his glance back at us from his rock-bound enclave, but he is also presented as sublimating in an expressly psychophysical sense, as a painter who converts his visual perception into the "soul" of which Fuller spoke during the high tide of American transcendentalism. The beckoning figure is the pivot of a genuinely artistic *sublimatio*—to use the alchemical term for a process of transmutation that is at once material and psychical[53]—and he stands in for the actual creator (i.e., the historical Thomas Cole, the man who in fact created this work) by acting as his self-projected deputy in the representational world of the work. In this way, Cole carries out (and shows himself carrying out) the very activity of artistic subliming of which *The Oxbow* is the exemplary outcome.[54]

As Thomas Cole's realization of natural sublimity brings under its encompassing aegis both the apocalyptic and the contemplative sublime, merging them in one work of art, so it also effects its own continuation into the beautiful, merging with this as well. In *The Oxbow,* the "unappropriated world"[55] of raw nature is doubly, indeed triply, sublated—that is, negated only to be preserved and carried forward at another level. First, it is taken into the work in an initial implacement by which the three primary places of the painting's pictorial space are determined: this is the sublation of natural scenery by which landscape as experienced becomes landscape as represented. In a second sublation, the sublime in one basic sense promulgated by both Burke and Kant, that is, as "terrible" and "fearful," is expressly infused into two of these primary places. And, in a third action of sublation, this very sublime—the sublime of danger and exposure, of precipice and abyss—is itself sublimed into the beautiful place, the pastoral paradise, into which the scene recedes in the river and plains below and the mountains beyond.

By this threefold sublation, the natural world in its savage stage has become fully appropriated, fully represented. This happens by its transformation into a set of re-implaced places, a reconstituted landscape that is the setting of soul as well as of matter. It has been transfigured into something that is at once sublime and beautiful. In becoming a painting under this double description, a piece of American scenery has become the represented world of *View from Mount Holyoke, Northampton, Massachusetts, after a Thunderstorm (The Oxbow).* And it is to *this* world—not the brute natural world or the sophisticated ideational world of the artist or spectator, the art critic or art historian—that beauty and sublimity finally belong, and belong *together,* in the most profound and lasting way.

CHAPTER 5

Representing a Region

East Anglia in the Eyes of John Constable

These scenes made me a painter.

> —John Constable, letter of 1821

Why, then, may not landscape painting be considered a branch of natural philosophy, of which pictures are but experiments?

> —John Constable, lecture of 1836

WHAT IS A REGION?

By "region" I mean a group of closely concatenated places that are (1) spatially contiguous with each other (i.e., between which there is no void space); (2) temporally coexistent and thus cohistorical—that is, possessing a shared history, whether or not this is recorded by human beings. A region in this sense is at once more constricted and more concrete than the "material region" posited by Husserl as the proper object of a "regional ontology."[1] Although for Husserl "landscape" would be merely one subset of "space"— itself just one of the many possible regions of regional ontology, such as "things," "souls," and "bodies"—in the practice of landscape painting region is a privileged, nonsubsumable domain in which natural presences, things and people and places, coinhere. It is privileged and nonsubsumable inasmuch as it is a primary way in which the natural world organizes itself in our experience. Whether as wild or as cultivated, this world presents itself to us as always already regionalized—in contrast with sites, which do not come with pregiven regional identities: instead, we organize and subdivide them for our own utilitarian purposes.

The perceived landscape, then, presents itself to us as regionalized from the beginning, not unlike Plato's model of primordial space, *chōra*, which manifests itself in four distinctive *chōrai*, or regions.[2] Less extensive than Nature as a spatial totality but more capacious than a particular place, a region is the coherent clustering of places within the openness of landscape—a clustering that both depends on and reflects the aforementioned properties of contiguity and coexistence. These properties furnish consistency and continuity in space and time to what might otherwise be a merely loose congeries of isolated "spots." Spots become genuinely placial only insofar as they become integral parts *of a region*: that is, places that belong together in and through the embrace of the same region. In this way, they belong together as members of the same "vicinage" or neighborhood, thus as essential to the identity of just *that* region. Despite very considerable differences in spatial and temporal extent—ranging from the tiny Canton Saint Gall to Siberia, and from the momentary scene of a spontaneous political demonstration in a certain city borough to a distinctive part of an ancient old-growth forest—regions hold their constituent places together in an intricate dovetailing of space and time. They act to individuate space and time, endowing them with a local habitation and a name: the name of a region in fact often reflects its spatiotemporal individuation, and its local habitation is based on the places that populate it and create a basis for shared experience.

Landscape painting can be regarded as the painting of regions in their placial specificity and felt vicinage. For this reason, landscape painting is not just another genre of painting, a merely formal classification, but something that plays a very important role in the appreciation of regionalized nature—indeed, often its very perception—and it affords a very special satisfaction; hence its extremely broad appeal in the West, in China and Japan, and in many other parts of the world.[3]

This is not to say that a region is easy to represent in landscape painting. The primary difficulty is that it cannot be adequately represented by the depiction of a single place. However emblematic a certain place may be—however much it embodies a region's essential traits—it cannot figure *as that region*. At most, it is a revealing but partial representation of the region. Painters such as the aptly named American "regionalists" Grant Wood, John Steuart Curry, and Thomas Hart Benton produced a copious body of paintings that represent particular parts of the Midwest; but for none of them did a single scene suffice to convey an entire region. The same is true of painters who painted several different regions in their work, for example, van Gogh in Holland, Paris, Arles, and Saint-Rémy, or Monet in Paris, Argenteuil, and Giverny.

The fact is that a considerable variety of paintings of a given region, whether painted by a single individual or by many painters, is required before regionality is properly imputed or recognized. This is not a merely contingent fact. The spatial and temporal complexity of a region calls for tenacity of effort on the part of painters who paint it as well as a dedicated subtlety in their means of representation. The nuance and persistence of these painters—their untiring commitment to the representation or, more exactly, the continual re-representation, of a particular region—corresponds to the overdetermined historicity of that region, while the sheer diversity of their renderings answers to its spatial density.[4] If at all successful, the paintings of a region capture that region's material essence, which they instantiate one by one and from particular points of view. They stand in for, and express, the region they take as their common theme, their shared topic, conveying a concrete sense of its identity by their own collective force and merit.

REMEMBERING THE REGION

If any landscape painter in the West deserves the appellation "regional," it is John Constable, whose paintings represent places located in his native East Anglia, England. Except for several brief trips to Derbyshire and the Lake Country—where he was not impressed by the places he saw—he remained in East Anglia until his marriage in 1816, and thereafter returned frequently to his homeplace for inspiration (staying in the very house in which he was born). He refused to go abroad, despite considerable courting from contemporary French painters, who greatly admired his work. As Michael Rosenthal has written, "Constable was unique in concentrating on a part of the country where his family lived and worked and which he knew so completely."[5]

To concentrate on "a part of the country" is to concentrate on a *region* of that country, that is to say, on the characteristic, and indeed idiosyncratic, features of the land and sky and water that inform its landscape. Constable was so taken by these features and so determined to put them into paint that he began early in his career to represent them in a comprehensive way: "by 1808 he began systematically to study the landscapes of the Stour Valley [of East Anglia] through oil sketches, a practice which was to be continued for the next ten or so years."[6] With an almost Linnean zeal, Constable scrutinized the local land for its most telling marks as well as for its trivial traits. The result of this persistent scrutiny of his region was that Constable became a painter who "*knew* his landscape, and more, he knew it both over time and from numerous angles."[7] To *know* a region is to be fully acquainted with it as it is spread out in space (i.e., as visible "from numerous angles") and as it is distended in time (i.e., as it changes "over time"). Such spreading out and distending undergird an intimate knowledge of a region, whose detailed and complex landforms call for continual return and reassessment of the very sort that Constable deliberately undertook in the early years of the nineteenth century—a practice not unlike Cézanne's continual exploration of the region around his native Aix-en-Provence in later decades of the same century.[8]

To know a region is also to be able to *remember* it. Rosenthal says of the East Anglian countryside that Constable "would have seen it change, and memory would have influenced this late conception."[9] Just as Constable was not tied to a single spot in the region he loved, he was not limited to a single moment of its perception. Unlike Monet, who strove to represent discrete and unrepeatable experiences of illumination (e.g., in his studies of the Rouen cathedral at different times of the day), Constable attempted to convey in his paintings the felt density of a region's past, its historical as well as its spatial depth, its very repeatability. And the past he knew was the past he remembered. As he wrote to his fiancée Maria in May 1812: "You know I have succeeded most with my native scenes[.] [T]hey have always charmed me & I hope they always will—*I wish not to forget early impressions*."[10] Why did Constable not want to forget his early perceptions of the region? Not just because they acted as a fund of details on which he could draw habitually and reliably—thanks to what Freud called "the pertinacity of early impressions"—or insofar as they would serve as a gauge of historical accuracy.[11] More important still is the fact that *only through memory is knowledge of an entire region sustained*. We encounter here a quite fundamental condition of representability, a condition of painting any landscape that conveys more than a momentary glimpse of its layout. Confronted with the spatially and temporally expansive character of a given region, a landscape painter draws upon his or her memory of numerous experiences of that region in order to hold together otherwise disparate experiences—and to be adequate to the complexity of the region itself.

Memory effects this critical holding action by means of a synopsis that re-members a region in a series of condensed but connected representations of it. The result is a peculiar topomnesiac power that is invaluable to landscape painters. Whereas the ordinary denizen of a region calls upon this power in the form of fleeting mental images and scantily descriptive words, the landscape artist is engaged in the difficult task of embodying the power in enduring images of what Constable liked to call "real scenes."[12] Just as the painter must transmute the perceived places of a region into the represented places of his composition, so he must specify his memories of these regional places in that same composition: the dematerializing of perception calls for the concretizing of memory in attaining the actuality of the work of art.

Now we can begin to understand one basic aspect of what Constable was doing in the assiduous landscape studies he made of his native region. He was building and improving upon a mass of memories from early childhood—and not just his own childhood but that of his contemporaries, for whom he was acting in effect as a stand-in *mnemon* (i.e., one who remembers for others). In this way he forged a topomnesiac core that is intersubjective in character and that answers to the collective essence, the shared historicity, of the region itself. It is crucial to notice, however, that Constable was not attempting merely to reactivate early memories as such, much less to represent them. His quest was not psychoanalytic: he was not seeking to reenact earlier experiences that are perceptually identical with a current experience. His desire "not to forget early impressions" was realized not by reinstating and analyzing these impressions but by enriching them through an infusion of *new* memories. These latter were what he acquired in his practice of repeatedly traversing his native region.

The project of grafting new memories onto older ones stemmed from the basic fact that in his return to East Anglia after 1802 he was returning *as a painter of its landscape*. He was perceiving it with the eyes of an experienced painter, not those of a child. He knew that his adult perceptions, occasioned by his return to familiar "rural scenery," would precipitate densely dovetailed memories of the same particular places that he had known as a child.[13] The first round of memories of a given place would thus be extended into other more recent rounds, and in such a way as to allow for the specifically painterly aspects of the landscape to color and deepen them anew. The process is that of continually compounding and complicating one and the same topomnesiacal resource.

It is not surprising that such memorial expansion bears affinities with that of Wordsworth in *The Prelude*. The two naturalists had met in the Lake Country, and in Wordsworth's great poem of 1805 the poet reconnects with his own early impressions of the natural world, showing them to be anticipations of his later, more articulate naturalism. Childhood experiences—more exactly, memories of them—constitute a prelude to his life as a poet by being intuitions of the same basic topics first taken up in *Lyrical Ballads* (1798). But the early memories are vaguer and less determinate in character than the adult remembrances: childhood memories absorb and modify nature through what Coleridge called "primary imagination," yet do not shape it actively for poetic ends by means of an acquired "secondary imagination." The childhood memories lack a *focus memorius* of the sort that the mature poet's continually regenerated re-rememberings will furnish.

It is this memorial focus that Constable also sought to provide by continually returning to his region of origin. He revisited it on foot as well as in images or words. Only by such an active bodily based regathering of his primary memories was he able to win through to a painterly reassessment of the region; for his rememorialization of East Anglia, especially East Bergholt and the Stour Valley, was not undertaken for the sake of memory itself; it was undertaken so as to engender a more adequate way of representing this region in paintings. Disdainful of the emphasis on sheer technique that he encountered while studying art in London—mere "bravura" he called such technique—he resolved in early 1802 to return to East Anglia to paint its land in as naturalistic a manner as possible. In a letter of May 29, 1802, he wrote to his friend John Dunthorne:

> For these few weeks past, I believe I have thought more seriously of my profession than at any other time of my life; of that which is the surest way to excellence . . . For the last two years I have been running after pictures, and seeking the truth at second hand. I have not endeavoured to represent nature with the same elevation of mind with which I set out, but have rather tried to make my performances look like the work of other men . . . I shall return to [East] Bergholt, where I shall endeavour to get a pure and unaffected manner of representing the scenes that may employ me . . . *There is room enough for a natural painture.*[14]

There was indeed room enough, but such painting was not to be achieved by Constable for some time. It was one thing for him to proclaim that "nature is the fountain's head, the source from whence all originally must spring,"[15] but it was quite another to return to this primal source in such a way as to *represent* nature convincingly—in Durand's strong sense of "represent"—while also being faithful to its regional essence.

Before this could happen, Constable cast about considerably in searching for an authentically naturalistic style of painting. Not only was he deeply influenced by Gainesborough and Lorraine as paradigmatic landscape painters,[16] but he adopted a style between 1802 and 1808 that can only be called "picturesque" in keeping with the term made popular by William Gilpin in his 1768 *Essay upon Prints; containing Remarks upon Picturesque Beauty.*[17] The picturesque is neither beautiful nor sublime, but instead charming and pleasing; it is very much on a human scale: country cottages were regarded as leading instances of the picturesque. Constable's very first representations of East Anglia upon his return there—seven or eight small oil paintings, all of them executed in 1802—are decidedly picturesque in their rather precious treatment of the places they depict. They also show the effects of Constable's simultaneous engagement in watercolor painting, in which he laid on the paint in thin washes that leave the texture of the paper exposed and visible. These inaugural works in oil and watercolor, even if not yet "pure and unaffected," faithfully represent scenes in East Anglia to which Constable will revert continually in years to come: Dedham Vale, the Stour Valley, Dedham seen from Langham, Fen Bridge Lane, and so on. Taken as a group, they already constitute a conspectus of places indigenous to the region and exemplary of it, places that will receive ever more complete representation "over time" and "from numerous angles."

At the end of this inaugural period, Constable had departed from the picturesque thanks to his absorption in the subjects of rural life—subjects that Gilpin had found "tolerable only at a distance."[18] Constable's intimate knowledge of that life was set forth by an increasingly bold technique that improvised with brilliant colors, often applied in sweeping brushstrokes of *impasto* thickness. It was as if his ever more ramified acquaintance with the region supported his growing painterly prowess; by drawing on his memorial resources and extending them, he could be bolder in paint, less beholden to the details of a given scene. Constable began to gain confidence in his own emerging mastery of the problems of landscape painting; by 1810, he was exhibiting paintings of East Bergholt for the first time, announcing to the world his commitment to representing his home region on his own (and *its*) terms.

The painting from this period of which Constable was proudest, *Dedham Vale, Morning* (1811) (see Plate 10), was sent to the annual Royal Academy exhibition with high hopes. Although accepted for the show, it went unappreciated. As a deliberately low-key scene from rural life, it fit into none of the reigning categories of the age. Not only was it nonpicturesque: it depicted the lackluster Stour Valley instead of the more conventionally attractive Orwell Valley only a few miles to the east.[19] It lacked the formal perfection that would have called for its designation as "beautiful" in the manner of Lorraine, and it did not aim at sublimity in any of its known variants. Yet Constable, if no one else, was aware of its revolutionary potential. Hence his concern over learning that the painting had been hung in what he considered to be an unfavorable position at the Royal Academy, as well his admission that "this picture cost him more anxiety than any work of his before or since that period in which it was painted."[20]

The painting, which not surprisingly went unsold, shows a very flat landscape (almost as flat as a Dutch marshscape) arranged in gently receding planes that move from a dark foreground of early morning shadow to an illuminated, pale green middle and far distance. No abysses or precipices here! One continuous scene! A silver birch at the far left recalls Lorraine's motif of the *repoussoir* (that leads the eye into the painting), but otherwise there is little evidence of the earlier painter's idealization of pastoral scenes. The cattle, horse, and the farmer next to the horse appear in the foreground as matter-of-fact beings whose compact presence anchors the scene at its visual fulcrum point. From their seemingly casual positions the look of the viewer can wander in two directions—either straight through a capacious and welcoming valley toward a delicately modulated horizon or, alternatively, to the right along the path that leads off the canvas (on this path a woman is shown in the middle distance and a man much further along to the right, underscoring the recession). The resulting biaxial directionality creates a pleasing spatial openness, a freedom of movement that is rarely encountered in Western landscape painting before this date. At the same time, the directness of the treatment is remarkable: natural objects are presented in such a forthright and unmannered way as to embody in their very pictorial quality rural life as it was lived in that particular region of early-nineteenth-century England. It is a convincing case of what Constable himself called "Landscape from nature."[21]

Thus it is an understatement to say that this painting "broke new ground for [Constable] by reflecting something of his own feeling for place."[22] More to the point, it reflected his rich memorial immersion in what he described to his fiancée as "this delightfull place . . . amongst my favourite Haunts."[23] The word *Haunt* is telling: it designates "a place of frequent resort" *(O.E.D.),* a place that possesses sufficient force to draw us back again and again, whether in reality or in memory. *Dedham Vale, Morning* encompasses several such haunts: the foreground scene of active rural life, the place of the winding Stour River (barely discernible above the three poplar trees), and a group of distant places (including the village of Dedham at the far left, recognizable by its prominent church tower). "This delightfull place" is in fact a group of places affiliated as partners of the same vicinage, designated simply as "Dedham Vale." Paintings bearing this particular place-name—and they are numerous in Constable's complete oeuvre—permit us to perceive the vale as seen from various vantage points. Two oil sketches survive from the period immediately preceding the artist's work on the finished painting of 1811 (see Figures 5.1 and 5.2). Each of these sketches shows Dedham Vale from a slightly different point of view.[24] The first brings the foreground into still closer proximity to the viewer, who is made to feel that he or she may be standing on the same ground as do the two figures depicted there; the second almost entirely excludes the foreground area in favor of a concentration on the middle and far distance. In both, the path that leads to the right, though suggested, is abruptly cut off by the edge of the sketch. In the second sketch, the presence of the tree as *repoussoir* is greatly reduced, producing an effect of open-endedness in the landscape. This effect—the same as that alluded to by Thomson in "The Seasons," wherein the poet invites the reader to "sweep / The boundless landscape"[25]—carries over into the completed painting (Plate 10), which manages to reintroduce the tree at the left without in any way crowding or limiting the landscape beyond: on the contrary, it opens it up in the most inviting way.[26]

The oil sketches, in other words, can be considered free variations on the painterly

Figure 5.1. John Constable, *Dedham Vale*, 1809–10. Oil on paper laid on panel. 8 ½ × 12 ½ inches. Collection Stephen Raphael.

Figure 5.2. John Constable, compositional study for *Dedham Vale, Morning,* 1810–11. Oil on panel, 6 × 10¾ inches. Whereabouts unknown. Reproduced from Michael Rosenthal, *Constable: The Painter and His Landscape* (New Haven: Yale University Press, 1983), Figure 54.

representation of Dedham Vale as seen from the same viewing place. Along with the finished painting, they put us into the vicinity of the vale, and are thus variations on its regional essence. Despite the manifest differences among these works, each of them affords an alternative take on Dedham Vale, whose characteristic features are resituated and reconfigured.

This continual re-implacement of aspects of a region in paintings answers to their re-representation in memories. Just as it is possible to remember the same thing but in different ways—for example, with varying emotional tonalities—so the painter of a region may render the leading features of its constituent places in various versions. These versions reinstate one and the same scene, a given haunt, as it presents itself in several acts of perceiving as well as in the history of the painter's remembrances. In the case before us, the oil sketches of Dedham Vale capture disparate moments of perceptual engagement, while the finished painting embeds these moments within a denser memorial history of being in that region. Constable ensured the pertinence of this history because "he generally painted places which he knew and with which he was completely familiar."[27] In fact, it is only on the basis of this kind of familiarity that he could create painterly variations so freely and fully in the evolution of a single painting such as *Dedham Vale, Morning.*

I am using the term "free variation" in Edmund Husserl's sense of the term: by freely varying a given item, that is, by exploring in perception or imagination or memory a series of settings in which it figures, we come to determine what is indispensable to its essence: what the item *cannot do without* and yet still *be* that item. What it cannot do without is discovered by noting the congruency or overlap *(Deckung)* between the variations themselves. *It is in the overlap that what is essential is to be found.* This is as true of simple tables and chairs as it is of complicated landscape regions. Just as I can detect what carries over from one instance of a chair to another, for example, its ability to support sitting human bodies, so I can attempt to discern what one view of a region shares with other views. The discovery of such a connective congruity amounts to grasping what is essential to a given region.

CAPTURING THE REGIONAL ESSENCE

Such essentiality, far from being a philosophical luxury, is basic to the apprehension of a region as it is represented in a landscape painting; for various paintings of a given region could not cohere as paintings *of that region* unless they possessed at least a minimal congruency with each other. They are in effect corepresentations, cohesive variants of a region regarded as a coherent collection of places. Moreover, they are variations not only on the region as such but *on each other* as well, as we see in the dialectic of sketches that led to the creation of the final painting, *Dedham Vale, Morning.* Not as strictly regulated as a well-ordered series of numbers or physical objects would be, these sketches nevertheless have important internal relations to each other, each being in effect an experiment vis-à-vis the others—yet all exhibiting a common adhesion to an immanent regional essence. The significance of such series becomes especially evident in Constable's practice of painting full-scale oil sketches preparatory to a finished work, a practice adopted immediately upon his return to East Anglia in 1802 when he decided to "make some laborious studies of nature."[28] Such sketches constitute an open-ended group of congruent exemplars. On their basis, the painter was able to achieve that "pure and unaffected representation" of the region of East Anglia which he so intensely wished to preserve and transmit, where *pure* points to something essential and *unaffected* signifies undefiled by merely contingent differences.

Constable conveyed the essence of a region by the very variation of his painterly representations of it. This variation was itself twofold in character. On the one hand, the choice of the particular place to be represented was varied—from Dedham Vale to Flatford Mill, from the East Bergholt Rectory to Willy Lott's farm. To change the choice of place is to change the venue of the immediate scene of action. Whether this chosen place fills the entire canvas or only a portion of it (in the latter case, it is cosituated with other places that do not lend their name to the painting's title, thereby constituting a vicinage of related places), it remains critical as a *focus memorius* for the artist who is in effect remembering it—and a *focus imaginarius* for viewers who do not know the region firsthand. Through the perceived particularity of the focal place—in its detailed filigree, as it were—the painter remembers the region and we in turn imagine it *as he perceives and remembers it.* Place is here a privileged conduit to region; it both embodies the essence of a region and refers us to it as something more encompassing than the place itself can be. In this capacity, a place both carries and condenses its region, singularizing what is necessarily generic at the level of region itself. Region is both instantiated and varied by place—by the various places of which it is composed.

On the other hand, the representation of a single given place can itself be varied. This occurs mainly by alteration of point of view. Either the point of view, the "command" or "prospect," as Constable called it, is shifted horizontally along the surface of the depicted ground—so that, for example, we see a certain lane near Dedham now from one side and now from another—or it is altered vertically: for example, in two paintings that set forth vistas of Dedham, in the first of which *(View of Dedham)* the spectator's point of view is elevated approximately sixty feet off the ground, while in the second *(Dedham Vale)* the viewer's vantage point is much closer to ground level and is more continuous with the foreground.

These two paintings illustrate three further ways in which the representation of a particular place can be effectively varied. The first is through the representation of the comparative proximity of the place, that is, its *apparent size* within the painting. Thus, the village of Dedham is seen as a diminutive island in pictorial space in the earlier work (Figure 5.3)—a place "way over there"—whereas in the later work (Figure 5.4) it emerges in the middle distance in much more detail (e.g., the stonework in its church tower is now visible). Second, the *position* of Dedham in pictorial space is significantly altered from being presented as perceptibly off-center in the first painting to being very much centered in the second painting, even though the position of the town is much the same in terms of the mathematics of the picture plane (in *these* terms, it is only slightly off-center in both cases). Third, the painting of 1814 is a horizontal rectangle measuring 21¾ × 30¾ inches, while the work of 1828 is a considerably larger standing rectangle (57⅛ × 48 inches). In

Figure 5.3. John Constable, *View of Dedham,* 1814. Oil on canvas, 21¾ × 30¾ inches. Courtesy Museum of Fine Arts, Boston. Copyright 2000 Museum of Fine Arts, Boston. All rights reserved. Reproduced with permission.

Figure 5.4. John Constable, *Dedham Vale,* 1828. Oil on canvas, 57⅛ × 48 inches. National Galleries of Scotland, Edinburgh. Reprinted with permission.

this last case, both the *shape* and the *physical size* of the canvas, that is, the material bases of what can be called the place-of-the-surface, differ, with important implications for the representation of regional place: the larger, upright canvas brings the village of Dedham dramatically closer to the viewer, confronting her with its forthright reality, while the smaller, sidewise canvas keeps the same village at bay and thus in effect farther away.[29]

The representation of a given place can be still further varied. For example, the manner of representing the physical objects that populate a place is significantly variable. Just as a region allows itself to be variegated through its constituent places, and as places are varied through multiple forms of approach and depiction, so the material things that populate places can be diversified as to number, arrangement, relation to place itself, and so on.

It will be noticed that the analysis of the sequence that goes from region to place to thing is essentially static. In such synchronic analysis, we concern ourselves with the sheer variability inherent in successive levels of increasing specificity without regard to the evolution of the representation of a region. Yet such evolution, being historical and memorial, entails its own variation of states of representation: it, too, contributes to the free variation of a region by a landscape artist. I have alluded to this diachronic dimension in my mention of the two oil sketches (Figures 5.1, 5.2) that preceded the finished painting, *Dedham Vale, Morning* (see Plate 10). Many other exemplary cases abound in the work of Constable, including two oil sketches and a final painting of *Flatford Mill from the Lock* (1812). But an even more pertinent instance is a sequence of three works, all composed in 1812 (or close to then) and consisting in an initial pencil drawing and two oil sketches (Figures 5.5, 5.6, 5.7). Each of these ostensibly preparatory works—in their case, no known finished painting caps the sequence—shows a sweeping view of Dedham seen from afar. In none of them is there any significant variation in the horizontal or vertical placement of the viewer, in the size and position of Dedham in pictorial space, or in the material objects by which Dedham is represented (the church tower alone, a standing synecdoche, is visible in all three works and is again visibly off-center in each). All of the works are rectilinear in a lengthwise direction and small in physical size (each is 12⅜ × 26¼ inches or smaller). Nonvarying in each of these ways, the works nevertheless introduce two new forms of significant variation: in dimension and medium. The *dimension* of the drawing is strikingly different from that of the two oil sketches, being much more elongated: its ratio of 2⅙:1 is to be compared to that of approximately 2:1⅛ and of 2:1¼.

Figure 5.5. John Constable, *Dedham from Langham,* c. 1812. Pencil with white heightening, 12⅜ × 26¼ inches. Statens Museum for Kunst, Copenhagen. Reprinted with permission.

Figure 5.6. John Constable, *Dedham from Langham,* 1812. Oil on canvas, 7½ × 12⅝ inches. Ashmolean Museum, Oxford. Reprinted with permission.

The effect is that of panorama, not unlike Thomas Cole's second drawing of the view from Mount Holyoke; each requires the concatenation of two separate pieces of paper. The difference in *medium,* from pencil drawing to oil sketch, brings with it a felt (indeed, a quite palpable) difference in the representation of place: the drawing is at once more delicate and more accurate, while the oil sketches are bolder and freer—in keeping with the fluidity of the medium. Taken together, these discrepancies in dimension and medium support the fact that the drawing is at once more sweeping and more topographic than the oil sketches, which are impressionistic in comparison.

In following the progression from the drawing to the first and then to the second oil sketch, we witness a development in Constable's representation of the region he is painting. By July 13, 1812 (the date of the first oil sketch, Figure 5.6), he has moved from an effort to convey the view of Dedham from Langham in exacting detail by precise graphic delineation to a forceful and daring representation in which the visual dynamics are based on the interaction of the dark and light masses created by brushstrokes. Attaining still one more variation by altering the image while staying in the same medium, his oil sketch of August 24, 1812 (Figure 5.7), becomes even more committed to the sheer dynamics of pictorial form and shape: only one recognizable figure is in the foreground (versus several in the earlier sketch) and tumultuous clouds have been introduced overhead to augment the drama of the scene. During the period in which Constable was making the two oil sketches, he wrote to Maria to say, "How much real delight have I had with the study of Landscape this summer, either I am myself much improved in *'the Art of seeing Nature'* (which Sir Joshua Reynolds calls painting) or Nature has unveiled

Figure 5.7. John Constable, *Dedham from Langham,* 1812. Oil on canvas laid on board, 5⅜ × 7½ inches. Tate Gallery, London, 2000. Reprinted with permission.

her beauties to me with a less fastidious hand."[30] The truth of the matter is that both of these alternatives holds true. Constable was indeed improving in "the Art of seeing Nature"—his perceptual powers were keeping pace with his growing fund of detailed memories of particular places—but Nature was also revealing herself to him more generously and openly. In direct response, his own hand became less scrupulous and more mobile, as we cannot help but notice when we follow his progress from the line drawing to the two oil sketches that follow upon it so brilliantly. Duration in time here colludes with variation in space to achieve a "pure and unaffected representation."

FROM DRAWING TO PAINTING

With John Constable we thus arrive at a quite different place from that occupied by someone such as Fitz Hugh Lane or Thomas Cole. All three artists make conscientious, meticulous drawings of a particular landscape to begin with; but from this point onward there is a critical parting of the ways. Lane directly transposes his drawing to canvas by a systematic transfer via grid lines. No intermediate stage is called for in this direct, part-by-part transposition. Even a second version of a finished painting alters only a few details and does not change the basic layout of the scene: we sense that we are being treated to an elegant, subtle variation of the same place. Both Cole and Constable, in contrast, eschew transposition and make no effort to carry over, detail by detail, the pictorial contents of preliminary drawings. Instead of moving straight to a finished painting from a drawing, both artists create intermediate works, oil sketches that are at once true to the drawing and yet bold departures from it. But where Cole transforms his oil sketch in turn into a painting, Constable undertakes no such step: his oil sketches are sufficiently satisfying to count as "complete" if not "finished." Constable here seems to be saying that if the spatial energetics of a sketch can become rewarding enough on their own terms, there is no need to proceed to a painting suitable for exhibition at the Royal Academy.

Indeed, for Constable an oil sketch and a finished painting converge in that each is a full statement of the landscape, albeit in different terms: one is more emotional and personal and spontaneous, the other is more formal, removed, and static.[31] But rarely in the history of painting has the sketch assumed such importance as it does in Constable; it is not merely preliminary, not just a stage on the way to a work to be exhibited. It exists in its own right and possesses its own aesthetic merit. Fresher and more vivid than the final product, it stands on a par with it. It does so not because it is a primus inter pares— as if the two kinds of landscape representation were involved in some kind of contest— but because it has virtues of its own that cannot be matched, or even carried on further, by the finished oil painting. Nor should they be! In Constable's competent hands, the sketch is not provisional or preliminary; it is already a work of art and to be judged on its own terms. As Ann Bermingham states, Constable's "oil sketches fall outside the territory mapped out for them by the academic tradition . . . Their relationship to the final work is not logical or programmatic but eccentric and coincidental."[32]

Another reason why oil sketches came to assume such importance for Constable and distinguish him so markedly from painters such as Lane and Cole (and from many other nineteenth-century landscape artists) was his practice of painting these sketches out of doors. They were created in *plein air*—in the very presence of an unfastidious Nature whose spontaneity inspired the painter's lack of inhibition. The most notable instances of such inspiration are Constable's celebrated "cloud studies" of 1819–22 in which the artist attempted to capture the ever-changing aspects of the sky from direct observation. Appearances to the contrary, these studies are not simple transcriptions of natural phenomena—it is now known that they were influenced by Alexander Cozens's earlier efforts and by the meteorological theories of Luke Howard[33]—but they remain well suited for expressing the freshness and changeability of natural phenomena such as clouds, whose very being consists in continual transmutation (see Figure 5.8).

A cloud study of this kind also endeavors to discover the invariant amid the variations, even if here the invariant is merely a recognizable shape of cloud formation. Back on the ground, the invariant for Constable belongs to the region of East Anglia construed as permanent and unchanging land whose variations occur as particular places and scenes, things and perspectives—all this under the ever-changing weather that hangs above it.[34] Further combinations of the invariant with the variant are possible: Constable's two oil sketches of *Dedham from Langham* (Figures 5.6, 5.7) set forth the sameness of a place in the very midst of a radically variable sky. In their unpretentious inventiveness, as open to the sky as they are immersed in the earth, these sketches represent a special twofold clearing in the natural world in which they were composed. Around this

Figure 5.8. John Constable, *Cloud Study,* September 5, 1822. Oil on canvas. V & A Picture Library; copyright The Board of Trustees of the Victoria and Albert Museum. Reprinted with permission.

clearing, the artist's memorially informed perception comes to concentrated expression, reflecting his evolving understanding of the region he knew so well and for so long.[35]

REGION RE-IMPLACED

Region is at once the most capacious unit and the most ramified content of landscape painting. It is the dynamic basis of such painting: that by which "natural painture" becomes genuine *representation* and not just representational. Thanks to the uncontainable energy, the undelimited synergy of a region, its lively painterly rendition is no mere replica of a natural scene. When painted in a way that captures its material essence, a region does not impose itself on the place that specifies the scene of a painting but is discovered within the place itself. Paintings of the kind created by John Constable stand in for an experienced place without merely substituting for it; they stand in its place without taking its place. Such paintings carry forward the placial complexion of a particular scene into a regional register, rendering it other than itself in their renovative pictorial space. This space is not a scene of mere *re*presentation, composed of depictive icons of secondary significance, but is primary in its forceful (and sometimes phantasmal) presence. In it and through it, re-presentation occurs and re-implacement arises.

The place at stake in regional re-implacement cannot be reduced to position or "simple location." Place in landscape painting is place of a higher order, place to the second power. It is a place-of-representation that offers space *for* representation. As such, it conveys the material essence rather than the complete inventory of a given natural scene. It offers the aesthetic concretion of this scene's undelimited circumambience, its literally incommensurable natural powers, its placial plenitude in the midst of regional richness.

The space of depiction, pictoriality as such, renders possible the event of representation: an event whose parameters are those of place and not of site, a scene from a felt and known region rather than the abstraction of a pinpointed position. The space of regional representation is configured as place—whether this occurs as a singular place, the contiguous places of a vicinage, or the memorially suffused places of a haunt.

THE SUBLIMING OF REGION

In preceding chapters, I have been tracing the origins of modern Western landscape painting to its inaugural efforts to reduce the abundance of landscape to the foreclosure of landskip. Landskip becomes full-fledged landscape painting when a desire to convey a perceptual totality (e.g., the full panorama of a place or region) or to embroider on it in decorative detail gives way to a passion for re-implacing a finite but intense experience of natural scenery. The aim shifts from transposition—a quasi-cartographic goal shared by topographic painting and by decoration, each thriving on exact repetition—to *transformation,* from strict transcription to free transmutation. (Photography, which I treated briefly in chapter 2, vacillates between these two extremes.) Displacement of site becomes transplacement of place; literal placement, *positioning,* becomes painterly re-implacement, that is to say, *placing again* (i.e., re-placing, not to be confused with replacing or substituting).

In this transformative movement from topography to topopoetry, the sublime arises from the very body of beauty as its porous threshold, its ethereal but powerful partner. Combining Heidegger with Kant, we can say that the sublime arises from the interaction of earth and world—and of sky and sea—with the mind (or better, psyche) in the creation of a work of art. Nature is not sublime in itself—with this much of Kant's theory we can agree—but it *becomes* sublime by a process of psychical internalization (in imagination and reason, memory and understanding, appreciation and evaluation) that leads in turn to reexteriorization (i.e., in the images of a painting), a final twist ignored by Kant. To the dialectic of repetition and representation is thus added the dialectic of psychical action and the natural world. The result is the sublime in various of its major avatars—contemplative, apocalyptic, and natural—that appear in the most disparate locales in the West: the Northeast and the Far West in America, as well as various places in England and France, Holland and Italy and Germany. That a painter such as Constable does not set forth a readily recognizable mode of the sublime (if anything, he appears to be "rustic" in contrast) does not mean that his work is any less sublime in the end.

The re-implacement at stake in landscape painting is intrinsically sublimatory, proceeding as it does right up to the threshold of its own powers of representation. But when it is genuinely sublime, in whatever mode, represented place is not a merely numinous spectacle. It does not evaporate into a mental mist; it is altogether concrete, albeit now as part of a painted rather than a perceived scene, for the sublime in landscape art is always situated in a particular place, which is itself embraced in the region it reflects. To capture the sublimity of a natural scene, the landscape painter must anchor experience and memory and imagination in representations that are as place-specific as they are region-reflective.

Representing landscape is a matter of *finding place* within the framework of a painting and of *proffering place* on its own surface; it is a matter of the rebirth of place under our very eyes and in those of the painter, whose hand guides us to the renewed scene of place. The circumambience of perceived and remembered and imagined region—region finally *understood*—must be brought into the insistent arena of represented place if the sublime is truly to suffuse the painted representation of landscape.

CHAPTER 6

Representing Place Elsewhere

Northern Sung Landscape Painting

Landscapes display the beauty of the Tao through their forms and humane men delight in this.

—Tsung Ping (A.D. 375–443)

People nowadays enjoy looking at landscape paintings as much as at the scenery itself.

— *The Mustard Seed Garden Manual* (A.D. 1679)

LANDSCAPE PAINTING IN ANCIENT CHINA

We have seen that the movement from landskip to fully committed landscape painting was remarkably prolonged in Western painting. For more than a millennium—from Pompeian murals to medieval manuscripts and Renaissance portraits and allegories—the representation of landscape was confined to the background or margin: it was merely, or mainly, scenery *for* an object or action that was seen as autotelic and self-sufficing, and thus independent of its landscape setting. This object or action could be situated virtually anywhere: hence the tendency to create an idealized landscape of the sort that beckons in the far distance behind, say, the Mona Lisa or through the window in many paintings of the early Italian Renaissance. Idealization of place turns place into Any Place, that is, no particular place at all.

Whether landscape is confined to accompanying scenery or to mere decoration, it fails to be a place of its own. Instead of being genuinely placiated, it bedecks a primary subject matter that is either literally placeless (as in certain abstract philosophical or religious themes) or, if offering an ostensible place (as in the case of a depicted dwelling), does so only thanks to the context of the work and not because of any intrinsic character of its own.

Not until topographic painting begins to be taken seriously do we find a respect for place in its particularity and uniqueness: as just *this* place and no other. Before the nineteenth century in the West, this respect was an exceptional occurrence: for example, El Greco's early-seventeenth-century *View of Toledo,* a painting at once mystical and topographic. By the early eighteenth century, topography had become a more pervasive concern, as we saw in the case of Smibert's *Vew of Boston.* This concern gave way in the course of the next century to an interest in landscape for its own sake, for example, in the early paintings of the Catskills by Cole or in the grandiose landscapes of Church and

Bierstadt. The poignant case of Lane reminds us, however, of how slowly and painfully this transition from a topographic to a sublime representation of nature may happen in individual cases—as we also see in the gradual development of Turner from a painstaking painter of appealing vistas to his later free-form landscapes and seascapes.

It would be ethnocentric in the extreme, however, to presume that this saga of the prolonged and often hesitant evolution of landscape painting out of descriptive and topographic origins holds true only for the West. In fact, an even longer progression is to be found in ancient China from pre-Han origins in the Chou era (c. 1030–256 B.C.) to the T'ang (A.D. 618–906) and Northern Sung epochs (A.D. 960–1126)—a period, all told, of almost two thousand years! Until the T'ang era, landscape figured mostly as background for stylized depictions of events that featured humans and animals—a situation quite parallel to later Western paradigms. In varying degrees, it furnished decorative detail to pottery cups and bases, "hill" jars and censers, painted shells, and other objects. In the case of bronze mirrors, for example, we witness painted processions of horses and humans in a setting of trees—trees that are rendered in fine ink lines that delineate delicate branches. A representative instance is the scene from the outer side of a painted lacquer *lien* excavated at Changsha in central China, dating from the Warring States period (481–221 B.C.) (Figure 6.1). In this miniature landscape, we see elegant, highly regularized representations of trees and hills, earth and sky (as signified by the two horizontal lines, respectively). These situate and orient the human figures, who ride over the hills as well as through a gate that stretches between heaven and earth. Of particular note is the circularity of the scene, which not only connotes an open, recurrent sense of time but also anticipates later scroll paintings with their evolving scenes and continuously changing points of view. Nevertheless, it would be presumptuous to designate this design a "landscape"; at most, it has landscape *elements*: elements that will become entirely abstract in certain subsequent developments. The tendency to transform natural motifs into purely decorative swirls and whorls—instead of integrating them into full-fledged landscape scenes—indicates that the earliest impulse in China, as in the West, is to confine landscape to scenery and scenery in turn to design.[1] Landscape in any full-bodied sense is simply not considered a suitable subject matter of pictorial art; it lies outside its proper boundaries.[2]

Indeed, painting itself is at first only a minor art, subordinate to poetry and calligraphy and even to architecture: "early painting in China generally played the subsidiary role of ornamenting objects for daily and ceremonial use or embellishing architectural elements or carvings."[3] Even when the landscape depiction is comparatively elaborate, its role remains instrumental or supportive—for example, in the case of Han tombs, where one encounters paintings in the form of strategic maps, as inventories of retainers, and as illustrations of calisthenics.[4] This pragmatic employment of elements of landscape still does not appreciate it for its own sake.

Despite these striking parallels in ancient Chinese art to early Western reductions of landscape to an ornamental or instrumental significance, there are two important differences from the comparable pre-landscape traditions of European and American art. The first is that various landscape motifs such as trees and hills and mountains, however partially they are presented and however much they tend to become abstracted, will not

Figure 6.1. Painted decoration around a cylindrical lacquer box. Chinese; Warring States period. Excavated at Changsha.

be forgotten in later developments in China. They constitute a repertoire of images, an abiding background of artistic knowledge, that will be drawn upon by painters of the T'ang and Sung periods who are fully engaged in landscape representation as such. There is thus a much more continuous overall tradition of such representation than one finds in the West. A second and still more crucial difference is that early Chinese artists were self-consciously thinking in terms of landscape and taking it into account, even if they did not commit themselves to its wholistic representation. At least this is so by the time we reach the Han dynasty (202 B.C.–A.D. 220). As Michael Sullivan observes:

> In this early period [i.e., of the Han dynasty] Chinese artists and craftsmen were already strongly inclined toward representing landscape wherever there was the least excuse to do so. Indeed, while the craftsman of classical Europe or ancient India instinctively filled a space to be decorated with human figures, animals, or plant forms, his Chinese counterpart thought first of a landscape and even when . . . his main preoccupation was a hunting scene, it was the setting rather than the hunt itself which determined the form of the design.[5]

But why was the artist or craftsperson thinking of landscape first? The answer to this question indicates a still more definitive difference between early Chinese and American or European artists. This has to do with their respective views of nature. Put in the crudest terms, the difference consists in the fact that for Western painters, whether those of the ancient world or Hellenistic times, the Middle Ages or the Renaissance, or the early modern period, nature is *without*: it is something external to conquer, subdue, and shape. The natural world is first of all something to *take over*—take over in its material otherness, its transcendence of the human realm—and then to *take in*: where to "take in" connotes both to internalize and to represent, and to do so by respecting its very otherness. Hence the sense of acute struggle between an external, dominating Nature (most concretely experienced in erupting volcanos, storms at sea, and more generally wilderness) and an internal counter-Nature that tries to master Nature in mainly human terms. This is the struggle of the dynamical sublime in Kant's sense of the term, and it reaches its artistic apogee in the Romantic period and throughout the nineteenth century in the West.

Deeply different is the view that nature is never strictly outside us—nor, in contrast, within us, in "human nature"—but *everywhere,* equally so and at all times. We are in nature, and nature is in us: "nature" in lowercase, no longer transcendent, wholly other, or altogether wild. It belongs to everyone, and everyone to it. Thus there is no imperative to struggle *against* nature, no reason to take it over or take it in. The natural world has always already made its appearance, and our task (and preeminently the task of painters) is to convey this pregiven immanence, this constant accessibility, this unstinting generosity of the natural realm. The effort (or, more appropriately, the noneffort) is to reverberate with a world already well known to us: to achieve "spirit resonance" *(ch'i-yün)* with all that is, including ourselves as an integral part of all that is. In this spirit, landscape is nothing other than the visibility of a ubiquitous nature, its manifest layout, its way of being present to us, not before us or against us, or even in us, but *with* us. Not just the "withness of the body" (in Whitehead's phrase) but the "withness of nature" is what is most at stake, both in everyday experience and in art. And the vehicle of this withness, the basis of any significant consonance with nature, is *ch'i,* the ever-circulating cosmic energy that inhabits human and nonhuman worlds at once, making them more same than different, *interanimated,* as it were.

This is a philosophical framework that was shared by poets and painters, connoisseurs and government officials, literary critics and everyday people. It was laid down early in Chinese civilization—at least as early as the *Tao-te ching* (fourth century B.C.)—and continued to influence all the arts (not only poetry and painting but ritual, music, archery, charioteering, and calligraphy) well into modern times. Within such a framework, landscape—and thus landscape painting—was bound to hold pride of place; for landscape encompasses all lesser things, artificial as well as natural, within its aegis. A painting of a landscape will therefore try to exhibit the same comprehensiveness as the landscape it represents. As Kuo Hsi put it in the eleventh century A.D.:

> There is a proper way to paint a landscape. When spread out on an ambitious scale it
> should still have nothing superfluous. [Even when] restricted to a small view it should

still lack nothing. There is also a proper way to look at landscapes. Look with a heart in tune with forest and stream, then you will value them highly . . . Approach [them] with the eyes of arrogance and extravagance, then you will value them but little. Landscapes are vast things. You should look at them from a distance. Only then will you see on one screen the sweep and atmosphere of mountain and water.[6]

Crucial in this claim is not only the sheer scope of landscape and its corresponding representation in a painting but its ability to induce a state of spirit resonance in which one can "look with a heart in tune with forest and stream." At that moment, the viewer of landscape realizes an abiding unity with the natural world—which has been forgotten in the distractions of daily life. This unity can be conveyed by a painting as well as by a direct experience of the natural world. As Kuo Hsi adds:

Without leaving your room you may sit to your heart's content among streams and valleys. The voices of apes and the calls of birds will fall on your ears faintly. The glow of the mountain and the color of the waters will dazzle your eyes glitteringly. Could this fail to quicken your interest and thoroughly capture your heart? This is the ultimate meaning behind the honor which the world accords to landscape painting.[7]

"The honor which the world accords to landscape painting": this was a hard-won recognition indeed, in the West as in the East. But in the East, the recognition was abetted by a regnant philosophy of nature that preceded the victory and made it possible. It also made it possible considerably earlier in human history than in the West: approximately one thousand years earlier and in a form that has lasted to this day.

The earliness and lastingness reflect what was hardest won of all: an iconography whereby particular features of the landscape world were given reliably similar, if not always identically the same, forms of representation. These forms—forms of mountains and rivers, hills and huts, trees and bushes, pathways and bridges—amounted to a virtual visual vocabulary, a coherent set of conventional symbols whose ready interpretability and recognizability could be counted on not only by a painter's own contemporaries but by many succeeding generations of viewers, at least until the end of the Northern Sung era (i.e., A.D. 1100). Although Chinese landscape painting does not represent anything transcendent, for example, in the manner of the apocalyptic sublime, it does embody a symbolic language of emblematic signs whose meaning, if not universal, is rarely difficult to decipher. As Sullivan remarks, these signs are "visual enough so that the forms that gave rise to [landscape painting] may be apprehended, conveyed, and recognized for what they are, yet abstract enough to confer upon the forms thus created the validity of a general, eternal truth."[8] General and eternal at least for all those for whom they are transparent in their significance, thanks to sharing the same philosophical and social framework—and meaningful for all others who can, even from afar, sense and savor this significance.

SCENERY VERSUS SCENE

In Europe and China alike, then, we witness a long march from decorative and utilitarian uses of landscape to its unrestrained painterly representation—a slow but insistent winning of the right to present it in its own light and in its own time and space. It was only toward the end of this long march that Ching Hao could say that "he who tries to express spirit [in landscape painting] through ornamental beauty will make dead things."[9] Ching Hao is writing in the tenth century A.D., fully seven hundred years before similar claims would be made in the West. In ancient China we witness the early triumph of landscape painting—early in terms of world history—and this is so, I have been suggesting, because of the subsumption of such painting within an embracing philosophy of nature. It was this philosophy that laid the ground for the triumph and assured its perdurance.

Of greatest interest to us, however, is not the overall effect of this philosophy but its specific implications for *place*. It is especially characteristic of the ancient Chinese that they construed major philosophical concepts in the most concrete manner rather than as material for sheer speculation. Take, for example, the celebrated instance of *yin* and *yang* as it was conceived in Late Chou thought (especially in the work of Chuang-tzu in the third century B.C.). Before it came to mean receptive and "female," *yin* first of all signified "the shady side of a hill" and thus implied a dark valley or other unilluminated low place; and *yang* originally meant "the sunny side of a mountain," long before the derivative senses of "male" and active accrued to it.[10] These primal senses of the two great metaphysical principles are *senses of place*: the shady side of a hill suggests a low-lying and protected place such as a glade or vale; the sunny side of a mountain invokes prominences of various sorts, such as buttes, mountaintops, and so on.[11] Rather than being mere illustrations of *yin* and *yang* construed as abstract principles, such basic kinds of place embody and prefigure these two principles at the level that matters most—that of the natural world. Hence the importance of the primal appearances, the *hsiang,* the "emblematic images," by which such places are conveyed. In the *I Ching* (The book of changes) we read that "The holy sages were able to survey all the confused diversities under heaven. They observed forms and phenomena, and made representations of things and their attributes. These were called the images *(hsiang)*."[12] Not only the holy sages but, still more tellingly, landscape painters were in a privileged position to "make representations of things and their attributes." In order to do so, the things must be represented in situ; otherwise, they will be ethereal, untethered, in effect no things, nothing at all. In particular, the artist-philosopher—the compound term is unavoidable in the ancient Chinese context—must select images of places that embody the power of such forces as *yin* and *yang.* The choice of these images is thus not an idle one: to paint mountains, or plains, or the sea is to paint the primordial appearances of things in their aboriginal places—in accordance with their *genii loci.*[13]

Such appearances-in-places fall into two great regions: heaven and earth. These cosmological terms came to prominence during the Chou dynasty, when heaven *(T'ien)* was said to exist in a conjugal relation with earth *(Ti),* both being the offspring of a primordial *ch'i.* Taken together as *t'ien-ti,* this divine couple gives birth to human beings, who are responsible for shaping their lives under the guidance of the divine father-mother. But the link between heaven and earth is also described in explicitly placial terms. Not surprisingly, as cosmic regions, *T'ien* and *Ti* are themselves collocations of places. Thus Kuo Hsi advises the prospective painter: "When planning to paint, you must first balance heaven and earth. What is meant by this? Supposing you have a foot and a half of material, you should leave heaven its due portion above and earth its place below, and then with ideas in mind set up the scenery in between."[14] We shall return to the role of scenery in a moment. For now, I want only to underline that heaven and earth, like *yin* and *yang,* are place-specific terms. As the Great Commentary of the *I Ching* puts it: "Heaven is high, the earth is low; thus . . . inferior and superior places are established."[15] It follows that both pairs of terms, heaven/earth and *yin/yang,* favor landscape painting—itself ineluctably a painting of places and regions—as their most auspicious mode of representation.

The four terms just examined are themselves further unified by something they share in common: *ch'i,* the cosmic energy that animates the universe as well as each particular thing and place in the universe. Just as concrete places serve to specify *yin* and *yang,* heaven and earth, so these various terms can be seen in turn as specifications of *ch'i.* Instead of being simply polar oppositions, the members of each pair are complementary and are ultimately the "manifestations of an undifferentiated whole, interacting in perfect equilibrium, and [owe] their origin to the power of *ch'i,* the breath of the universe."[16] Like nature itself, *ch'i* is at once within the individual—organically, as "life-breath," yet also as "consciousness" *(chih-chüeh)* and as "spirit" *(ling-ch'i)*—and around the individual, as material nature, as "matter but not necessarily with [determinate] form. Solid, hard, and tangible condensation of *ch'i* is termed substance *(chih)*."[17] *Ch'i* connotes vital force, but its scope is so cosmic that it can be expressed as "matter" as well as "spirit." And as matter, it cannot exist unimplaced: it cannot exist without place and still *be* such matter. The ancient Chinese are here in accord with an equally ancient Greek notion, namely, what I have elsewhere called the Archytian axiom: *to be (at all) is to be in place.*[18]

Ch'i as sheer cosmic energy is not only all-pervasive—*pandemic,* always already part of every thing and every place—but the basis of all relationships between things and their respective places. And when it takes the specific form of *ch'i-yün,* "spirit resonance" or "spirit consonance," it is the source of the appreciation of places by human beings, who resonate with the places they inhabit or explore. This resonance is heightened and intensified in a landscape painting, which, by its exemplary *ch'i-yün,* reverberates in its viewer even when the viewer cannot actually visit the places represented in the painting. Spirit resonance is therefore a source of ramifying connectiveness between persons and places, whether these latter be perceived or painted.

For the ancient Chinese, spirit resonance is the most important factor in painting, and it is even more significant than representational likeness, which is considered to be an effect of such resonance.[19] The likeness at stake in spirit resonance is not that between disparate things, but that of a basic resemblance *of kind* whereby things of the same character resonate with each other. The *yün* of *ch'i-yün* connotes rhyme or assonance (in poetry), membership of the notes of the same key (in music), as well as bells ringing among themselves.[20] *Ch'i* figures into *ch'i-yün* in that, as Alexander Soper comments, "The painter must see to it that the *ch'i* of everything animate within his picture shall be able to find and respond to its like, not merely elsewhere on the silk but by infinite extension throughout all the universe."[21] Doubtless for this reason, spirit resonance is placed at the top of the list of the celebrated "Six Essentials [or Laws] of Painting." In Hsieh Ho's classic formulation of the fifth century A.D.:

The first [of the Six Laws of Painting] is called "animation through spirit resonance."
The second is called "structural method in use of the brush."
The third is called "fidelity to the object in portraying forms."
The fourth is called "conformity to kind in applying colors."
The fifth is called "proper planning in placing [of elements]."
The sixth is called "transmission [of the experience of the past] through copying."[22]

In contrast with the other five Laws, as Kuo Jo-hsü comments,

> "spirit resonance" necessarily involves an innate knowledge; it assuredly cannot be secured through cleverness or close application, nor will time aid its attainment. It is an unspoken accord, a spiritual communion; "something that happens without one's knowing how" . . . If spirit resonance is already lofty, animation cannot but be achieved . . . In general, a painting must be complete in spirit resonance to be hailed as a treasure of the age.[23]

Place is very much part of the Six Essentials—not as "proper planning in placing" (this has to do with composition, not with place per se) but as *scene*. In Ching Hao's more practical version of the Essentials, scene *(ching)* is singled out as the fourth factor, along with spirit *(ch'i)*, resonance *(yün)*, thought *(ssu)*, brush *(pi)*, and ink *(mo)*. "Scene is obtained," he adds, "when you study the laws of nature and the different faces of time [different times of the day or seasons of the year], look for the sublime, and recreate it with reality."[24] "Reality" is here to be distinguished from mere "lifelikeness." The latter bears only on outward appearances, on the *show* of life, rather than on the reality of place, which is scene proper. Scene embodies spirit *(ch'i)*, and it is with scene in all its temporal and seasonal density that the painter needs to have resonance *(yün)*. When the scene is part of the painting of a place that captures its spirit, it can be said to possess *ch'i-yün*.

Kuo Hsi remarks that the task of the landscape painter is to learn "how to pluck a scene from beyond misty reaches."[25] In order to do this, the landscape painter must attend to the essence of the scene depicted—rather than to its outward trappings, which are mere "scenery."[26] Only scenes are worthy of genuine landscapes, being integral parts of them. Scenery, on the other hand, is the degradation of landscape, its superficialization. Li Ch'eng-sou (c. 1150–after 1221), a Southern Sung writer looking back over the brilliant achievements of the Northern Sung, has this to say: "The younger generation does not understand technique. They draw minutely and add wash cleverly in order to charm the uncultivated eye. *Such works are just figures and scenery assembled in a patchwork, definitely not landscapes.*"[27] A genuine scene—a complex of places—constitutes the essence of a landscape; scenery is its mere appearance. The fact that the difference between essential scene and inessential scenery is marked as such by Northern and Southern Sung painters and writers indicates how far landscape painting has come from a period in which scenery qua ornament was the only form in which landscape was permitted to appear in art. Kuo Hsi, the outstanding theorist of painting in the Northern Sung dynasty, explicitly relates scene with essence as well as with *ch'i*:

> Each scene in a painting, regardless of size or complexity, must be unified through attention to its essence. If the essence is missed, the spirit *(ch'i)* will lose its integrity. It

must be completed with the spirit in every part, otherwise the essence will not be clear. The artist must accord his work overriding respect, otherwise the thought will have no depth. Fastidious attention must be given throughout, otherwise the scene will seem incomplete.[28]

To attend to the essence of the scene is to capture the spirit of the place. Thanks to the intimate bond between essence and spirit, the role of place is far from secondary: the fourth Essential is closely allied with the first Essential, and any rank order among the Essentials begins to dissolve. Spirit is in every part of a painting—in every scene it presents—and it is the enspirited whole, the entire place-world of the work, to which the artist must accord "overriding respect." Without this respect, the painter will risk being merely superficial; he will lack resonance with the reality of the place he is painting: he will deliver scenery rather than presenting the essence of the scene.

GETTING INTO DEPTH

Kuo Hsi was also an accomplished painter of the Northern Sung period, indeed, a special favorite of the emperor. Some consider him to be the greatest painter of the time. His work exhibits an impressive monumentality for which it is justly famous. It gives us truly a *scene in depth* rather than the depthless, literally superficial surfaces of mere scenery, as is evident in his exemplary painting *Early Spring* (Figure 6.2). Far from being a mere setting, the landscape in this painting is the true subject matter.[29] The situation obtaining in earlier times has been reversed: there, landscape features served as mere backdrop to the primary action of human beings (e.g., that of the figures on horseback and in a chariot as painted on the cylindrical lacquer box of Figure 6.1); here, tiny human figures are barely discernible to the right and left of the promontory in the near distance, being dwarfed by the high rocks and mountains that tower over them. On the lacquer box, humans and animals leap over hills; on the scroll of *Early Spring,* hills and mountains leap over humans. What was mere scenery has become a massive scene. Our eye is immediately drawn into the dark twisted area of land, whose convolutions it follows with pleasure and a sense of ever-heightening drama.[30] This is where the action is—in the active shaping of the land, its synergy—and we are invited to experience an intense spirit resonance with this convulsive action of the landscape. Rocks, trees, water (including strategically placed waterfalls) literally configurate to make the near sphere of this work into a complex representation of depth.

Depth is no simple matter, as perceptual psychologists will testify. It is not merely the "third dimension" along with height and width. If anything, as Merleau-Ponty avers, it is the very *first dimension,* indeed, the dimension of dimensions, being the basis of "the reversibility of [all] dimensions."[31] We see such reversibility at work in Kuo Hsi's painting when we realize that the height of the upper, overtowering mountain is not a matter of sheer verticality but *of height in depth*: we are led to grasp this height not just from our own viewing point outside the painting, but more particularly from the position provided by the rock formations below the mountain and in the foreground of the painting. These formations have their own delimited height-in-depth, which prefigures the extreme heightened depth of the mountain at the top: the way the lower mass turns away

Figure 6.2. Kuo Hsi, *Early Spring*, 1072. Northern Sung. Hanging scroll, ink and light colors on silk. 62¼ × 42⅝ inches. National Palace Museum, Taipei.

from us in its own self-enclosed depth adumbrates the way in which the remote tall mass turns away from it in turn. This signifies that the near and the far are much more closely implicated than would be the case on the conventional view expressed by Jao Tzu-jan in the middle of the fourteenth century: "In painting a landscape, you must first divide the near and the distant . . . you must first divide the far from the immediate, making the high and low, and the large and small, all suitable."[32] But in the interinvolvement, the reversibility, effected by depth as it is actually experienced in landscape—and as it is set forth by Kuo Hsi in *Early Spring*—one cannot keep the near separate from the far; the two mesh intimately, mirroring each other rather than standing apart; so much so that there is no place for middle distance in a work such as this, which exhibits a dynamic dialectic of the near and the far—of the rocks below and the mountain above and behind it. Reversible as it is in terms of particular features (e.g., the lines that curve away from the vertical in opposed but symmetrical ways), this dialectic is not, however, an altogether egalitarian affair. The high mountain, though farther away and painted in lighter values, is the main determinant of depth. Kuo Hsi was quite aware of this. He distinguished three kinds of depth—considered as modes of "distance"—all of which exist in relation to the mountain presence:

> Mountains have three types of distance. Looking up to the mountain's peak from its foot is called the high distance. From in front of the mountain looking past it to beyond is called deep distance. Looking from a nearby mountain at those more distant is called the level distance. High distance appears clear and bright; deep distance becomes steadily more obscure; level distance combines both qualities. The appearance of high distance is of lofty grandness. The idea of deep distance is of repeated layering. The idea of level distance is of spreading forth to merge into mistiness and indistinctness.[33]

Early Spring presents all three kinds of depth/distance. High distance is exhibited in the way in which the loftiest peak seems seen from far below, at its foot, yet is "clear and bright"; deep distance is represented by the horizontally arranged "repeated layering" at the far left, in an area that seems to go *around* the mountains and to disappear altogether into obscurity; level distance is suggested by the areas of mountain below the peak—areas whose very vacuity invokes mist. Kuo Hsi knows whereof he speaks! His painting and his text illustrate each other, as is only apt in view of the fact that calligraphy is for him a branch of painting and of equal importance.[34]

Moreover, the multidimensionality of depth and the intercalation of far and near are both supported by a system of implicit orthogonals (i.e., diagonal structures) that form a grid for the pictorial content while affording multiple points of view on the scene as a whole.[35] Despite all this structure, a basic mystery still inheres in a scene such as *Early Spring* presents—a mystery that never ceased to astonish early Chinese painters and cognoscenti. This is that such expansive spaciousness as we here witness can be displayed in such a confined actual space as a painting provides. "In the few inches of a painting," exclaims Ching Hao, "a hundred thousand miles of scenery may be drawn!"[36] Han Cho remarks similarly that "it is as if a thousand miles were right before one's eyes; there is an abundant spirit that is almost tangible."[37] And Shen Kua adds that "the divine grace of creation, the light or darkness of *yin* and *yang,* a distance of a thousand miles—all this can be captured in the space of a foot."[38] Each of these writers is struck by the capacity of landscape painting to miniaturize vast geographical distance within the close confines of the painted work.

How does this happen? The key lies in what can be called the principle of multiple differentiation—the skillful refusal to let objects standing at a distance from each other be confused with one another. This principle underlies Li Ch'eng's advice to the painter: "If there are a thousand cliffs and myriad valleys, they must be low and high, clustered and scattered, and not identical. Multiple ranges and layered peaks must rise and fall, lofty and high, and yet each [must be] different."[39] Li Ch'eng has struck upon a crucial aspect of the representation of distance of any kind: there has to be overlap between objects stationed at various points in the distance, yet not to such an extent that they replace each other, much less become confused with one another. Merleau-Ponty, discussing the enigma of depth, formulates the matter thus: "The enigma consists in the fact that I see things, each one in its own place, precisely because they eclipse one another [i.e., overlap], and that they are rivals before my sight precisely because each one is in its own place. Their exteriority is known in their envelopment and their mutual dependence in their autonomy."[40] A particularly illustrative instance of the principle of multiple differentiation is found in the painting *Tall Pines in a Level View*, attributed to the same Li Ch'eng cited just above (Figure 6.3). Here each object, whether rock or tree or water, is in its determinate place, and yet each eclipses another object or region, preventing its full perception. The branches of the several trees closely intertwine, and the rocks overlap in a decorously receding manner—as do the mountains, which occlude not only each other but the far sky beyond. At the same time, the rocks in the near sphere surround the water in the foreground, and the tall pines shelter the rocks in turn; the mountains in the far space embrace the plain in the middle distance, while the sky encompasses the whole scene. The resulting intimate interplay between exteriority and envelopment guides our gaze from the active foreground into the quiescent middle region and thence into the gentle mountains beyond. In this case, middle distance is especially important because the view of the mountains is literally a level view in which our implicit vantage point as onlookers of this scene gives us the impression that we are standing opposite, and on even keel with, the far peaks. In conveying this view, the painting as a whole illustrates what Kuo Hsi had said of "level distance" in particular: it is a "spreading forth to merge into mistiness and indistinctness." It is from out of this very indistinctness that a clear succession of particular places is garnered for our viewing: for example, the water place at the center of the foreground, the various rock places surrounding it, the broad place of the plain, the far place of the mountains beyond. These places, each filled with its own appropriate objects, recede in depth together in a forceful and elegant way.

How different is a third painting, also of the Northern Sung and this time of a truly monumental mountainous mass. If Kuo Hsi's *Early Spring* (Figure 6.2) had given equal prominence to all three kinds of depth, in Fan K'uan's masterpiece *Travelers among Streams and Mountains* (Figure 6.4) there can be no dispute as to the supremacy of high distance—which is such as to obliterate "deep distance" almost altogether ("almost," given that there is a bare vestige of a side view at the left), as well as to diminish the effect of level distance (in contrast with its centrality in Li Ch'eng's painting). Everything in Fan K'uan's work is subordinate to the overwhelming peaks, which together constitute one enormous vertical plane, tilted slightly away and back so as to increase the awesomeness of the spectacle. The tiny animals and humans below, however difficult to discern, allow the central massif to assume its full monumentality; in fact, a "leaping scale" takes one's eye from these diminutive figures to the trees and lower rocks and thence to the

Figure 6.3. Attributed to Li Ch'eng, *Tall Pines in a Level View,* tenth century. Northern Sung. Hanging scroll, ink on silk. 205.6 × 126.1 cm. Chōkaidō Bunko, Yokkaichi.

mountains above in progressively more extensive units.[41] This is in accordance with Kuo Hsi's dictum:

> In landscape painting there are three degrees of magnitude: a mountain, which is larger than a tree, which is larger than a human figure. If the mountains are not piled up by the score, and if they are no larger than the trees, they will not look imposing. And if the trees are no larger than the human figures, they will not look large.[42]

Figure 6.4. Fan K'uan, *Travelers among Streams and Mountains,* early eleventh century. Northern Sung. Hanging scroll, ink and color on silk. 81¼ × 40¾ inches. National Palace Museum, Taipei.

Except for the glimpse of lake at the base of the overwhelming rocks, there is virtually no middle distance, whose very absence is marked by the mist rising between the near world below and the peaks up above—a mist that sets off the high mountains all the more effectively.[43] Of particular note is the fact that there is very little recession in depth in this painting, despite its grandiose scale: instead of eclipsing or enveloping each other, the various objects and places that populate it seem to be compressed together onto the same picture plane. The effect is to bring us right up to the outer face of the mountain—to confront us with its unmitigated frontality. By propelling us *there,* into that sheerly vertical plane where there are no footholds, we are made to feel the precipitousness of the mountain even more forcefully.

This is a painting in which a mountain stands before us as "a major object," as Kuo Hsi called it—in which it "seem[s] to gaze down from its eminence and survey the ground below."[44] A mountain, the epitome of a megathing, dominates the already "large thing" that is landscape, as if the mountain were itself the condensed emblem, the *hsiang,* of landscape in general. The very word for landscape painting in Chinese, *shan-shui-hua,* means "paintings of mountains and rivers." And in fact mountains constitute the quintessential element in Northern Sung painting; they are the "bones," the very structure, of the earth, hence of nature overall. "A single mountain," insists Kuo Hsi, "combines in itself several thousand appearances."[45] This is owing not only to its unique power of inducing various kinds of depth perception, but also to its equally unique instantiation of metaphysical forces: "the different appearances of mountains and streams are produced by the combinations of vital energy *(ch'i)* and dynamic configuration *(shih)*."[46]

The dynamic configuration is that of the depths of a scene, and it is enhanced wherever one can distinguish between a "guest" and a "host" mountain—as is clearly the case in Fan K'uan's painting, which places the humble guest mountain below and to the left, as if paying homage to the central "master peak."[47] The dynamism is also that of *yin* and *yang.* A first-century A.D. text says: "Why must [the sacrifice be offered] on Mount T'ai? It is the place where the ten thousand things originate and where [*yin* and *yang*] alternate."[48] It is also the place where *li,* constant rightness, is enacted.[49] More dramatically and completely than any other natural thing, a mountain is therefore a place for the interplay of such basic cosmic forces as *ch'i* and *li, yin* and *yang.* It is also the primary place of authentic wilderness. In short, a mountain is a *place of places*—a place from which other places originate: the originating host of all the guest places in a given landscape. No wonder, then, that "to the Chinese all mountains are sacred . . . since remote times, the Chinese have held that the cosmic forces, the energy, harmony, and ceaseless renewal of the universe, are in some way made manifest in them."[50]

No wonder, either, that the Chinese valued so highly wandering in the mountains. Such wandering was both an act of meditative acknowledgment by the traveler[51] and something essential to the vocation of the painter, who is enjoined to journey among the mountains at length before attempting to paint them. Kuo Hsi thus admonishes: "In learning to paint a landscape . . . you must go in person to the countryside to discover it. The significant aspects of the landscape will then be apparent."[52] Such a journey is not to be confused with either sightseeing or dwelling. In wandering, one loosens one's clothing—a deeply symbolic gesture for the ancient Chinese—in order to absorb the essence of the landscape. Only when this essence has been assimilated should one attempt to paint it in an image that allows the viewer to travel in turn amid its wonders: hence the point of Fan K'uan's title: *Travelers among Streams and Mountains*.[53] In contrast with the open-endedness of wandering, which is ideally suited for coming to know what is essential to a landscape by the very indirection of walking, traveling connotes an activity with a definite aim in mind, for example, getting to a destination or gazing at a painting in which one's look is guided by the artist.

Where traveling usually obeys a schedule of some sort, wandering is free from the imposition of calendrically determined units. Thus it can take years before the painter is ready to put down on virgin silk what he has learned of the landscape by slow exposure; the time required is the time during which the "natural order, which is but a visible manifestation of the cosmic order, may reveal itself to him."[54] And not only must the natural order, the order of essence, be manifested to him; it must also be *incorporated into him* before creative work can occur.

Such incorporation is crucial. Without it, there can only be superficial work—only "figures and scenery assembled in a patchwork"; in other words, work that is not worked through, work that stays only at the surfaces of things rather than coming from the depths of place. (These depths may, however, appear *on the surface,* as happens in Fan K'uan's painting, where the drama of the depth of high places is played out in the sheer frontality of the central mountains.) But in order to be adequately borne out in painting, these placial depths must first come to inhere within the artist who has wandered in their midst.

IDENTIFYING WITH THE LANDSCAPE

One of the most striking convergences between Western and Eastern (specifically, Chinese) thought occurs at this very point. The convergence concerns the process of identification, to which allusion was made in an earlier chapter. Freud, inspired by the thinking of Karl Abraham, speculated that in the circumstance of mourning—and more generally in the formation of character—an identification with the lost other occurs. This other, no longer present in one's present perceptual world, is introjected into one's psyche, where it merges with the ego: an identification that allows the other to survive as an integral part of one's own self, as an aspect of one's characterological equipment, as it were. If this incorporated presence is not worked through in mourning, it will have pathological consequences such as melancholia.[55]

What the ancient Chinese painter undergoes is formally parallel—yet revealingly different in context, substance, and outcome. To begin with, the ancient painter identifies not with a person but with the landscape through which he wanders for years in preparation for painting; the purpose of the wandering is not just to note and observe but to make himself one with what he has moved through. Kuo Hsi observes that "an artist should identify himself with the landscape and watch it until its signification is revealed to him."[56] Moreover, this identification involves a strong factor of internalization, even to the point of pathology: the serious painters of previous centuries, says Shen Kua (A.D. 1031–95), "had streams and rocks in their vitals, and clouds and mists as a chronic illness."[57] But the internalization of the landscape, whatever its risks, remains necessary, so far as painting is concerned. This is clear from the claim of T'ang Hou (active c. 1320–30):

> Landscape is a thing naturally endowed with Creation's refinements. Whether cloudy or sunny, dark or gloomy, clear or rainy, cold or not, and in morning or evening, day or night, as one rambles and strolls [i.e., wanders], there are inexhaustible subtleties. *Unless there are hills and valleys in your heart as expansive as immeasurable waves, it will not be easy to depict it.*[58]

In a Western way of putting this, one must internalize and identify with the hills and valleys if one is to paint them with the flair they deserve and the *ch'i* they embody. At this very moment, however, the ancient Chinese paradigm begins to diverge markedly from that of Freud. Not only is *what* is internalized significantly different—landscape features instead of personal characteristics—but still more important, its valorization is quite distinct. Rather than being a burden within (as is the incorporated but not worked-through others who remain entirely "encrypted" in one's psyche), the contents of the landscape with which the painter identifies are singularly positive presences.[59] They are the indispensable basis for a representation that resonates with the spirit of the land, that captures its true forms and conveys them to the viewer of the painting.

In terms of outcome as well, there is a significant difference in the two instances. In the case of interpersonal identification, the best one can hope for is a nonproblematic, nonpathological relation to the incorporated other, who after identification can become redifferentiated as a nonambivalently remembered person.[60] But, in the case of painting, the material taken within the artist is deployed very differently; it is there not to be worked through and redifferentiated but to be expressed in the work of art—which means, in turn, to be *transmitted* to others, who can then enjoy the same natural scenes and even move in them as in an image-world. Tung Yu, an author of many colophons on paintings, is explicit about this:

> Yen Chung-mu painted for his pleasure and was extremely good at making true forms in landscapes, but he never painted without due preparation. He would travel, searching everywhere, and when his spirit was aroused by what he saw, he would store his exhilaration in his heart; hence the immensity in his breast naturally produced hills and valleys. Only when his inclinations had taken an appropriate turn, did he let them come forth. If one investigates his paintings from the point of view of forms, they are all completely what he saw; he could not paint his thoughts into them. He designated himself as "able to transmit," and only real landscapes went into his paintings, so that all his pictures were made from what he had seen.[61]

"Exhilaration in his heart," "immensity in his breast": these describe the internal emotions and perceptions that fairly bristle to be set forth, freed, by the painter. So much so that he has the sense that they are speaking spontaneously to him—that his body is a mere conduit, or else like "dry wood," waiting for the spark of these interior contents to ignite. Others stress the involuntary and nonconscious state of the creating body: "If all is ordered in your bosom, your eye will not see the silk and your hand will be unaware of brush and ink, and through the immensity and vastness [of your internal stock of images] everything will become your own painting."[62]

The body is thus a carrier for a burgeoning set of "mind-heart-imprints," as the ancient Chinese called these internalized images. What matters is not the sheer fact that the artist has identified with the content of these images, or even that he expresses them, but their transmission to others in paintings. In the quote by Tung Yu above, he is citing Chuang-tzu when he uses the phrase "able to transmit," a phrase that can also be translated as "transference of power": "When the body and vital power suffer no diminution, we have what may be called the transference of power. From the vital force there comes another more vital, and man returns to be the assistant of Heaven."[63] These words of Chuang-tzu remind us that in transmitting the landscape the painter is at the same time transmitting forces and powers that cannot be reduced to the idiosyncrasies of the occupants of the landscape—yet that are experienced intensely by these occupants; for "the spirit of the landscape" *(ch'uan-shen)* dwells in them, and the artist's task is to convey that spirit to others through his work.[64] And this is possible only insofar as "in his breast there naturally are hills and valleys, and they issue forth and appear as shapes."[65] These shapes are the forms made available in the painted work, where "forms" include colors and values as well as lines and volumes.

If hills and valleys can thus inhabit the painter's breast, it follows by the chiasmatic logic of reversibility that the painter can inhabit the landscape: "the painter takes his body with him," as Valéry said.[66] On the more general scale with which the early Chinese were wont to think of landscape as set forth in paintings, the place-world of the painter is itself to be conceived in expressly bodily terms. "In popular belief," says Michael Sullivan, "the mountain is the body of the cosmic being, the rocks its bones, the water the blood that gushes through its veins, the trees and grasses its hair, the clouds and mists the vapor of its breath."[67] Body and landscape here merge in a cosmic equation that exceeds the scope of any psychoanalytic paradigm—even if it shares with that paradigm the recognition of the importance of the self's identification with what is other than self.

BETWEEN VERTICAL AND HORIZONTAL

One of the very virtues of ancient Chinese painting, especially of the Northern Sung era, is that it continually reminds us of the importance of beginning *and* ending with the larger natural world—even if the internal region of mind-heart must be traversed along the way.[68] The painter starts with impressions garnered on his travels: even if there is no *plein-air* tradition in this era, experiences of wandering in the mountains are highly valorized. Memories of such experiences are held within the artist's interiority, becoming thereby a rich resource for eventual landscape painting. But this painting also draws upon the more delimited natural world that is specifically *bodily experience*. Beginning with walking and ending in the physical action of painting, from sensing natural phenomena outright to perceiving their painted representations—in all such bivalent actions, the body plays an essential role. Thanks to its effective agency, the artist moves in the very world he will paint; he will paint the vision he has gained while wandering; thanks to the body as well, the viewer of the painting receives with his or her own eyes this vision in representational form. Indeed, the painted work bears the direct imprint of the artist's bodily action—above all, in the expressive brushwork that is characteristic of Chinese painting and that brings this painting so close to calligraphy.[69]

But beyond these particular uses of the body in painting, there is another, more subtle employment. This has to do with its role in the generation and maintenance of the basic landscape schematism of horizontality and verticality. This schematism underlies the depth and distance that we have seen to be an abiding preoccupation of Northern Sung painters and theorists; for the three kinds of "distance" as distinguished by Kuo Hsi—interpretable as modes of depth on Merleau-Ponty's model of envelopement/exclusion—all depend on a very basic fact of human perception. This is that the upright human body stands up in the landscape and looks out to the horizon. Only from that stance and in that looking are high, deep, and level distances possible. The body is always on *this* side of such distances or depths—acting as their silent sentinel—while the visible horizon is always on the far side as their ultimate outpost.

The verticality of the body is perpendicular to the horizontality of the horizon, the two together thus crossing in a way that, were it to be formalized, would ironically resemble the intersection of the two axes of Cartesian analytic geometry. But unlike the purely formal intersection of x and y axes, the intersection here at play is altogether material and dynamic. It is the intersection of the lived body and the perceived landscape, their intertangled conjunction within the circulating energy, the force field, of *ch'i*. This same conjunction subtends the *mysterium coniunctionis* of heaven and earth, which also relate to each other as vertical and horizontal—as do *yang* and *yin* in their own manner. Making these mysterious metaphysical relationships possible in every concrete case is the unmysterious, yet by no means simple, relationship of body and horizon.

In the case of early Chinese landscape painting, this basic relationship—so basic that it is often overlooked in discussions of such painting—obtains both for the painter (as wandering and then painting) and for the viewer of the painting (as he or she looks at the finished work). In every instance, an upright body intermeshes with a set of places whose outer perimeter is that of the horizon. The body and the horizon, the near and the far, limit as well as encompass the places that make up the landscape as such: they are the effective epicenters of this force field. The resulting scene is distinctively triadic:

BODY–PLACES–HORIZON

In each of the paintings we have examined from the Northern Sung period, this triad, with its inherent horizontality/verticality, has been an important ingredient. In Li Ch'eng's *Tall Pines in a Level View* (Figure 6.3), for example, it is especially evident. The graphic horizontals of the plain in the middle region create a horizon *under* the mountains—a hypohorizon that undergirds the distant horizon outlined by the mountainous mass—while the tall pines on the left intersect with the horizontals of the plain. This intersection establishes a grid within which local variations can occur, notably in the form of the diagonal tree trunks and convoluted rock formations in the foreground. Here a vertical/horizontal grid is explicitly presented, and the sense of the upright body is conveyed directly by means of the level view—a view that signifies that the standing body is situated such that it can look straight into the far distance.

Even in the more complicated topography of Kuo Hsi's *Early Spring* (Figure 6.2), the schematism of horizontal and vertical is powerfully operative. The vertical axis is by no means a perfectly straight line, but there is an unmistakable upward movement in the center of the painting from the standing pines in the near promontory to the crevice and waterfall just to their right and then to the ascending edge of the high central mountain, an edge that leads in turn to the very top of the host mountain. The axis thereby traced is no less effective for being somewhat circuitously sketched (it seems to assume a vertical motion upwards)—and we identify with it as standing bodies that are themselves less than ramrod in their own uprightness. At the same time, two horizontal bands cut across the monumental vertical axis: one is formed from the level of the water in the lower left part of the painting—a level reinforced by the presence of human beings in a boat and at water's edge—and the other by the open and misty space just above the center of the painting, on top of the bulbous central mass of rocks.

In both of these paintings, particular places are located and oriented in relation to the biaxial systems I have described. In *Tall Pines in a Level View,* they arise in the continual recession that goes from the foreground (where a group of rocks shelters an appealing pool of water) to the open plain (itself a field of places), and on to the mountains (a group of far-places with little internal distinction). In *Early Spring* a string of localities literally clings to the sinuous vertical axis while others, paler and more diffuse, are found along the two horizontal planes. Similarly, in Fan K'uan's *Travelers among Streams and Mountains* (Figure 6.4), places are sheltered amid the trees and rocks in the near sphere—a barely visible temple at the upper edge of the forest, a passageway for beasts of burden at the lower right, several water places, many rock places, and so on. Not to mention the precarious places on the massive vertical mountain, whose uprightness matches that of the standing human body: even here there are locations for vegetation and for freely falling water.

Perhaps the most telling conjunction of horizontality and verticality—with places configured in between—is found in yet another masterpiece of the Northern Sung period, *Summer Mountains* (Figure 6.5). This is an altogether remarkable example of the biaxial schematism at work in early Chinese painting—a schematism that unites body and place in a common enterprise of landscape representation—and all the more remarkable in that this is a hand scroll that reveals quasi-narrative episodes: the return of fishing boats to their port, two lonely travelers crossing the first bridge on the right (hardly visible in this reproduction), another bridge leading to residences and places of retreat in the center and on the left. These episodes are specified, however, less in terms of time than of place—as my descriptive words *port, bridge, residence* suggest. The action that is adumbrated in the lower half of the painting, instead of being a drama of singular events, is that of a continuous, implicit *journey between places.* These places themselves inhabit the interstices of the scene as a whole—a scene whose name, *Summer Mountains,* evokes a place-world viewed at a certain seasonal moment, immediately after a summer shower, when "the verdant mountain peaks and dense foliage are suffused within a cool mist."[70] In other words, the scene is a scene of complex interconnection between aqueous, terrestrial, and mountainous places. The purely placial aspect of the painting is most evident in its upper half, where mountains link up and overlap in ways that convey at once recession in depth and broad outlay in space, thereby allowing adequate room for a multitude of particular places at each step. The artist's technique supports this ambitious placialization of space by its subtle and innovative use of lines and washes to draw us toward an infinite horizon located somewhere *beyond* the mountains.[71]

Figure 6.5. Attributed to Ch'ü Ting, *Summer Mountains,* early eleventh century. Northern Sung. Handscroll, ink and light color on silk. 17⅞ × 45⅜ inches. The Metropolitan Museum of Art, Gift of the Dillon Fund, 1973 (1973.120.1).

In its vastness of view—considerably more extensive in its spatial outreach than any of the other three paintings we have considered—*Summer Mountains* offers many points of entry. Instead of restricting us to a single place of access located squarely in the foreground (as in the other works, however differently presented), this painting invites our perceptual participation at many points throughout: most conspicuously at the coastal inlets, but also in the hidden pathways that quietly traverse the entire wooded region and in the numerous mist-bound ravines and valleys just above this lower region. A generous and open spirit prevails, as if to say, "Come and enjoy yourself! You too can enter this scene!"[72] But the hospitality—the easy availability of particular places within undelimited space—is reinforced by a delicately balanced juxtaposition of horizontal and vertical axes. Instead of being drawn straight upward into the sheer high distance of Fan K'uan's *Travelers among Streams and Mountains* or far outward into the beckoning level distance of Li Ch'eng's *Tall Pines in a Level View,* we are given *both* options in Ch'ü Ting's masterpiece. Not only do the horizontal and vertical axes subtend the organization of the painting overall (this much is true of all the paintings here under analysis), but they are made perspicuously present in their very equipoise with each other. Beyond the predominance of length over height in the shape of the work as a whole, horizontality is reinforced—given places to be—in the spread-out sea, the mists that cling at approximately the same level above the vegetation, and the sweep of the mountaintops as they march across the upper part of the scroll. But at every point there is an answering verticality as well: the masts of the boats on the lower right, the tall trees on the far left, the ascending hills and mountains in the middle distance. These latter, in contrast with the altitudinous mountains in *Early Spring* and *Travelers among Streams and Mountains,* do not overwhelm our perception but rise gracefully in tandem with each other, gently guiding our vision upward. The technique of overlapping planes, which are presented "frontally and additively,"[73] is much the same here as in Fan K'uan's painting; but the planes are more supple, thanks to the technique of parallel brushstrokes that merge into ink wash (rather than, as in Fan's case, being rendered by an intense stippling that gives to the central mountain mass its imperial independence).[74]

The concrete result of this two-way painting—painting that is sensitive to both horizontal and vertical axes—is a work to which our lived body can instantly relate. Perceiving it, we are not awed or stunned, as is so often the case when we savor works by Ch'ü Ting's immediate predecessors; instead, we are charmed and bemused by the accommodating layout of the entire landscape. Not only do the mountains forms "read from right to left, and front to back, in a part-by-part, sequential manner,"[75] but the whole painting reads in accordance with the two axes here mainly at stake: right to left in the horizontal direction, and front to back in the vertical direction. The former is the basis for narrative, that is, temporal depth, the latter for recession, that is, spatial depth in the sense of "deep distance."[76] The skillful use of diagonals to designate the mountain slopes softens the harshness of a strictly perpendicular relationship between horizontal and vertical.[77]

Thus a work such as *Summer Mountains,* despite its vast vista, does not present itself as "a conceptual vision of the macrocosmic universe"[78]—as in Fan Ku'an's highly structured gigantism, the graphic embodiment of the very idea of height—but as a measured vision of a habitable region, which asks us to join it in any of several possible places. Between the diachrony of the horizontal and the synchrony of the vertical, a landscape opens in *Summer Mountains* in which our viewing body can easily and pleasantly find places to explore. In fact, with the gentle and poetic vistas of this landscape, we have taken a decisive step away from Northern Sung monumentalism and toward Southern Sung lyricism—that is, toward a minimalist, subtractive painting no longer under the sway of biaxial directionality and more fully engaged with the aesthetic subtleties of place and its meditative possibilities. If the rectangular format suits Northern Sung painters more completely (the hanging scroll reinforcing verticality, the hand scroll promoting horizontality), the square and especially oval formats that became increasingly popular among Southern Sung artists ushered in an era of more intimate relationship to the landscape in which a single given moment or a particular isolated place, rather than a vast vista of collocated places, is emphasized.[79]

THE SPIRIT OF THE PLACE

Whatever came to happen subsequently, the Northern Sung artist undeniably accomplished two major things. On the one hand, he imbued his depictions of landscape with an inherent *order*; whether tacit or explicit, this order is another kind or level of *ku,* of "bone," that strengthens the world of his work. On the other hand, by employing the schematism of vertical/horizontal, the Northern Sung painter afforded *orientation* to the viewer's body, which is able to take its bearings in the representation from the very beginning of its acquaintance with the work.

But something else is also at stake in the landscape paintings of the Northern Sung epoch: namely, *re-implacement.* Not only are places the mediatrix between body and horizon, that is, their creative common matrix—this holds as much for ordinary perception as for aesthetic experience—but in the context of landscape painting they facilitate the ingression of perceived and undergone realities into depicted ones. At the same time, landscape paintings are by no means restricted to representing such depicted places (this way lies the rigorous route of exact description); they also *transmit* them—first to the surface of the representational vehicle, then to our perception as viewers.

All too often in the West, the transmission at stake in art is thought to be a conveyance of images and emotions, sensations and memories, as is signified in Wordsworth's phrase "emotion recollected in tranquillity." These various items constitute an indispensable experiential milieu—without them, the artist's scope would not extend to the expansive point required by creative work—but they are also the medium *for* the content that matters most in landscape painting. I have been arguing in this and previous chapters that this content is *place* as it is configured in landscape. Beyond the particulars with which a given landscape is populated (rocks, trees, bushes, pathways, people, etc.), *place is what is primarily transmitted in landscape painting*. A painting done in the Northern Sung spirit offers to us a set of places, whether these be open places (e.g., the middle distance of Li Ch'eng's *Tall Pines in a Level View* or the coastal waters in Ch'ü Ting's *Summer Mountains*) or emphatic foreground-places (as in Kuo Hsi's *Early Spring*) or precipitous mountain-places (as in Fan K'uan's *Travelers among Streams and Mountains*).[80] These variations do not begin to exhaust the rich placefulness of landscape painting done in the Northern Sung style, which often presents to us a complex congeries of places—including "guest" and "host" places, concatenated and disconnected places, land versus sea places, inhabited versus wild places, and so on.

The painter wanders in the midst of natural places so that he (or she: in other traditions, other times) may take these places in, identify with them, and then transmit them to others. What is for Freud an entirely psychical process that has human beings for its exclusive focus is for Northern Sung painters a literally parapsychical process—a process "alongside the psyche"—in which places and things rather than human beings are the true substance of the identificatory experience. It is these things-in-places that become integral to the artist's internal repertoire, part of his parapsychical being, that which he must transmit to others who hunger after the same landscape presences. These others take delight in what he has suffered from—and then releases into paintings that occasion the delight. The artist's disburdenment, unlike that of the psychoanalytic patient, does not take place in terms of words and memories, or even reexperienced emotions. The disburdenment is of place itself; for it is place that fills the painter's breast to the point of bursting and that he must relay to others by means of the agile bodily actions required by painting itself. He lends his body to place so that the reinstatement of place in his work can be lent to others in turn.

Such a transmission is a process not of purely psychical redifferentiation but of cosmic re-implacement: putting places somewhere other than where they first were (whether in the natural world in which the artist wanders or in his vagrant imagination), or even were secondly (in his heart or chest: his "memory"). *It is a matter of putting places into paintings.* This means putting them into the place-world of the artwork, making them its primary content: not its mere *Inhalt* (i.e., what is additively contained there) but its true *Gehalt* (its "substance," what holds it together as one work). When this is done, re-implacement has happened; and when it is done well, it effects *ch'i-yün*, an expansive spirit resonance that reflects the expansiveness of the world held within, while also embodying the *ch'i* and *li,* the *yin* and *yang,* the heaven and earth, of the place-world without.

Re-implacement is a means of attaining artistic truth. It is not to be confused with achieving mere "lifelikeness," that is, formal similarity. The critique of lifelikeness holds for places as well as for things. Ni Tsan asked himself, "Why should I worry whether my [painting of] bamboo shows likeness or not? . . . Others may call it hemp or reed."[81] The same holds true for painting places. Numerous are the reported instances of "wild painters" who, when drunk or in a frenzy, would throw paint against a wall and afterwards transform the spatterings into a painting of beautifully rendered places. Kuo Hsi himself instructed a plasterer to leave a wall in its fresh rough condition so that he could paint it directly as a landscape.[82]

Northern Sung painters and literati are in accord on this basic point. Ching Hao cites an elder who says: "Lifelikeness means to achieve the form of the object but to leave out its spirit. Reality [i.e., truth] means that both spirit and substance are strong. Furthermore, if spirit is conveyed only through the outward appearance and not through the image in its totality, the image is dead."[83] "The image in its totality" is the enspirited place-world, and the aim of painting is to grasp not the external form of something but its essence or "true image," as Ching Hao calls it.[84] Such a true image is tantamount to "true form." "Yen Chung-mu painted for his pleasure," remarks Tung Yu, "and was extremely good at making true forms in landscapes."[85] Such forms are not just true to objects, or even true to oneself[86]—though they are also this latter, because the self holds the forms in inner trust before painting them—but *true to place.* This is the truth of *t'ien-chen,* of being "true to nature."[87]

The crux of the matter is whether the artist wishes to transmit the formal likeness of a place—"verisimilitude," as is said in the West, as if to mock the original sense of this term (i.e., true-similarity), or *hsing-ssu* ("formlikeness"), as it is said more directly in Chinese—or the spirit of that place. For the ancient Chinese, the answer is unequivocal: the task of the painter is to transmit this spirit by re-implacing it into his painting. This is what Fan K'uan, among others, did so eminently well in the Northern Sung era. Fan was described thus in a royal catalog of his time:

> [He] lived among the crags and coves and wooded hills in the Chung-nan and T'ai-hang mountain ranges, daily observing the clouds, mist, the melancholic ways of wind and moon, and the indescribable effects of the darkening and clearing skies. Silently his spirit communicated with them . . . Such were his thousand cliffs and ten thousand gorges that they instantly made one feel as if walking along one of those shaded mountain paths, and even in the middle of summer, one felt chilled and wished to be bundled up in a cover. Thus it was said that K'uan was able to transmit the spirit of the mountains.[88]

The spirit of a place, including a mountain-place, may be transmitted in a representation that is literally *pu-ssu,* "not resembling" or "unlike." Yang Wei-chen (A.D. 1296–1370) puts the choice in this telling way: "when one judges the high or low quality of painting, there is either the 'transmission of likeness' or the 'transmission of spirit.'"[89] By the latter, adds Yang Wei-chen, is meant "spirit resonance [hence] life-movement."[90] The Chinese words for this last statement are transliterated as *ch'i-yün sheng-tung.* These are the same four words that compose Hsieh Ho's First Law of painting, which can also be translated as "animation through spirit resonance."[91] The transmission of spirit sought by the painter is *the transmission of the spirit that inheres in a place* with which one is fully resonant—hence that is continually in motion, given that resonance *(yün)* and motion *(tung)* are inseparably allied.

What the painter transmits is thus an enspirited place or set of places, the detotalized totality of the *genii loci* that make up a landscape and keep it moving—and that keep its representation in a painting from becoming a "dead image," a diagram or mere map of its literal contents.[92]

"By all means," counsels Han Cho, "follow rules and standards, but base your painting on spontaneous spirit resonance *(ch'i-yün)*. By doing so you will certainly attain a sense of life *(sheng-i)* in the painting."[93] To this we need only add: and you will certainly attain as well a sense of place, of its animating and indwelling spirit.

Interlude

Material Conditions of Representing Place
in Landscape Painting

I

In this interlude I shall consider particular material conditions of representing places in landscape painting (and, by implication at least, in maps). In Part I, I have taken these conditions for granted in my concern for how places are painted in their beauty and sublimity. Now I shall try to spell out several of these conditions, which serve to make the representation of place possible in landscape painting.

Place-of-exhibition. From one place to another: the representation of a natural place in a work of art—what I have called the place-of-representation—itself comes to reside in a quite different place, whether this be the artist's studio, an exposition hall, a museum, or a living room. This place of manifestation and residence is not incidental to the place-of-representation. Indeed, the latter may be so deeply affected by it that we find ourselves experiencing "a different picture" as a single work of art is moved around from place to place—something that anyone who has hung paintings in his or her own home knows about. Turner was so sensitive to the effect of exhibition space on his paintings that he was known to repaint them extensively in the annual Royal Academy exhibition after they were already hanging on the wall! This is only an extreme instance of what anyone might be inclined to do once he or she had perceived the striking difference that a change in immediate surroundings makes on a painting: a change in the way its representation of place, its literal "topic," reflects the very place in which it is situated.

The place-of-exhibition expressly concerns the interaction between architectural space and painting. Here the relationship is two-way: vividly so. On the one hand, the building, and most particularly the room and the wall for exhibiting a landscape painting, influence the viewer's perception (as well as the comprehension) of a painting. We are aware of this from extreme cases of overcrowded exhibition space—in which one painting distracts us from another, discouraging focused attention—or from the opposite circumstance in which a painting is seemingly lost in the void of too much space, diminishing its effective presence to us. Ideally, the exhibition place is sufficiently subdued and sufficiently capacious to let a representation of a place of nature create its own clearing within the architectural setting.

On the other hand, this setting will itself be affected by the presence of the painting; it will be enriched and enlivened—or deadened and weighed down. Some paintings, by their vivacity and strength of outlook, seem to act as apertures in the walls on which they are set: to be "windows" onto the represented scene of nature quasi-glimpsed through these walls. Capitalizing on this wall-piercing property, Frederic Church constructed for the opening exhibition of his *Heart of the Andes* (Plate 6) an enormous neo-Renaissance frame of presentation—thirteen feet high and fourteen feet wide—replete with velvet curtains; it was viewed in this dramatic setting by thousands of paying patrons when it was first shown in New York City in 1859. Church said expressly that his masterwork was intended to be seen as through "a palatial window or castle terrace [giving] upon an actual scene of picturesque mountains, tropical vegetation, light and loveliness."[1] Other, quite different paintings consolidate wall space, congealing it, as it were, and making it cohere even more than if it had no painting whatsoever on it (as occurs in the almost depthless garden landscapes of medieval tapestry such as those now in the Cluny Museum in Paris). Either way, whether they complicate or consolidate, there can be no doubt as to the considerable impact of landscape paintings on their own place of presentation, including a table on which a Chinese landscape painting might be unscrolled or a showcase in which a fragile work is exhibited. As this place affects their own appearance, they reconfigure it in turn, creating a dialectic of placing and being placed that must be taken into account in any thorough consideration of landscape art. It is this dialectic that takes the situation as a whole beyond that of pure decoration and toward a circumstance in which architecture and painting meet in the place-of-exhibition as in a common ground.

Place-of-the-surface. Another important material condition of landscape painting is the place afforded by the physical canvas (or paper, or whatever other surface) on which the representation of landscape appears. Of this we are normally even less aware than we are of the exhibition place. Much of the art of painting—especially in its classical and neoclassical forms—consists in concealing the presence of the surface to which paint is applied; this includes incorporating the surface, even as unpainted, into the larger totality of the work, in which it then plays an integral role. This latter strategy is deftly employed by many traditional Chinese landscape painters, as well as by Cézanne, who in numerous works makes such subtle use of untouched canvas that it becomes essential to the overall effect he wishes to achieve: he was fond of observing that a painting should be considered complete at every stage, even if it is never finally finished. Completeness can very well include untouched canvas or paper. Another strategy consists in letting the texture of the painted surface show through, allowing us to glimpse the texture through the paint. This occurs prominently in landscapes done in ink washes or watercolor, which at once stains and reveals the grain of the paper on which it is laid down. In such cases the surface is present by adumbration, that is, by a literal "shadowing-forth." Whether the texture is obscured by sheer overlay or allowed to emerge through the painterly medium or left unpainted altogether, its presence is of critical importance in the representation of landscape.

There is also a close relation between the place-of-the-surface and the size of a given painting. Even when the same subject matter is represented in the same way in two paintings, their respective sizes make a considerable difference in how we regard them. No matter how colossal the subject of a painting may be in actual perceived reality and no matter how great the distances implied by this subject, a diminutive rendition will act to undercut its monumentality: size crucially affects scale, even though the two factors are in principle independent of each other. This is why the spatial schema of a preliminary sketch for a painting often has to be modified if the size of the sketch differs significantly from the painting for which it serves as a model. As we have seen, Cole's preliminary studies for *The Oxbow* underwent important changes before the final schema for the painting itself was determined. Lane resorted to a grid by which he could transfer a sketch of Norman's Woe to a full-size painting: the positioning of particular parts of the sketched scene by grid lines facilitated their literal transposition to the larger painting, which thus maintained the same scale. In contrast, Constable insisted on making oil sketches of precisely the same size as the finished painting—for example, in the case of *Salisbury Cathedral from the Bishop's Grounds*—thereby avoiding the complexities of projective distortion. In this respect, the Chinese take one further step by often making the sketch coincide with the painting itself.

Beyond texture and size, the surface of a painting possesses visual and tactile and kinesthetic properties that contribute to its status as a place of a very particular sort. The surface may even attempt to present itself as a synesthetic whole—for example, in Red Grooms's three-dimensional reconstructions of *Chicago* or *New York,* or, more recently, *Nashville.* In such extreme cases as these, a departure from two-dimensional canvas or paper allows for the creation of a distinctive cityscape that condenses the feeling and style of an entire urban way of life. Here the surface of the work has gained a third dimension that we as spectators are encouraged to enter by literally walking through the artist's assemblage. The representation of a city as a particular place elicits the full resources of the representing medium, including size, texture, color, and shape—all of which, in Grooms's case, are combined in an expressive and emblematic three-dimensional simulacrum of that city. The result is something like a twentieth-century equivalent of the nineteenth-century American panoramas or fourth-century Chinese scroll paintings. In each instance, the physicality of the representing work, and more particularly the sensuous quality of its surface, is made to subserve the task of representation.

Much the same is true, albeit less dramatically and grippingly, of two-dimensional landscape paintings. Such paintings provide place not just by expressly pictorial means—for example, by drawn and colored forms that relate in some specifiable way to the place they represent. More than this: the represented place is conveyed *through another place*: the place of the painting itself. And this latter place is constituted primarily by its sheer surface. The surface is the bond, the tenuous tissue, that ties together representing and represented places. If the recognizable content of a painting establishes the place *of* what it represents, surface establishes place *for* the painting itself: its own proper place. Any given painting is realized *in terms of* the surface that is its *Träger* or "bearer." (The very word *sur-face* connotes bearing something *on* the *face*.) The properties of this surface influence, as from a gravitational ground, the qualities of the pictorial space that appears by its means. The pictorial space, configured as represented place, is not so much superimposed on this ground as allowed to arise from it.

The place constituted by a painting's surface is thus essential to the artwork considered as providing a place-of-representation. Such a surface place is a third thing existing between the actuality of landscape and this landscape as it is represented in the painting.

Although it is entirely physical, the place-of-the surface is ingredient in represented place, which is pictorial and not physical in status. Without a surface place, represented places could not enter into our aesthetic perception; they would be merely imagined in the mind's eye. To be given in such perception, represented places must find a proper surface of inscription. This is the surface of the painting in which they appear—a surface that constitutes its own kind of place.

Quasi-place-of-the-frame. Situated between a painting's surface and its exhibiting wall, the physical frame of a painting is a *parergon,* a "by-work" in the literal sense of existing at the edge, *as* the edge. As *parergon*—"work" *(ergon)* "alongside" *(para)*—it assumes the ambiguous status to which landscape painting itself was consigned in Western art before its own intrinsic interest was recognized. Regarded as mere scenery, landscape painting was in fact sometimes termed "by-work." Although the marginalization of landscape painting has been decisively overcome, the status of the frame remains for the most part "external": marginal, at the edge of the action. From this sideline position, the physical frame takes on a number of seemingly contradictory roles. For example, it acts both to *include* (the entirety of the painting's presented surface and the represented things that appear on it) and to *exclude* (the same surface and objects from the surrounding surface and the space of the exhibition wall). By the same token, the frame of a painting is at once *inside* and *outside*: snugly inside the place allotted to it by the place-of-exhibition and yet just outside both the place-of-the-surface and the place-of-representation belonging to the painting proper. It is, in Derrida's phrase, "a hybrid of outside and inside."[2]

Compounding these ambiguities is the fact that the external frame is at once a place and a nonplace (hence I call it a "quasi-place"). Considered as a literal physical entity, a frame is placelike in that its own surface affords room for something purely decorative to appear on it.[3] Even if it does not exist in the manner of a classical frame with its elaborate ornamentation—witness the bare, minimal frames of Mondrian and many others in his wake—the outer frame is, in any case, the place where a given painting *comes to an end,* literally terminating there. And yet, at the same time, the frame is just as decidedly a nonplace insofar as it is discontinuous with a painting's place-of-representation. Contiguous with the physical painting as it is, its spatiality cannot be conflated with that of the pictorial or represented items of that same painting, or even with its surface of presentation.

It is undeniable that many frames (especially if they also include mats) do take up considerable space, sometimes even more than the painting they surround. But sheer amount of space is not to be confused with importance of place, and it remains the case that the frame is generally subordinate to the place-world constituted by the painting it surrounds.[4] As Mikel Dufrenne puts it, it is "annexed" by this world.[5] This is all the more the case if the frame is reduced to an absolute minimum. Nevertheless, even a "frameless" painting around which no wood or glass has been placed is still perceived *as framed*—framed by its limiting edges and by the surface of the wall on which it is hung. In this case, the frame is quite literally *no place*; it is nowhere to be found in two- or three-dimensional space; and yet a factor of enframement, of "frame-work" if not of formal framing as such, is still experienced by the viewer. A framing effect of some significant sort—albeit merely in the form of a dim backdrop—is always present.

The frame, then, is a kind of no-man's-land of landscape painting. Neither natural place (i.e., the actual place of the environing natural world) nor represented place (i.e., the place set forth by the painting), it is a zone of indeterminacy that is nonetheless determinable, an arena of nonaction that remains active in our perception of the painting for which it serves as the frame. Thanks to this deeply divided role, the quasi-place of the frame is more pointedly discontinuous with natural place and its representation than is either the place-of-exhibition or the place-of-the-surface. The exhibiting wall exists in tandem with its architectural housing, and this latter with the natural space outside, in a closely concatenated set of increasingly inclusive containers—or better, backgrounds.[6] Even the surface place shares with natural place properties of size, volume, and texture. The frame, in contrast, is not reducible to the natural properties it shares with the environing natural world; like a Gestalt ground, it cannot be exhaustively analyzed in terms of sensory properties. Nor does it partake of the forms of represented place that appear within the work. Although represented place is at least coextensive with the surface place of its presentation (being projected, or envisioned, on or at this surface), it lacks common ground with its frame. Thanks to its deeply ambiguous being, then, the frame disrupts the profound continuum that exists between places that exhibit (on walls), present (on surfaces), or represent (in pictorial space).

The frame of a painting is thus disparate both from naturality and from pictoriality. It is a third term that, instead of mediating, serves to disconnect. It effects what Derrida calls "double separation."[7] In the complex procedure of representing place in painting, it is the odd man out. Unlike a place-of-the-surface, a frame does not bring together representing and represented places. Instead, it holds them apart by emphasizing the irreducible difference between the illusory or virtual nature of represented place and the real or actual character of the place represented.

This decided oddness, this literal idiosyncrasy, is to be contrasted with such seemingly similar things as boundaries or horizons. A *boundary* (in contrast with a strict border or perimeter in a legal or real-estate sense) allows for the interfusion of both sides, the inside and the outside, of the place or region that it nevertheless serves to delimit. These sides *meet* in the boundary, which thereby participates in what is located on both sides. "Good fences make good neighbors" not because (as in the popular interpretation of Frost's famous line) they hold neighbors apart as distinct individuals but precisely because a stone wall in its irregularity and imperfection is at once precarious and porous: it is an interplace that brings together otherwise separated neighbors.[8] A *horizon* is even more fully a place of meeting: it is *where* sky and earth (or sea) meet, a "point of astonishment" (in Emerson's phrase). As such, it is a full-fledged place, a crossroads of the elements in which the elements themselves commingle.

Boundaries and horizons alike serve to bring together regions *of the same genus,* such as neighboring places or juxtaposed elements. The outer frame of a painting, in contrast, acts to keep apart quite disparate things: painting and wall, surface place and exhibition place, represented place and representing place, the place of depiction and the place of perception, and so on. This provides yet another reason for calling such a frame a "quasi"-place. Unlike a landscape place, or its representation in a painting, it does not serve to contain and connect—to be a link or, still better, "a link of links" (in Bruno's phrase), concatenating what is already concatenated to some degree, above all, various discriminable places in a given natural scene. In this respect, the frame of a painting is an empty interplace, a place between places rather than a bona fide place.

Apparent exceptions to this general rule are easy to locate: Seurat's or Bonnard's or Hartley's painted frames are continuous thematically and colorwise with the paintings they present, and Marin's handcrafted frames are indispensable to the dynamism of his vivid watercolors. Such frames continue their paintings "by other means." But that is just the point: even these painterly and painted frames remain *other* to the works they surround. Such bold experiments do not undermine the general rule, which is that the external frame of a painting is a by-work that acts in tension with the painting it encloses. Even if this frame "is never completely isolated"[9]—and in the cases just noted it certainly achieves a minimum of isolation—it nevertheless serves to disconnect the painting from its immediate setting. As a *parergon* par excellence, the frame "inscribes something which comes as an extra, *exterior* to the proper field . . . It is lacking *in* something and it is lacking *from itself.*"[10] What it lacks is precisely place, that is to say, "the proper field" of painting itself. At most, it has, or better *is,* a quasi-place in the activity of representation in which all painting, and landscape painting in particular, is engaged.

II

We are not done with paradox and perplexity yet. Despite its literally ambiguous status and its quasi valence, the frame can be even more significant in the painted representation of the "natural presences" of landscape than the place-of-exhibition or the surface of the painting. Like them, it is a material condition of possibility for such representation; but it can have an especially determinative effect on the final outcome. To understand how this happens, however, we must pursue further the very meaning of *frame.*

A concrete example is called for. A painter, perceiving a shed in its own natural environment (e.g., as part of the backyard of a friend's house), is suddenly struck by it and makes a sketch of it with the intention of including it in an eventual landscape painting. In the painter's previous perceptual experience, it had not been conspicuous and had seemed, for example, unrelated to the horizon of land lying behind it. But once promoted into the sketch and then into the painting evolved from the sketch, its roofline becomes continuous with the horizon, which it reinforces by its own concerted horizontality (it is a long, low shed). What has happened in this transformation of an ordinary object into a central presence in the painting? One of the main changes wrought is the inclusion of the shed, formerly a freestanding physical entity, within the *internal frame* constituted by the horizon in represented space. The mere fact of this inclusion has sufficed to bring about a significant alteration in the fate of the shed—a change that may surprise the artist as much as it does the spectator.

What is happening in this circumstance is not at all literal framing of the usual sort, that is, sheer containment, but what can be called "frame-working." In frame-working, a frame *sets the artwork in motion.* The bare perceptibility of the pertinent frame (here, a represented horizon acting as internal frame) contributes to the highlighting of an otherwise inconspicuous object in pictorial space (i.e., the shed, especially its roofline), bringing this object into an unexpected prominence whereby it becomes a figure in its own right—in contrast with the way an external frame, for example, the decorative frame of the painting itself, forms a figure against the wall as background.

In general, the work of frame-working may be said to consist in converting the sheer surface of a painting into a charged "energy map."[11] By this latter term is meant nothing cartographic, much less topographic. It refers to the way in which the forms and qualities of things represented in pictorial space are not located in this space as so many inert entities positioned in an indifferent field of presentation. They have their own life there, a visual (and often also tactile and kinesthetic and synesthetic) élan vital that turns them into dynamic presences in vivid interaction with each other. In such an energized circumstance, the most diminutive object can become a critical, determinative force—as we witness so strikingly in the case of a paltry shed that becomes a major motif in a landscape painting, or sometimes even a single spot of paint that resonates throughout the work.

No wonder panoramic painting of the sort undertaken in Henry Lewis's *Mammoth Panorama of the Mississippi River* was foredoomed from the start. A landscape painting that purports to be literally panoramic bases itself on a refusal of the external frame—or, more exactly, it keeps rolling the frame back (literally so, in the case of a Chinese scroll painting). It tries to ignore, or to repress, the meaning and value of the fixed framing of a view or set of views, while downplaying the energetic effects generated by the internal dynamics of the painting. In this way, the panorama attempts to be faithful to the circumambience of the landscape as this is grasped by the freely orbiting head of the human subject—and, more ambitiously still, in the movement of this subject's whole body through a long stretch of the landscape such as might be apprehended while traveling down the Mississippi River on a riverboat or across a whole region of China in the Northern Sung era. But place as represented does not require any such panoramic presentation—any more than it needs precise cartographic localization. What it does require is an external frame of some sort. This frame is not merely a negation that cuts off a complete view of the landscape; by its very action of exclusion, it encloses a field of force that a painting needs for the dynamics of representation. Such representation, being less than complete, will fail to satisfy the totalizing passion of panoramic or topographic painting. But precisely as detotalized—as offering partial views only—it concentrates and holds the energy called for by the demands of representing places in a convincing and lively manner.

The external frame, by its bivalent action of excluding and enclosing, helps to generate a field of remarkable force within the delimitation of what is technically termed the "picture plane" (i.e., the two-dimensional physical surface on which representation is realized). Were there no such enclosure, the pictographic energy would be so diffuse as to ultimately dissipate. As a quasi-place, the frame contributes to the ongoing implacement and re-implacement of represented objects in the painting. These objects find place—or, more exactly, they *take place*—by dint of frame-working. Assembled in one composition, the resultant places open up and punctuate the internal spatiality of the painting, creating its aesthetic extensity. Integral to this spatiality in the case of landscape painting are such internal framing factors as the horizon and other lines of force.

Frame-working constitutes an energy map, and it helps to convert a landskip into a landscape, a by-work into an art-work. It creates the clarified openness of a landscape painting. It clears the painting in both senses of the term: it opens it from within (as one might clear a space in a forest) and it exceeds it from without (as we say that one object "clears" another by going around it). In such a *Lichtung,* "earth" and "world" (in Heidegger's sense of the terms) come forward to convert the flatness of the picture plane into the varied strata of represented three-dimensionality.[12] Earth and world come to a stand in the pictorial space cleared and complicated by the energizing effects of the frame-work.

This is not to deny other energetic aspects of landscape painting. One of these, for instance, arises from the interaction between the picture plane and pictorial space (i.e., the place-of-representation). This interaction has its own dynamism, as we realize when we focus on the relationship between the geometric center of a painting and the pictorial space of the represented places in the painting: their center may not correspond to the geometric center at all. In Cole's *Oxbow* (see Plate 9), for example, the geometric center is found at precisely the point where the dark and threatening storm at the upper left gives way to the pastel-colored cloud, a cloud from which no rain can fall and which gives way to the halcyon clarity of the right half of the picture. But the cynosure of the pictorial space is found in the oxbow of the river, the extraordinary curve that provides both theme and title to the painting. This curve is situated just below and to the right of the geometrically central cloud. Partly in shadow and partly in the light, the oxbow is as much a transition between the threatening and the benevolent areas of the painting as is the pacific cloud hovering above it at an indefinite distance. Thanks to a gentle tension of the elements—a tension between light/sky and earth/water—the two centripetal areas, geometric-surface and pictorial-representational, respectively, create a pitched vortex into which the spectator's eye and, indeed, her entire virtual body are powerfully drawn. The resulting vortical force subtends and strengthens the tri-partite placial movement from foreground to abyss to pastoral distance in a rapidly receding depth that opens out the picture plane from within.

But this animated interplay of depicting surface and depicted depth exists only within the frame-work of the painting as a whole, whose dynamism draws upon the fund of energy ordained and released by frame-working. In the end, the *energeia,* the dynamic actuality of the painting, is embodied and exhibited in the form of the particular places of its own representation.

III

In this cursory discussion, three major material conditions of the representability of place have been identified. Each of these contributes in a distinctive way to landscape painting, whether Western or Eastern in provenance. Both the place-of-exhibition and the place-of-the-surface bear on the *presentation* of such painting; they constitute its two main forms of physical givenness, within which the particularities of visual representation can be displayed. This is so even though the exhibition place is intrinsically changeable (the same painting can be exhibited in a diversity of settings, albeit with the effects noted earlier), while the surface place is strictly unalterable: once the surface of a given painting has been painted, it is forever the surface of just that painting and no other.

(This holds even when, as in Constable's case, two or more paintings have the same subject matter and the same exact size: however closely resembling the surfaces of these paintings may be, they retain their separate identities. This is a fact on which Andy Warhol capitalized by creating whole series of works having the same size and the same image, yet differently colored and highlighted in each case.)

The external frame of a landscape painting surrounds its represented content: it is the outside of this content, whose inside is constituted by the contents of the scene represented within its pictorial space. The circumambient frame is a zone where representation ceases; it is, as it were, a materialized void: a nonplacial place that is the very condition of possibility for the scene represented on the surface of the painting. (In its paradoxical power, the void of the frame is to be compared to the power of stillness possessed by the empty spaces in Northern Sung paintings.) The outer frame thus mediates between three apprehended plena: the representational world of the painting, its painted surface, and the surface of the wall on which the painting is hung. As the boundary between all three of these plena, this frame has the status of something "quasi." It is *as if* it were the frame of each, even though each is an entirely different kind of thing. In this very vacillation, the external frame reveals its status as a threshold, a *limen,* an indeterminate yet decisive interplace.

It must be emphasized that the variety of possible frames is considerable, not just in style (e.g., ornate, wooden, painted) but also in regard to whether a given frame is an adventitious or an integral part of the painting: Can you remember the frame of the *Mona Lisa*? More to the point: can you *forget* the rope frame Picasso placed around his elliptical *Still Life with Chair Caning* (1912)? Magritte engaged in an especially ingenious variation in framing—or, more exactly, in *represented* framing—by depicting framed paintings set *within* a represented landscape (*The Fair Captive* [1947] and *The Waterfall* [1961]) as well as unframed paintings that are shown as directly continuous with the landscape or cityscape that surrounds them (*The Human Condition* [1933], *Euclidean Walks* [1955]). More recently, Harry Hodgkins spread the boldly painted motif of his paintings directly onto the frame, which is thereby subordinated to the major motif while still being discernible beneath the actually perceived strokes.

Such vicissitudes of framing (including internal frames of represented boundaries and horizons) are fully compatible with the stability of the regional essence of a landscape painting's content. This essence, always unique in status (no two regions of the earth are interchangeable), informs the painting from within as its own nonexchangeable core of sameness. Constable provides the most compelling case in point, but we could have found much the same stability of regional essence in the case of any painter, East or West, who wishes to remain faithful to his or her region (whether this region is urban or rural or both: this latter being saliently the case in the ambidextrous instances of Edward Hopper and John Marin). We have seen (in chapter 5) that the sheer variation in representations of one and the same regional essence only ensures a more complete grasp of this essence *in its very uniqueness*: here it is a matter not just of the many-in-one but of the one-by-the-many. "All my life," remarks Charles Olson, "I've heard one makes many"[13]—to which the poet might have added: "and many makes one."

The representability of place, then, depends on a set of at least three indispensable material factors, each of which makes a decisive contribution to landscape painting as a whole. I say "at least" in order to indicate that other factors are doubtless also in play: image, line, color, medium, apparent size, depth, and so on. But the factors I have singled out are the predominant parameters of landscape painting when this latter is considered with regard to place, whether this be the place *of* the painting (concerning which the exhibition place and the frame place are most determinative) or the place represented *in* the painting (for which the surface place is the most important).

These same factors are also operative in the constitution of maps, which have their own place-of-exhibition, place-of-the-surface, and quasi-frame—however differently realized these may be in given cases. Maps also convey variant representations of the same region; indeed, this variation is one of their primary virtues given that these representations convey to us alternative ways of grasping the spatiality of that region. But rather than entering into a detailed discussion of their distinctive modes of enframement, let us instead come to terms with maps themselves as forms of representation. This is what we shall now proceed to do in Part II, where we shall explore how maps of many kinds represent the many places that make up the surface of the earth.

PART II

Mapping the Land

The difficulties that we run into are like those we would have with the geography of a country for which we have no map, or only a map of isolated places . . . We may freely wander about within the country, but when we are compelled to make up a map, we get lost. The map will show different roads which lead through the same country and of which we could take any one at all, but not two . . .

—Wittgenstein, from a Cambridge lecture of 1933–34

CHAPTER 7

First Considerations

Geographie is the imitation, and description of the face, and picture of th'earth.

—William Cunningham

The science of Geography . . . is, quite as much as any other science, a concern of the philosopher.

—Strabo, *Geography*

In Christ's name . . . I placed the said island in the present chart in its proper place where it ought to be.

—Francesco Beccari, "Address to the Reader," portolan chart of 1403

PREHISTORIC MAPPING

The earliest known maps are petroglyphs—rock carvings that represent landscape in some significant manner. From the Upper Paleolithic period (c. 40,000 B.C.) through the end of the Neolithic era (about 2000 B.C.) and well into the Bronze and Iron Ages (ending in the fourth century B.C.), rock art portrays various features of the local landscape: mountains, rivers, paths, dwellings, cultivated fields, and so on. These carved representations on durable stone constitute a primitive "cartography"—a word that signifies, literally, the tracing of a map. For example, the Bronze Age rock drawings on Mont Bégo in the Ligurian Alps, discovered early in the twentieth century by Clarence Bicknell, appear to represent pathways that link huts to each other and to enclosures in the larger landscape.[1] "On his many journeys up and down the valleys to Mont Bégo," remarks a contemporary geographer, "[Bicknell] repeatedly observed the striking likeness of the carved combinations of solid rectangles, subcircular forms, pecked surfaces, and irregularly interconnecting lines to features in the landscape when these are viewed from above—seen in plan, that is, from a vantage point high up the mountainside."[2] One of these rock carvings, located near Mont Bégo in the Fontanalba Valley, is shown in Figure 7.1. Another petroglyph, from the same valley, is shown in Figure 7.2. Both of these figures qualify as what historians of cartography call "simple maps," given that each is composed of at least six cartographic signs. Figure 7.1 has two hut signs (the dark areas), four or five land enclosures, and three land-use signs (the stippled areas); Figure 7.2 has two enclosure signs,

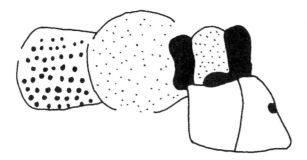

Figure 7.1. Petroglyphic map from Val Fontanalba, Mont Bégo.

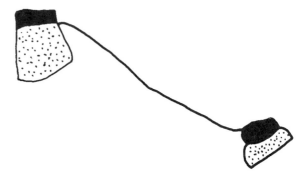

Figure 7.2. Petroglyphic map from Val Fontanalba, Mont Bégo.

Figure 7.3. Neolithic wall painting, Çatal Hüyük, Turkey.

two hut signs, one pathway sign (connecting the two main units), and two land-use signs.[3] As Bicknell himself wrote, "here people who stood on the [high] terraces might have looked down at the ploughing in the flat land of the valley, or on other terraces beneath them, and seen the operation from above as it seems to be depicted on the rocks of the higher regions."[4] On these rocks, then, we find maps—or at least "protocartographic images"[5]—that show condensed, schematic views of huts, pathways, and fields as glimpsed from positions higher up on Mont Bégo. Such views in plan are to be contrasted with perspective or elevation views, in which the profiles of objects are seen from the side, as for instance in a notable Neolithic wall painting from Çatal Hüyük, Turkey (Figure 7.3). This remarkable image is in fact a hybrid figure, composed partly of pictographs (i.e., the double-peaked volcano, Hasan Dag, seen in profile) and partly of views in plan (i.e., the living units of the settlement below the volcano). The pictographs are visual images of identifiable objects—no one can mistake *this* mountain for a molehill!—while the views in plan are so highly schematized that they can stand for any number of things: houses with inner courtyards, planted plots, and so on. The same is true of another petroglyphic carving, this one from Bedolina in Capo di Ponte of the Valcamonica region of northern Italy and dating from circa 1500 B.C. (Figure 7.4). Here too we see a commixture of "picture-map" and "symbol-map,"[6] the pictorial elements consisting in the readily recognizable stick drawings of humans and animals and in the sketches of houses as seen in profile, the symbolic factors being the roughly square fields and circuitous pathways as seen from above, including the stipling (perhaps standing for crop planting) and the circles with a dot in the middle (maybe representing supply wells or springs).

In the case of the Bedolina map, which is considered "the earliest work of its kind yet known,"[7] we know that the pictographic elements were not inscribed at the same time as the schematic features: some of the pictographs preceded the latter, and some were added later. But the collective result reads, nevertheless, as one more or less coherent composition, despite the fact that the two kinds of element call for two different kinds of perspective viewing.[8] Because of this complex coherence, it becomes dubious to

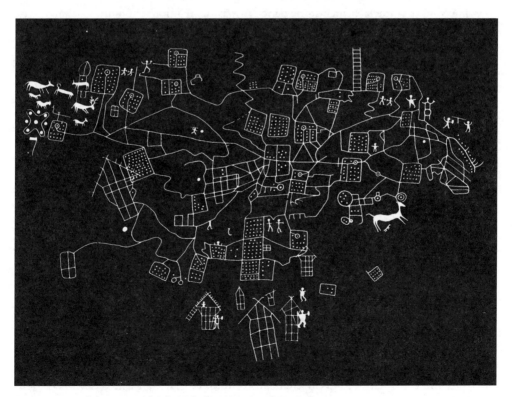

Figure 7.4. Bedolina petroglyph, Valcamonica, northern Italy.

maintain that there is always a straightforward "progression" (as one scholar puts it) from symbolic to pictographic maps, since instances that are undeniably both at once (and both quite successfully) can be found from Çatal Hüyük (late in the seventh millennium B.C.) to Bedolina (middle of the second millennium).[9]

It is a revealing fact that two leading scholars of prehistoric maps offer completely opposed views on the evolution of mapping in preliterate societies, even though they base themselves on the very same maps we have just been considering! The dilemma of interpretation that they present takes the form of this question: how important are *images,* and in particular *images of landscape,* for maps in general? For Catherine Delano Smith, images in the form of pictographs are of questionable value in prehistoric cartography; in her view, they serve mainly to distract. Thus she prefers a frankly bowdlerized version of the Bedolina map, from which all pictographs have been eliminated, remarking that in this case "the earlier figures and later additions have been removed to reveal a complex topographical map."[10] The distinct implication is that this map is more essentially a *map,* more effectively cartographic, in the absence of pictographic elements. Not only this, but their removal is seen as a step forward in cartography:

> The topographical maps [of the Mont Bégo and Valcamonica areas], including the simple ones, seem to demonstrate a concept of graphic representation distinct from that represented in picture-maps, namely the depiction of all features in plan . . . This new viewpoint must have constituted a cartographic step every bit as significant in the context of the later prehistoric period as was the reintroduction of the ichnographic city plan [i.e., ground plan] in the sixteenth century A.D. . . . . It is arguable that what appears to be a new (or perhaps increased) incidence of maps drawn in plan reflects prehistoric man's recognition that depiction in plan provided a more effective means of recording a spatial distribution than did a pictorial map."[11]

To drive the point home, Smith analyzes the nearly two hundred separate markings of the Bedolina map into just three elements of genuine cartographic significance: rectangles that enclose series of points, circles with central points, and irregular connecting lines.[12] Given Smith's overall definition of maps as "permanent graphic images epitomizing the spatial distribution of objects and events,"[13] any change (such as that from the pictorial to the schematic) that would enhance the representation of such distribution is *eo ipso* an amelioration. It is not only "new," but *better* as well. Not only does cartography not require pictography: it prospers in its very absence.

This view stands in stark contrast with that of P. D. A. Harvey, who sees the reverse movement from symbolic to pictorial maps as cartographic progress. His model of a "symbol-map" is taken from the early Bronze Age (c. 2000 B.C.) from Seradina in the Valcamonica valley (Figure 7.5). Harvey juxtaposes this nonpictographic map in plan with the mixed map of Bedolina (located less than half a mile uphill!) in order to observe that the Seradina map "could hardly have been recognized for what [it is] were it not that later carvings in the same tradition [such as that of Bedolina] show houses and other features pictorially, in simple elevation [i.e., in profile]."[14] At stake here is a twofold claim: first, that there is a natural development from nonpictorial ("abstract," "symbolic") to pictorial (i.e., in the literal sense of *giving a picture of*); second, that the pictorial is supe-

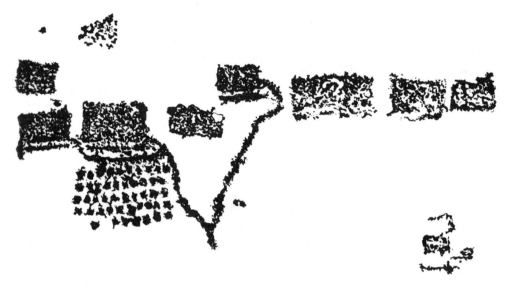

Figure 7.5. Early Bronze Age map, Seradina, Valcamonica.

rior to the nonpictorial, that is, to the schematic or symbolic. It is held to be superior inasmuch as in pictographic maps "representation becomes more realistic"[15]—a change that "may have come about through a simple advance in artistic methods and concepts among the carvers of the valley."[16] In a burst of assertiveness, Harvey sums up his position with the statement that "the evidence of Valcamonica is very clear: we see the symbol-map giving way to the picture-map as part of a wider artistic or even cultural change."[17]

But the situation itself is far from clear. This is that two experts in prehistoric mapping, both analyzing the same maps of the Valcamonica region, come up with quite disparate judgments: in the one case, a perception of increasing confusion and regression as we move from the simpler Seradina map to the Bedolina hybrid; in the other case, a conviction of a more evolved and sophisticated mode of mapping to the extent that pictographs gain prominence. How are we to adjudicate a situation like this, where the evidence, seemingly so unambiguous at first glance, gives rise to such deeply divergent interpretations?

We should begin by examining the primary premise of each interpretation. On the one hand, Smith holds that an increase in recognizable pictoriality cannot be considered an unalloyed good. Thus she maintains that "in the absence of substantial indications to the contrary, it would be more appropriate to attribute these maps to the primarily symbolic purpose behind much prehistoric rock art, in which artistic or visual significance was subordinate to a now unknown abstract context or message."[18] "Visual significance" is another way of articulating the idea of pictorial representation, which Smith regards as secondary to a symbolic purpose such as that latent in an early religion or cosmology. Prehistoric maps mainly ultimately serve the interests of the latter, and therefore their value as precise pictorial descriptions of the surrounding landscape—as mere "factual records"[19] in her implicitly disparaging term—is considerably less than their value as religious or cosmographic, that is, "symbolic," representations of this landscape in its sacred significance.

On the other hand, Harvey maintains that pictoriality per se is an intrinsic feature of good mapping: the more of it, the more adequate the map. His entire book, *The History of Topographical Maps: Symbols, Pictures, and Surveys,* is structured around the movement from "Symbols" (the title of Part I) to "Pictures" (Part II). This movement is regarded as a movement from more rudimentary to more advanced. As he says explicitly, revealing his own primary presupposition, "when the most primitive peoples turned their hands to mapmaking, the result would normally be a symbol-map, whereas a more advanced society would tend to represent features of landscape by pictures."[20]

The conflict between Smith and Harvey admits of one ready resolution. If our interest is in clarity of reading and practical utility, then it is plausible to consider pictoriality an advance over symbolism. If, however, we are more concerned with the representation of the sacred dimension of everyday life, then it will be equally plausible to think of abstract and schematic maps as expressing this dimension and thus as superior forms of mapping. The conflict will then become that of two interpretations of what mattered most to late-Neolithic and Bronze Age peoples: topographic transparency or symbolic numinosity.[21]

However we come down on this last matter, and even if it helps to explain the glaring differences between Smith and Harvey, one basic issue remains unresolved: How important are *images*—especially images of landscape—for the construction of maps in general? Are such images (i.e., in the express form of pictographs) indifferent to maps, perhaps even detrimental to them, and therefore dispensable, as on Smith's literally iconoclastic interpretation? Or are they in fact essential to maps that have evolved beyond a primitive stage, as Harvey holds? The issue is by no means confined to prehistoric maps—it will continue to be posed in much later stages of cartography—nor is it limited to maps as such. It is an issue that concerns virtually all models of the known world (of which maps form one distinctive, but hardly exhaustive, subset). In regard to world-models of every kind, those used in science as well as those employed in art and cartography, we must ask whether pictoriality is adventitious or required.

In pursuing this question, we will continue to witness the emergence of the same two poles of opinion: everywhere, the antipictorialists wage war with the pro-imagists. Max Black, a representative of the former group, asserts that "to make good use of a model, we usually need intuitive grasp ('Gestalt knowledge') of its capacities, but so long as we can freely draw inferences from the model, *its picturability is of no importance.*"[22] Jean Piaget, holding out for the latter position, argues that images, as a primary means of coming to know the world, are indispensable to all models of knowledge, save the very most formal.[23]

Closely related to this basic debate is the question of the desirability of isomorphism in mapping. Is it true, as Alfred Korbzybski claims in a much-cited statement, that "a map is *not* the territory it represents, but, if correct, it has a *similar structure* to the territory, which accounts for its usefulness"?[24] Or is it rather the case, as Amando Cortesào declares, that "we can never be sure that a map corresponds exactly to the portion of the Earth's surface that it purports to represent, even if it be mathematically correct"?[25] It is difficult to say what is the right response to this question of isomorphism in mapping. The difficulty is indicated by the fact that Catherine Delano Smith, whom we have seen to be in principle an antipictorialist in her theory of maps—and thus a trenchant critic of the requirement of cartographic isomorphism—can nevertheless claim, at the end of her immensely illuminating essay "Cartography in the Prehistoric Period in the Old World," that "what is urgently needed as conclusive evidence for the identification of map images such as those of Mont Bégo and Valcamonica . . . is a reconstruction of the real-world localities to which at least some maps may refer . . . The primary task [is that] of recovering the contemporary [i.e., the original] local landscape."[26] But to ask for the recovery of the original landscape to which prehistoric maps may "refer" is to ask for the reinstatement of the "real-world" landscape to which a map supposedly approximates by sheer similarity of structure. But must a map, in order to be *authentically* a map, approximate to the real geographic world on which it is purportedly based? Certainly not, at least not for those maps that are cosmological or religious in import, as Smith herself holds to be the case in most prehistoric maps! Indeed, does any map, even the most scientifically exact, ever truly *approximate* the world it purports to represent? How can two such different things as a map (pictorial or symbolic) and the landscape it describes become so much the same as to achieve anything like strict similarity? These are questions to which we shall continually return as we consider what it means to map a landscape.

For now, let us only agree that a map may be *true to* its landscape of origin—its "contemporary" world—without necessarily being *true about* that primal scene. Being "true to" a landscape means conveying the sense of that landscape, its material essence; being "true about" it signifies being informative of it in detailed (e.g., pictorial) fact. Prehistoric maps are true to the terrain they represent. They are reasonably accurate, as virtually every scholar of the subject attests,[27] but to claim this is not to have to claim that they reproduce in fine the particulars of the original landscape—as would be the case if they were exact and complete projections of these particulars onto the lithic surfaces on which they are almost always inscribed (allowing, of course, for differences of scale between the original and its projection: were there no such differences, we would be in the absurd situation parodied by Borges wherein the cartographers of a certain country attempted to construct "a map of the Empire whose size was that of the Empire, and which coincided point for point with it").[28] Nor is it even to hold that the map and the landscape possess the same formal structure: that they are iso-morphic in a literal sense. It is only to claim that there is sufficient *likeness* between the two for the map to serve as a recognizable representation of the landscape it sets forth. This likeness need not be isomorphic, a matter of sheer similarity; it need only convey the gist of what it represents, its morphological essence, as it were. As Husserl avers, such an essence can be quite vague in content and yet still be forceful in effect.[29] In other words, even if a map is schematic and manifestly fails the test of isomorphic pictoriality, it can still achieve *veracity* in its representation of a given landscape.

This is precisely what happens in the instances we have examined from the Mont Bégo and Valcamonica regions of southern Europe. Whether they make use of pictorial images or not, they are all true to their local landscapes; all involve a successful and telling "reconstruction of the real-world localities to which [they] refer." But the mode of reconstruction, as well as the type of reference, varies from case to case, sometimes being more "symbolic" and sometimes more "pictorial," and often being both at once without any diminution of the *verisimilitude* thereby realized. Contributing to their truth-in-likeness (as "veri-similitude" literally signifies) is not only the precise linearity, the exact graphism, of these maps—the chiseled-out lines that enclose by their very undulation as well as the pounded-in points that specify in their very delimitation—but such other factors as the texture and curvature of their surface of presentation (i.e., the place-of-the-surface), the consistency and hardness of the stone on which they are inscribed, and their own location in the earth. Short of any strict isomorphism, all of these factors contribute to the map's representation of its larger landscape, its feel for it, its sense of it: its being there *for it*—by being *of it.*

Location in the earth is especially important in the case of the maps we have been considering: its *own* implacement in the ground allows a petroglyphic map to act as an effective agency of cartographic re-implacement. The surrounding soil serves as the map's quasi-frame. This is perhaps most strikingly true of the Bedolina map, both in terms of its actual locus (stationed high in the mountain slope that rises from the flatlands it represents) and of its situation vis-à-vis the valley and river it overlooks, the mountains opposite (which provide an external frame) and the near mountain to which it clings. Multiply implaced in the larger landscape that encloses it as well as in the particular valley region it expressly represents, the Bedolina boulder map re-implaces this landscape and this region on its perduring surface: the landscape by projection from the outer edges of the map, the region by the incisions that constitute the map itself. The overall effect is to bring actual places *over* there and *down* there over *here,* onto this glacier-polished slab—to make them part of its very being. This set of images, inscribed on the stone, holds in thrall the primary features of the landscape around it—puts them in petroglyphic trust, enabling an entire future of viewing (including my own contemporary viewing, 3,500 years after they were first set down on the stone).

As in the case of landscape painting, so in prehistoric mapping the re-implacement of landscape and region rather than the reinstatement of form and position is the central action of representation, the motor of the movement that makes representation possible in the first place. Just how this is so is something we shall now have to discern in other contexts, other ways.

PICTURE/WORD/SYMBOL

We cannot reduce the complexity of prehistoric mapping to a simple choice between the priority of pictorial images versus the priority of symbols. P. D. A. Harvey himself has proposed that *between* symbol-maps and picture-maps there is a third genre, which takes the form either of three-dimensional "reliefs" (e.g., in the Han dynasty in China, in Inca maps as observed by the Spanish conquistadores, and in Moroccan maps of the fourteenth century A.D.) or of "picture-stories" of the sort found in American Indian artifacts. "Picture-stories and three-dimensional models," writes Harvey, "may have been ways in which the symbol-map developed into the picture-map."[30] Of particular interest to us is the picture-story. Beyond image and symbol it introduces the factor of the *word.* Words need not be expressly included on the map; but they are implicitly present as the vehicles of the narration by which cartographic signs are to be interpreted. Take, for example, the picture-story of a chief of the Tortoise clan of the Delaware Indians (constructed in the late eighteenth century) (Figure 7.6). The turtle at the upper left (1) identifies the clan; the triangular pictograms (4–7) represent scalps and prisoners taken; and the rectangular objects (8–11) show Fort Pitt and Detroit in plan. We are presented in effect with the ingredients of a heroic story of the Delaware chief—a story which, if actually told, would detail his exploits within the world of the white man. From such a picture-story, it is indeed "a very easy transition to a picture-map,"[31] for example, of the very sort that we have observed in the case of Bedolina or of Çatal Hüyük (both of which combine explicit pictographs with schematic images in plan).

In still other cases, the story is placed alongside the map as its explanation or "legend." Here the verbal element is present as such, and is combined with the pictorial element: a combination that is as ancient as a clay tablet map from Nuzi of 2500–2300 B.C. and as recent as a current road map. In the case of the Nuzi map, the words are inscribed in cuneiform directly on and around the map's pictograms (Figure 7.7).[32] Here the juxtaposition of word and image is smoothly integrated, the text being internal to the map itself: it names the three towns signified by the larger circles, and it gives the size of an estate

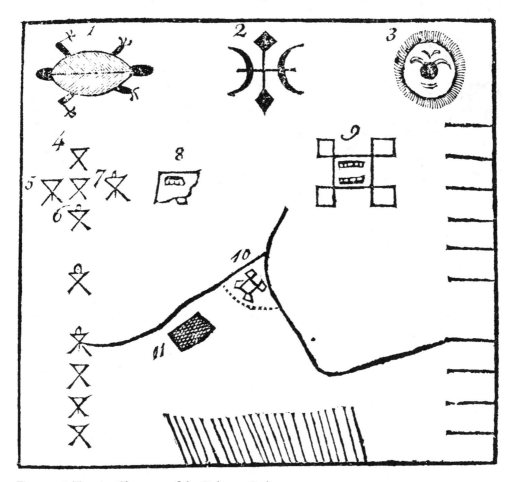

Figure 7.6. Tortoise Clan map of the Delaware Indians.

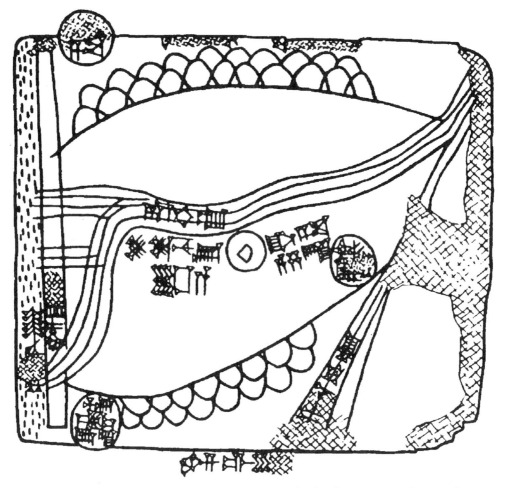

Figure 7.7. Clay tablet map from Nuzi (northeast Iraq), third millennium B.C. Courtesy Semitic Museum, Harvard University, SM 4172. Reprinted with permission.

located in the middle of the map. At the same time, ovoid pictograms designate two rows of mountains and the undulating continuous lines indicate a central river (with branches) between them. Thus, in the oldest known topographical map, the pictorial and the verbal coexist equally.

It might be thought that the verbal factor exists merely to supplement or to specify the pictorial—for example, to literally *name* the places that are given pictorial (or, alternatively, symbolic) representation in a map such as this. But the matter is more complex than this. On the one hand, it can be argued that words, far from being merely supplementary and therefore in principle replaceable, are needed in a fully informative map. They are needed not only for place-names—which are, after all, sheerly ostensive identifying references—but for explaining such things as the scale of a given map (i.e., its manner of representing distances between places) and for the exact meaning of pictograms themselves. Hence the importance of the "legend" on modern maps. On the other hand, words can never replace images altogether: words need images as much as the reverse. "It can be difficult," notes O. A. W. Dilke, "to list places accurately without some pictorial basis,"[33] and he gives as an example the Ravenna Cosmography, a hodgepodge listing of place-names of the eighth century A.D., which is difficult to regard as any kind of map even though it contains considerable geographical information. A few carefully selected images, situated in an appropriate spatial expanse, would have concretized and oriented these place-names in a maplike manner. Such images would have shown the proper positions of their referents in geographic space.

Among other early maps in the West to combine pictograms and words, both being inscribed on the cartographic representation of regions and territories, are the official maps of the Roman Empire as these were first ordered by Augustus Caesar and initiated by his son-in-law Marcus Agrippa in the first century B.C. One of the only surviving direct copies of such maps was discovered by Konrad Peutinger in 1508. Existing on eleven narrow strips of parchment (themselves probably the precipitate of an original scroll), the Peutinger Table shows towns and other notable architectural sites interconnected with roads that bear the inscription of distances.[34] The representation of roads is highly schematic, with no differences marked between major and minor thoroughfares; the placement of towns is indicated either by a mere crook in the road or by a condensed image of a town, typically a facade with two towers. In addition, temples and watering places, granaries and lighthouses, altars and officers' headquarters, are specified by highly stylized pictograms. Important harbors are represented by semicircular structures, and the very largest towns (Rome, Constantinople, and Antioch) receive personified images: the goddess Roma, with scepter, orb, shield, and crown is seen seated on a throne. Landscape features are quite cursory in form; larger rivers are represented as green lines and mountains as pale brown areas.

A section of the map, that depicting Rome and north-central Italy, is shown in Figure 7.8. Such a "strip map," as it may be called, offers an elegantly economic representation of information in symbols, words, and images. Most important, it brings together in one continuous cartographic space all three stages in Harvey's narrative of the evolution of maps from symbols through stories to pictures.

HOW TO MAKE A MAP: THE REPRESENTATION OF RELIEF

It is a remarkable fact that contemporary maps continue to do the same thing. Far from leaving symbolism or words altogether behind in order to move to pictorial images— a development one might expect on Harvey's progressivist model—such maps make emphatic use of all three cartographic vehicles.[35] Open up any road atlas and you will see that this is so. For example, Rand McNally's 1985 *Road Atlas* for the United States, Canada, and Mexico sets forth, on its very first page, in addition to a Table of Contents, an Explanation of Map Symbols. The symbols are presented seriatim in a column on the left side of the page; to each symbol, reproduced in color, is attached a brief verbal description: for an evenly segmented blue line we read "[road] under construction," a rectangular yellow strip symbolizes "urbanized area," a slash signifies a dam, a black line with regular marks on it indicates a railroad, and so on. In short, we witness a direct

Figure 7.8. Section of the Peutinger Table, showing Rome and environs, believed to be an eleventh- or twelfth-century copy of a third-century map. General Research Division, The New York Public Library; Astor, Lenox, and Tilden Foundations. Reprinted with permission.

juxtaposition of pictogram and verbal account—of *eikon* and *logos*. Or, more exactly still, we perceive three kinds of cartographic sign: words, pictograms (i.e., images formed on a relation of likeness, however minimal, to the geographic entity), and symbols proper (i.e., signs that are conventional or cultural in character).[36] Nevertheless, the three kinds of sign fall into two major groupings. On the one hand, pictograms exhibit some degree of resemblance with that which they depict: thus, national forests and grasslands are represented in the *Road Atlas* by a cartographic sign that is doubly pictographic: by virtue of color (green) and by virtue of texture (a stippled effect). Words and symbols proper, on the other hand, are arbitrarily related to what they designate: no relation of resemblance links the sunburst figure in the *Atlas* and "great river roads," nor is there any isomorphism between orangish pink and the "historic sites and monuments" for which areas printed in this color stand.

Perusing the *Road Atlas* also reveals that the signs for landscape features—notably for mountain peaks, forests and grasslands, and parks with (and without) camping facilities—are all pictographic in status, whereas the signs for roads and urban areas (and their features) tend to be conventional and nonpictorial. We may wonder why this is so. What is it about natural landscape that calls for readily recognizable pictorial images requiring minimal verbal interpretation? Does this mean that such images are cartographic signs par excellence, as would follow from Erwin Raisz's axiom that "a good symbol is one which can be recognized *without a legend*"?[37] Why does urbanity, in contrast, call for conventional designations that necessitate explanations in words? It would appear that conventionality itself—that is, the fact that such designations are established wholly by historical and interpersonal agreement—reflects the deeply acculturated character of the urban environment. This latter fact facilitates abstract and nonpictographic relations between the concerns of a particular city culture and the signs that designate these concerns, whether these be in the form of traffic signs, advertising signs, monuments, or the insignia at play in the map of a given city.

What, then, is it about natural landscape scenes that invites outright pictographic depiction? As we have seen in the history of American landscape painting, such scenes seem to call for precise pictorial representation from the very start. "Landskips," after all, are the painterly equivalents of pictographically rendered maps; as topographically precise, they were in fact used in a maplike way for purposes of orientation and knowledge. Maps of landscape features may be said to make a virtue of what in painting can become—and did indeed become in the late eighteenth century in the United States—a delimiting and even stultifying presence. Where painters such as Allston and Cole felt impelled to rebel against the narrowly topographic representations of earlier engraver-painters such as Smibert and the anonymous decorators of colonial homes, American mapmakers continued to seek ever more adequate pictographic representations of landscape features.[38]

In fact, by the end of the nineteenth century, the situation had gotten out of hand. Mapmakers in the United States and in Europe had so proliferated representations of the landscape that a virtual cartographic Babel ensued. As Lloyd Brown notes:

> In 1880, there were more than a thousand different kinds of geographical features
> shown on maps, starting with cities and towns, roads, beaches, and rambling on

through rivers, bridges, and ferries. There were other symbols that represented or-
chards and vineyards, and still others that represented manufacturing centers and
military installations. When it came to such important things as boundary lines,
compass points and the slope of the land every map maker had his own ideas. Map
makers needed to get together.[39]

And get together they did! At the Third International Geographical Congress in Venice
in 1881, standard measures were finally adopted, including the standardizing of place-
names and of units of distance measurement. Of course, this left open the possibility of
instigating new pictograms and symbols for large-scale (i.e., close-up) maps of particular
places—for example, in the case of Liverpool, England, one map of which created special
symbols for boat builders, ropers, coopers, and sailmakers.[40] But by and large, the direction was toward simplification of representation—as we see one century later in the
Rand McNally *Road Atlas,* which lists forty-nine cartographic signs in its Explanation of
Map Symbols, only nine of them directly related to landscape features.[41] Most of the
standard signs have been devised within the last century, but two are of very ancient
vintage: circles for cities and triangles for mountains, both of which are already present
in the earliest Babylonian maps.[42] It is as if *one* symbol of what is paradigmatically man-
made—the city, whether this be Babylon or Baltimore—and *one* symbol of what is just
as paradigmatically prehuman in the landscape, that is, mountains, have survived. All of
the rest have been subject to the vagaries of the history of cartographic representation.

If signs for boundaries between states or nations (typically composed of a series
of alternating long and short dashes overlaid by a wide band of a solid color, neither of
which bears any resemblance to natural traits of the landscape) are the epitome of conventionally established symbols, signs for the relief of the natural landscape are exemplary of iconic pictograms. By "relief" I mean not just elevation but the "lay of the land,"
its seen and felt and at least partially known outline, the way it presents itself to us in its
unique configurational character. The question of how to represent relief adequately is
basic to the representation of all landscape—as painters know by dint of long practice
(what else was Cézanne doing in his many representations of Mont-Sainte-Victoire than
depicting the contoured complexity of its height and depth?) and as mapmakers have
known from time immemorial: "The representation, recognition, and description of relief features [of a given landscape] from their contour patterns, ranging from simple examples such as concave slopes, spurs, and cols to complex land-forms, provides much of
the content of map-work as generally understood."[43] There are many means of representing relief, ranging from the mere designation of "spot-heights" (i.e., relative to mean
sea level) to full-fledged landscape color maps in two or three dimensions.[44] Let us consider briefly three of the outstanding manners in which the configured mass of a landscape is given representation in maps: hachuring, contour lines, physiography.

Hachuring. This is a technique that was developed at the end of the eighteenth century, especially for military use, and that then became widespread (more than eighty
forms of hachuring were devised in the nineteenth century). It consists in the drawing of
"hachures"—the French word for short, controlled lines—in such a way that the form of
a hill, mountain, or other protuberance is exhibited. The lines are drawn in the direction

of the steepest gradient, that is, in the direction of water flow down a slope. Thanks to their individual density or comparative closeness as a group, they represent the slope in its inclination: "by the way in which the cartographer chose to draw these hachures thick or thin, to pack them together, slant them, shorten or lengthen them, he could not only depict the contour of the hill but also show the varying slopes of its sides."[45] The result of this carefully crafted but often tedious effort is a representation that is at once detailed and accurate. Where earlier two-dimensional maps had shown relief by merely adding lines of shading to landforms—suggesting shadows cast by an imaginary sun—the technique of hachuring produced startlingly realistic representations in the hands of a master cartographer such as G. H. Dufour, whose atlas of Switzerland (1833–63) shows Swiss mountains and valleys in a dramatic and informative display (Figures 7.9 and 7.10). The advantage of this method of representing landscape is that it allows comparatively obscure features such as small rills or ridges to be given representation as parts of a slope whose degree of inclination can be determined with exactitude.[46] The distinct disadvantage, however, is that any "absolute information" (e.g., of height or distance) is not represented and has to be added by the superimposition of numbers to the map. Moreover, hachuring is less useful for low, rolling country and for small-scale maps, in which the landscape is seen from a considerable distance.

Contour lines. This is perhaps the most common of all techniques for representing the relief of a landscape in a map. First developed in Holland in the early eighteenth century for purposes of navigating rivers—yet another innovative Dutch contribution to the representation of landscape!—it is now employed routinely by almost every country as the basis for its official topographic maps. Contour lines give accurate information about absolute height and angle of slope by the relatively simple method of drawing lines that connect all points located at the same distance from a standard "datum" point, typically mean sea level. The extent of the interval between contours is critical; the narrower the interval, the more fully will the contour reflect the actual lay of the land, as is evident in

Figure 7.9. Portion of map from G. H. Dufour's atlas of Switzerland, 1833–63.

Figure 7.10. Dufour "hachure metre," showing number of hachures per inch for each five degrees of slope.

the contour map in Figure 7.11. The primary virtue of the contour map is that it allows a view of a given landscape from any vantage point to be represented adequately on a map. Indeed, so imagistically suggestive are contour maps that one can in principle reconstruct an entire landscape as seen in profile from the map, even though the map itself represents the same landscape in plan (i.e., from above) (Figure 7.12). It is evident from this diagram that some of the most basic elements of landscape painting—for example, the position of the horizon and the role of the picture plane—are easily projectable from a contour-line map. We are put in mind of Fitz Hugh Lane's effort to "map out" the cove of Norman's Woe in a careful drawing and to transcribe this latter, section by section, onto the two paintings of the cove he finished in the early 1860s. Extending his efforts, Lane could easily have drawn an actual contour-line map of the same cove that he sketched in profile. Indeed, the collusion between profile views and contour maps is close; this is because "a profile gives the most easily conceivable expression of slope along a certain line."[47] The most telling such line is the contour line.

The disadvantages of contour-line maps are twofold. On the one hand, because the placement of particular contour lines depends on the selection of certain crucial points in the landscape, the choice is left up to the individual cartographer's whim as to what counts as important—what functions as a genuine "landmark."[48] On the other hand, the

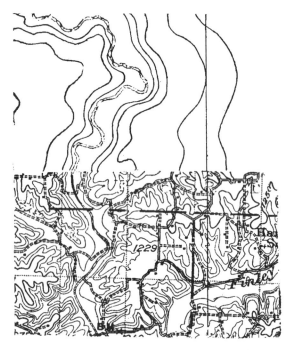

Figure 7.11. Two adjacent areas in the same scale and contour interval. From Erwin Raisz, *General Cartography* (New York: McGraw-Hill, 1948), p. 109. Reprinted with permission from The McGraw-Hill Companies.

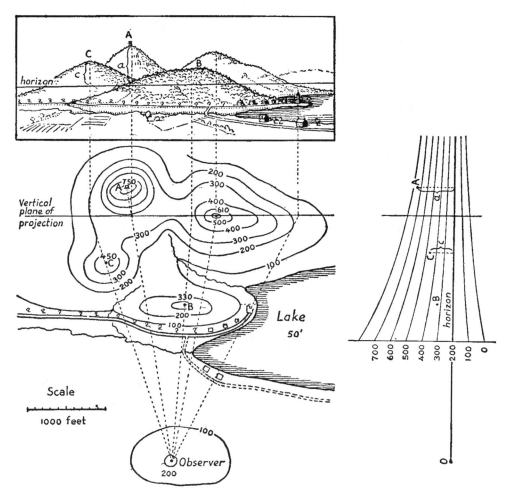

Figure 7.12. The landscape visible from any point as projected from a contour-line map. The vertical scale consists of hyperbolic curves. From Erwin Raisz, *General Cartography* (New York: McGraw-Hill, 1948), p. 113. Reprinted with permission from The McGraw-Hill Companies.

curvaceous character of contour lines tends to bypass any precipitous or angular feature of the landscape, including any feature that lies *between* two contour lines.[49] Thus the pleasing pictoriality of contour-line maps—the very feature that allies them with landscape drawing and painting—also acts as a limitation on its strictly cartographic merit.

Physiography. Also reminiscent of landscape art is the physiographic map—sometimes termed a "landform" or "morphographic" map.[50] Proposed by Raisz in a 1931 paper revealingly titled "The Physiographic Method of Representing Scenery on Maps,"[51] physiography offers a vivid and quasi-three-dimensional representation of a given landscape, including a landscape that is seen from afar in a small-scale map. All views in a physiographic map are oblique; that is, the land mass is seen in profile and from an angle that is always less than 90 degrees in relation to the surface of the area being mapped and typically from an angle of about 45 degrees (to be seen from precisely 90 degrees is to be seen in plan). Thanks to the special angle of view thus produced, the physiographic map, despite its recent official origin, harks back to the ancient pictographic maps of Çatal Hüyük and Bedolina, both of which display items seen obliquely. [52]

Raisz himself comes up with a list of some forty pictographic images of landscape features that can serve as elements of representation for physiographic maps; these images include iconic renderings of tundra, bush, savanna, selva, mesas, glaciated mountains, peneplains, volcanoes, and moraines.[53] To each image is attached an identifying word or descriptive expression, thereby constituting coherent image-*cum*-word units.[54] Of course, we are not restricted to these forty exemplary icons: "special symbols can be devised for [other] particular landforms."[55] The fact that this is possible indicates that the icons themselves need to be convincingly representative of given features—that is, they must obey a fairly stringent criterion of pictorial resemblance to be persuasive in this kind of map. As Raisz puts it, the physiographic map "suggests actual country and enables [one] to see the land instead of reading an abstract diagram [of it]."[56] This can be seen from one of Raisz's own maps of China (Figure 7.13). The very accessibility and pleasing aesthetic quality of such a map, however, bring with them the danger that it neglects the representation of certain important aspects of the landscape—for example, its exact elevation.[57] Furthermore, the physiographic method is not satisfactory for large-scale maps, where the displacement caused by oblique views becomes an obstacle to full representation: close-up images of large landmasses such as mountains literally block the representation of those features that are located behind them, something that does not happen in the case of hachuring and contour-line maps because of their top-down viewpoint.

The importance of such methods of representing the relief of a landscape, its variegated shape and implicit mass, cannot be underestimated. In the 1985 Rand McNally *Road Atlas*—a publication intended for the most general and efficient use by millions of travelers—there is attached to each map of an American state a miniature relief map rendered in green. This tiny relief chart is in effect an *icon of an icon*: it gives the relief in plan of the state's unique geomorphology. One can see at a glance, roughly and quite approximately, what kind of terrain the state possesses—landforms that are excluded from the larger map, presumably from a concern that the latter not become too cluttered. But the presence of two adjacent maps of the same region also reflects a recognition that the *kind* of representation is significantly different in each case. In the flat-out primary map, there is a representation of such (mainly conventional) things as state and city boundaries, roads and highways, state or national parks, and the location of cities; as for natural entities, only major water masses (ocean, lakes, rivers) are explicitly represented. As a consequence, the role of words and nonnaturalistic pictograms is massive. In the relief map, on the other hand, there is a minimum of conventional representations—only a black star marks the location of the unnamed capital city, and the initials of the state itself are superadded to the relief representation. This map presents itself as a condensed image of the state's configured landscape. If the major map represents the state in a maximally informative format by the use of largely conventional cartographic signs, the supplementary map provides a laconic representation of the same state in its three-

Figure 7.13. Physiographic map of China. From Erwin Raisz, *General Cartography* (New York: McGraw-Hill, 1948), p. 1113. Reprinted with permission from The McGraw-Hill Companies.

dimensional landscape being. Linking the two, however, is the shared image of the overall shape of the state—a shape that, even if conventional (i.e., historical-political) in origin, is nevertheless *perceived* as a pure pictogram in cartographic space.

ART, MAPS, AND PROJECTION IN DEPTH

To speak of the representation of relief in modern mapping is to suggest two basic themes of special pertinence to this book as a whole. The first theme is the *close collusion between art and maps.* This collusion has been strikingly evident from the prehistoric period down to the present day. Just as it is difficult to separate what is artistic from what is cartographic in a predynastic Egyptian bowl map of the mid-fourth millennium B.C.— or in the Egyptian "Book of the Two Ways" of circa 2000 B.C.[58]—so these two factors coexist in equally close embrace in medieval *mappae mundi,* in sixteenth-century portolan charts, and in seventeenth-century Dutch world maps.[59] Even if it is true that, by the end of the eighteenth century, the purely decorative parts of maps had come to be restricted to the cartouche and to border designs—the empty, unexplored areas of the earth, formerly offering space for uninhibited artistic expression in the margins of maps, had by then been filled in with detailed cartographic data—and even if, in the latter half of the twentieth century, any trace of decoration is absent from such maps as are contained in the Rand McNally *Road Atlas,* this evolution in no way undermines a continuing alliance between art and maps. The hachure maps of Dufour, many subtle contour maps, and the remarkable physiographic maps of Raisz still exhibit indisputable artistic merits; what they lack in pure decoration or in pure fantasy they make up for in an elegant representational realism. In fact, they combine accuracy of cartographic representation—an ever-increasing ability to depict the *form of the land,* the exact configuration of its "lay"— with a perspicuous and pleasing representational format. As much as in the paintings of Cole or Lane, Constable or Cézanne, we find a transfiguration of the place of landscape, its re-implacement and sublimation in the realm of representation.

The precision of representation so characteristic of modern maps should therefore not lead us to think that the collusion between art and maps has been lost—any more than Cole's concern for topographic accuracy in painting *The Oxbow* should make us think that he was less than fully artistic in his representation of the Connecticut River at that particular place. Ptolemy declared that "nobody can be a cartographer if he does not know how to paint."[60] Given that painting is an emblematic means of representing landscape, Ptolemy's words remain true today, when it is still the case that, as Leo Bagrow has said, "maps are works of art."[61] To agree with Ptolemy and Bagrow is to concur that mapping the landscape is not merely an *alternative* to painting it, as if it were a matter of a determinate choice between two distinctly different means of representation. It is to envision the possibility (and to admit to the frequently achieved actuality) of an active collaboration between these two approaches to representing the form of the earth in its endemic landscape features.[62]

A second theme to which the representation of landscape relief leads us is that of *depth.* Here we must consider that *every* representation of the contour of a given landscape is in effect a representation of its depth—or, more exactly, its depths. For if depth in its full significance encompasses the dimensions of height and width, as well as such other factors as the visible horizon and the panoramic sweep of open space—not to mention the three kinds of depth at play in Northern Sung landscape painting—then relief maps can be considered privileged means of representing the diverse depths of a landscape. Whether such depths are indicated by a set of hachure marks, by a series of non-intersecting contour lines, or by a concatenation of physiographic images, they are effectively conveyed to the map reader—as effectively as if they had been painted by the most accomplished landscape painter. To gaze at a Dufour map of a mountainous Swiss canton is to experience a version of the drama of depth that is so evident in Turner's celebrated painting of the Gothard Pass: in both cases, we are initiated into the precipitous depths of the Swiss Alps in all their frightening abyssal glory. If Turner's painting displays a high Romanticism, a melodrama that invites storytelling, Dufour's lithograph, by the dramatic chiaroscuro of its oblique illumination, also suggests tales of dangerous traversals. Both exemplify Harvey's thesis that the "picture story" comes between the sheerly symbolic and the purely pictorial map; in their case, the story is of how depths are perceived, explored, remembered, and represented in an image. That this image is so openly and freely painterly in the one case, and so carefully cartographic in the other, matters less than the fact that painting and map alike represent the peculiar depths that the Alps so distinctively embody.

Consider, moreover, that in J. J. Gibson's view the depth of ordinary perception is structured upon (indeed, *like*) a grid, a set of regularly decreasing spaces that recede from the perceiver even as they situate him or her in relation to these spaces. Renaissance paintings often make explicit use of such a grid. Maps, too, are gridlike; in fact, in virtually all Western maps constructed since the time of Dicaerchus and Eratosthenes, the grid or "graticule" has been its basic unit.[63] Just as the gridded schematization of perceptual space structures the environing world on the Gibsonian model of depth—a depth that is the basis of much Renaissance, and indeed, post-Renaissance, painting—so the division of the earth into quadrilaterals bounded by parallels and meridians serves to structure its surface as this is represented in comprehensive maps. This is not to mention the use of the grid in the division of urban space ever since the eighteenth century in the West.[64] In each instance, the depth that we perceive and experience is given a coherent and consistent representation; it is set forth in such a way that we can count on (and literally *count*) the regularity of its recession away from a given vantage point, whether this be at a particular intersection between coordinates of latitude and longitude in the case of a map, an actual street corner in a gridded city center such as we find in Philadelphia, or the implicit position outside the frame of a painting from which the spectator views its representational content. In these various ways, the depths of landscape (or cityscape) are at once *shaped* and *presented* for our viewing; they are "projected" before us in a clarified and modeled manner.

I put the word *projected* into quotes because of its special status in the history of cartography. Projection, construed not in the technical sense of an "orderly system of parallels and meridians on which a map can be drawn"[65] but as any effort to represent three-dimensional things in two dimensions, is indispensable for representing depth (or more exactly, relief-in-depth) insofar as all maps that convey the topography of landforms attempt to project depth, or the ground for depth, in one form or another. This has been the case from the very beginning: "since the earliest days of map-making, the depiction of relief has been one of the major problems of cartographers, for it involves the representation of three dimensions upon a plane surface."[66] To represent three dimensions in two dimensions is to condense depth into its planar representation; it is to *project* it there. Such projection is the common fate of all maps that attempt to be faithful to the earth's surface (except for those unwieldy "relief maps" that are themselves three-dimensional replicas of a much larger three-dimensional landscape). As Cortesào has written:

> A geographical map is the graphic representation of the earth's surface, or of a part of it, on a plane, through some system of projection and by reduction to a given scale, with the features symbolically represented. Even the earliest, mostly utilitarian, maps of Babylonians and Egyptians, drawn on clay tablets or papyri, those of the earlier

Greeks, mostly based on philosophical speculation, and those of many primitive peoples, even [those created] in modern times, imply some kind of projection (not necessarily in the strict mathematical sense), because, like any drawing or painting, they endeavor to represent, however inaccurately, three-dimensional forms on two-dimensional planes.[67]

Once more the affinity with painting is striking: just as a landscape painter may be said to project a fully configured landscape onto the flat surface of the canvas, so the map-maker projects the earth's variable landforms, their peculiar heights and depths, onto the equally flat surface of the drawn or printed map. It is not surprising to learn that map-makers were once called "Plattmakers"![68] For it is their established fate—a fate they share with landscape painters—to be involved in the perplexing project of representing depth, by definition a nonflat phenomenon, onto a flat surface of presentation.

Maps in particular are engaged in an unending struggle of "a plane trying to imitate a sphere."[69] As many theorists of maps like to point out, this is a losing battle. That which is the only strictly adequate model of the earth (its *paradigma,* as the Greeks would have said), namely, a spherical globe, is precisely what most fiercely resists representation in two dimensions. If it is true that "only a [constructed] globe can give a valid picture of the earth as a whole,"[70] it is also true that we cannot rely on globes themselves as maps (they are too cumbersome and expensive, and do not allow a single full view of the entire earth). This is not to say that any flat representation of a globe is adequate either. As David Greenhood has remarked sardonically, "we must accept the inevitable: that there is no way of flattening out a global surface and keeping intact all the useful information we wish to have."[71] In such flattening, there is a trade-off between shape and area: we achieve constancy of shape in a two-dimensional representation while sacrificing equality of area or vice versa: "no projection can be at the same time equal-area and orthomorphic."[72] So great are the necessary effects of distortion that we cannot even maintain sameness of a single geomorphic shape over any considerable region of a flat map. All in all, there are four major kinds of deformation entailed by the transition from a sphere to a flat surface: not only disfigurations of area and shape but distortions of distance and direction as well.[73]

Location alone is preserved across these distortions—that is to say, sameness of position as defined by the intersection of lines of latitude and longitude. This is crucial in that "we may distort shape or sacrifice area; we may fudge on distances and connive about directions . . . but one function of a map is indispensable: showing *true location.*"[74] It is indispensable because otherwise we would be lost in cartographic space and thus lose the one unprescindable character of any functional map: to specify our present location vis-à-vis other possible locations, *to indicate where we are in relation to where we have been or might be.*

Where we *are* is decidedly *on the earth*: on the body of Gaea, as the Greeks might have said. Just as it has been claimed that "the limitations of the globe [as a model of the earth] gave rise to cartography,"[75] so we may say that the earth itself gives rise to its own mapping and, in particular, to the projections that represent it in two dimensions. Indeed, cartographic projections are altogether dependent on the earth from which they spring; like Antaeus, they "derive their strength from Gaea (*geo*-graphy!) their mother, and where they lose touch with her they weaken."[76] Just as the "world" of the work of art in Heidegger's conception exists only in relation to the "earth" as its ultimate self-secluding ground, so the world of maps (and maps of the world!) arise only in relation to the earth that is their unique source: a source so reclusive that its own most apt geometric figure, the sphere, is incapable of complete representation in the flat world of drawn and printed maps. This is so even though there are in principle an infinite number of possible projections of the earth onto maps, just as there are also innumerable possible worlds projectable from the earth into works of art.[77]

The middle term that links earth to map, and world to earth, is *landscape*. This is what all four items possess in common. It is their shared integument, their double-sided flesh. Like flesh, landscape is the living and lived surface of a body—the earth's body. It is how earth appears to the gaze and the touch, how it *surfaces* to view and grasp. It is also what is projected, complexly, into the two-dimensionality of maps and taken up, multiply, in the n-dimensional worlds of works of art. The relief that is represented in maps belongs ultimately to the fleshlike surface of landscape, just as the dark recesses of earth that are brought forward in painting belong to the depths of this same surface. If maps for the most part represent overt landforms in detail and in all possible verisimilitude, paintings represent the land's surface in its implicit and concealed character. Both accomplish a truth that is based not on isomorphism but on bringing the reclusiveness of earth into the openness and light of world.

CHAPTER 8

Cartography and Chorography

Geographia imitatio est picturae totius partis terrae cognitae.

—Ptolemy, *Geography*

There is in Dutch painting a pleasure taken in description that is akin to what we find in the world of maps.

—Svetlana Alpers, *The Art of Describing*

CARTOGRAPHY VERSUS TOPOGRAPHY

The reader will recall how often the issue of topography has arisen during discussions of the representation of landscape in painting. It arose at first in an effort to trace the arduous birth of landscape painting in the United States. Such painting, I suggested, had to win itself *away from* considerations of topography (as also of interior decoration) if it was to establish itself as an art in its own right. But the question of topography, that is, of the precise representation of landscape features, would not go away. It kept arising—not just in the expected form of the analogy between painting and photography (a link that, if anything, liberated painting more completely for nontopographic aims) or in the consideration of painting as panorama (i.e., the impossible project of representing a vast totality of places in a single continuous sweep), but within the discussion of landscape painting itself. Early in this book we witnessed the depth of Fitz Hugh Lane's topographic concerns. Lane, like so many other artists of his generation, had been an engraver called upon to provide topographic "views" of his native Gloucester. It was a natural step for him to continue to furnish such views even after he had left engraving as his principal occupation. The fact that the views in which he was most interested in his later years were not those of a townscape but of unpopulated parts and stretches of the surrounding landscape and seascape did not mean that he was any less topographic in his approach; on the contrary, he remained rigorously committed to a topographic technique of exact transposition from drawing to painting.

Lane is an extreme case, and we were led to contrast his contemplative sublimity with the apocalyptic sublime in Bierstadt and Church. But these latter painters, for all their visionary prowess, still engaged in topographic transcription—most notably in the case of Church, who produced minute and precise pencil drawings and oil sketches of every aspect of his subject matter.[1] Even Cole, the first fully accomplished American landscape painter, based his breakthrough oil paintings of the late 1820s on careful sketches that he made in the Catskills in 1825. Moreover, as if to continue the earlier gesture of Smibert's frankly topographic *Vew of Boston* of 1738, Cole painted his own panoramic view of the same city a century later; and in *The Oxbow*, as we have seen, he painted a vista that he had earlier drawn in pencil at least twice. Only in an oil sketch of the same subject matter did Cole transcend limited topographic considerations and realize Durand's ideal of pursuing a subject to the point that the resulting work "*represents the model*—[and] not . . . merely resembles it."

Nor is Constable, whom we have taken to be paradigmatic of such unconstrained representation of landscape as Durand prescribed, by any means beyond topographic concerns. His pencil drawing *Dedham from Langham* of circa 1812 (Figure 5.5) unashamedly seeks resemblance to the subject matter of its view, even if the oil sketches done just afterwards are much less bound to this ambition. Indeed, Constable's cloud studies of the early 1820s (e.g., Figure 5.8) can be considered topographic delineations; if the sky is (in Constable's own words) the "chief organ of sentiment," his drawings and paintings of its changing face are done in the spirit of accuracy, of truth-about, as I have termed the topographic ideal. Still more important, Constable's lifelong effort to depict the Stour Valley amounts to a virtual mapping of his home region. In the course of discussing two 1815 paintings of views from Constable's father's house in East Bergholt, Michael Rosenthal comments that

> Constable was concerned with as naturalistic a representation of these scenes as possible, a quality which extends beyond reproducing local forms or colours in paint, for if the pictures are checked against the 1817 East Bergholt Enclosure Map . . . they will be found to follow the topography very closely. Field boundaries match, and buildings are where they ought to be.[2]

To illustrate this point as graphically as possible, Rosenthal presents, along with a detailed Ordnance Survey Map of 1805, the Enclosure Map to which he has just referred.[3] A perusal of this map shows that Constable's two paintings—and doubtless many others painted after he returned to the Stour Valley region in 1802—in fact "follow the topography very closely."

Not only will a topographic concern not go away easily; it remains of central significance in the paintings of several of the greatest of modern landscape artists. Further, it retains such significance in a manner that not merely allows but fosters *a conspicuous alliance with maps*. This is most obviously true of seventeenth-century Holland, to which we shall turn in the section titled "Pictographic Propensities" in this chapter. It is also true of England at the time of Constable: "The rise in landscape painting in Britain during the eighteenth and early nineteenth centuries coincides with a renaissance in British cartography."[4] For our purposes, the crucial questions are: What are we to make of the link between topography and cartography within a form of landscape painting that is in no way devoted to a slavish imitation of nature? Is the link merely contingent, or does it point to something more profound?

Topography exists *between* cartography and landscape painting. With cartography—

that is, "the drawing of maps and charts" *(O.E.D.)*—topography shares a double concern for *(a) drawing* or *tracing* certain physical features of natural places (as the common stem, *graphia,* indicates by its allusion to graphic representation of any kind); and *(b) describing* these places in ways that are true about them in terms of their identity, history, and various qualitative characteristics. It is evident that "describing" possesses considerable scope in that it involves inscribing or writing as well as drawing. Descriptive writing is by no means foreign to cartography, as we see in the case of maps in which the sheer linear image of a country is supplemented with place-names, a key to distances, an explanation of pictograms, or an identifying cartouche. Topography itself, literally, "the tracing of a place," is defined in the *O.E.D* as "the science or practice of describing a particular place, city, town, manor, parish, or tract of land; the accurate and detailed description or delineation of a locality." It remains that if cartography stresses the tracing of lines more than describing in words—there can be, after all, a map *without* words, difficult as it may be to interpret, as we saw in the Bedolina and Seradina maps (Figures 7.4, 7.5)—topography characteristically attempts to transcend the linear as such (e.g., in the form of boundary lines between regions) so as to achieve a pictorial representation that is descriptive in character. The description may not be itself verbal, but it invites verbal commentary—as occurs in Constable's cloud studies, to each of which the artist affixed in writing the precise time, place, and weather condition of the sky scene he was sketching.

Furthermore, whereas cartography can map a region without representing any of its local features (it need only trace the correct overall shape and the correct relative size in order to be considered minimally adequate, for example, in most conventional maps of the United States), topography aims at setting forth precisely those local features that comprehensive cartographic maps assume they can afford to neglect. This is why the very conception of a "topographic map"—which brings together in a single technical term both cartography and topography—is that of a map in which such local items as hills, highways, ponds, and other constituents of the landscape are given concrete representation. In particular, such a map tends to represent in detail the configuration of the terrain, including its comparative elevation and the exact shape of its parts: in short, the relief of the landscape.[5] Topographic maps of a specifically physiographic sort provide us with a representation of a landscape *in its depth*: they are two-dimensional pictorial analogues of a landscape's three-dimensional features. Not only do they furnish the perimeters that are the stock-in-trade of a purely cartographic representation of a land mass; they also de-scribe—literally, "draw a line around"—the local features that make a place that *particular* place and no other, that make it "a detailed description or delineation of the features of a locality" (in the reformulated first definition given in the *O.E.D.*).

Locality is a key word, because it designates not an indifferent site (as is the case with the featureless space of a sheerly cartographic map), but a quite unique place not to be found elsewhere on earth. A group of related localities make up a "region" as we have used this term with reference to John Constable's paintings of East Anglia. These paintings do not so much "follow the topography" of the Stour Valley—this phrase of Rosenthal's risks confusing the structure of the landscape with the representation of that structure—as *constitute its topography in painterly terms*; for, if topography is also definable as "the features of a region or locality [taken] collectively" (*O.E.D.*; second main definition), Constable's paintings indeed compose a topography of his native region, its collective representation in paint.

The example of Constable demonstrates that just as cartography and topography share certain basic concerns and traits, so there is also considerable overlap between topography and landscape painting. What the latter share is a common concern with the particularity of given places—with what makes these places the unique localities they are. Indispensable to both is the idea of the nonexchangeability of represented place. Although the representational space of a map could in principle be occupied by any geographical entity—that is to say, any of several countries (or parts of countries) could fit into a given grid as determined by latitude and longitude lines on a modern conventional map—such is not the case for what I have called the place-of-representation in a topographic map or a landscape painting. Onto a genuinely topographic map only *this* set of represented features can fit, for the primary determinants of such a map are not the outlines of rectilinear grids or the legal borders between states or nations but the inlines of discrete and unrepeatable shapes. These shapes belong to the land as its intrinsic contours: they help to make it one particular part of the earth rather than another. The fact that they are merely drawn or traced—as are the outlines of pure cartography—should not mislead us. Each contour traced by a topographic inline conveys the concrete *morphē* or form of the particular landform that is represented, the figure of its configuration. As thus de*lineated,* the landform is also de*scribed.*[6]

This is not to deny important differences between painterly and properly topographic representations of landscape features. Topographic representations in maps are immensely abbreviated and may even be quite abstract or symbolic. Typically, however, basic items of the landscape such as "perennial streams," "small rapids," "foreshore flat," "disappearing stream," and so on,[7] may be represented on a topographic map by extremely simplified icons or pictograms that at once reflect the contour of the terrain by virtue of their linear flexibility and yet are descriptively self-contained by virtue of their constant, coded meaning. Even pure contour lines are coded by their "contour interval," which gives the unit of measurement between successive contours. Otherwise put, topographic maps may rely on iconic and symbolic signs to relay information about the landscape they represent, though both are employed in a much more pictorial mode than is found in the use of such signs in nontopographic maps. Paintings, in contrast, characteristically dispense with icons and symbols altogether. This is just what Constable does in his drawings, oil sketches, and full-scale paintings of the Stour Valley: here even the cursory verbal descriptions of his cloud studies give way to fully imagistic representations. Such representations doubtless employ various schemata in their organization of pictorial space.[8] Yet these drawn, sketched, and painted images bear the burden of representing place *by themselves alone.* They do so by picking out various features of a place—not necessarily the same features as a topographic map might select—and by representing them in a way that renders them recognizable as features of that very place.

TOPOGRAPHY AND DESCRIPTION AS MIDDLE TERMS

Topography, then, finds itself implicated with two otherwise unlikely partners, cartography and landscape painting. As paired with cartography, it belongs to the larger study of "geography," the drawing and describing of the earth as a whole. As paired with landscape painting, it belongs to "chorography," the tracing or describing of particular regions: *chōrai,* Plato's term in the *Timaeus.* Ptolemy, who first proposed the distinction between geography and chorography—a distinction that continued to be respected through the eighteenth century in the West—linked the pursuit of geography with science and chorography with art.[9]

As affine with both geography and chorography, topography has predictably suffered from an identity crisis. Also foreseeable is the fact that it has played an ambiguous role in the history of landscape painting, sometimes exalted as an ideal for such painting and sometimes condemned as a debased and delimited effort. No wonder we have found it to have such a complex and checkered career in the evolution of landscape painting in the United States and England: as middleman between the cartographic and the painterly, here too it is often the odd man out.

There was a time when things were different—when cartography and painting were intimately allied and thus were far from being antithetical ventures calling for mediation by a third term such as topography. I refer to the seventeenth century in the Netherlands, above all in Holland. More than a hundred years before landscape painting was to arise in Britain and the United States in full dress, Dutch landscape painters were already working at full tilt. From earlier beginnings in Jan and Hubert van Eyck and in Pieter Brueghel (both father and son) and Jan Brueghel, landscape painting had become a central genre by the first half of the seventeenth century in the Low Countries. It featured such stellar practitioners as Jan Van Goyen and Albert Cuyp, Saloman and Jacob van Ruisdael, Hercules Seghers and Philips Koninck, and (in a delimited part of their work) Rembrandt and Vermeer. It was, in short, a glorious age for landscape painting, which came into its own more completely in Holland than anywhere else in Europe.

It was also a glorious age and place for mapmaking. A survey of maps and mapmakers proclaims that "The Low Countries for roughly a century, from 1570 to 1670, produced in some respects the greatest map makers of the world . . . For accuracy according to the knowledge of their time, magnificence of presentation and richness of decoration, the Dutch maps of this period have never been surpassed."[10] Such mapmakers as Abraham Ortelius, Gerard De Jode, Gerard Mercator, Jodocus Hondius, Willem Janszoon Blaeu, and Jan Jansson were cartographers of a stature comparable to that of the landscape painters of the period.[11] Their works, inaugurated with a flourish by the publication in 1570 of Abraham Ortelius's *Theatrum Orbis Terrarum* (Atlas of the whole world), ushered in a century of remarkable activity that was no less creative and significant than the ferment of landscape painters in the same century. In fact, there was an active liaison between the two groups. Maps were produced in such sumptuous formats and were so richly decorated that they hung on the walls of prosperous burghers along with paintings of landscape, still life, common life, and portraits. Painters themselves were often employed in the production of these maps, especially for the drawing and coloring of such integral elements as the cartouche and insets along the sides. These insets included everything from tableaux featuring local inhabitants of the mapped region to topographical views of cities in that region. Not only was there a complex mixing of genres in this fashion, but the real possibility of confusing genres also arose. Thus certain maps presented themselves as quasi-topographical views of their subject matter, complete with a spread-out vista and a horizon. For example, Pieter Saenredam's engraving titled *The Siege of Haarlem* defies classification, because it is at once a map (in its precision of locations and in its prominent explanatory cartouche) *and* a landscape view of Haarlem in a state of siege that includes a foreground military camp and a middle-ground group of barracks, with Haarlem and the sea in the far distance. Saenredam himself is usually regarded as a painter, but in fact he was a serious cartographer as well.[12]

This was an extraordinary situation indeed, unprecedented both in the history of art and in the history of mapmaking. Cartographers and landscape painters freely exchanged roles, and in any case influenced each other profoundly. Vermeer's celebrated *View of Delft,* often assumed to be a masterpiece sui generis,[13] was in fact modeled on specially situated topographical views of cities that were part of map collections such as Braun and Hogenberg's *Civitates Orbis Terrarum* (1587–1617), to which Vermeer almost certainly had access.[14] In this respect, Vermeer's painting is as much a mapping of Delft as a painting of it; it is in fact *both at once.* Cartography and landscape painting here join forces in such a way as to raise topography to a new level of accomplishment. *View of Delft* is certainly topographic in the customary sense of a detailed rendering of a particular place. But it is also something more. Beyond giving us precisely delineated cartographic certitude, this painting also proffers the clearing or open vista of landscape (here in the guise of cityscape); it is at once a chart of Delft and a portrait of it. "In Vermeer's *View of Delft,*" remarks Alpers, "mapping itself becomes a mode of praise."[15]

Just as revealing is another painting of Vermeer's, *The Art of Painting* (see plate 11A). In this interior view, landscape is confined to the large wall map that hangs behind the female model: but what a magnificent confinement is this! The map is not only centrally situated in the pictorial space, occupying a full fifth of this space; but it is more finely detailed than, and as richly colored as, any other part of the painting. Here indeed mapping is a mode of praise: the panegyric is as much to the map as to the model who poses demurely before it holding a long trumpet and a book. The intimate and enclosed domestic space of the room, illuminated from a window hidden on the left, is juxtaposed with the open cartographic space of the map, which depicts the Netherlands in its major central portion while offering topographic views of notable cities of the Low Countries on either side. Mixed in with these two primary pictorial modes of the map—cartographic and topographic, respectively—are identifying labels that designate particular places (including city views) that are depicted on several cartouches as well as on the strip at the top and bottom of the map. At the bottom, there are the words "I Ver-Meer" and on the top, among other words, the significant word *Descriptio* (see Plate 11B).

Image and word conjoin in the very notion of *descriptio.* It is by description that mapping and painting commingle as they do in Dutch art of the seventeenth century. These otherwise antithetical enterprises meet in a common venture of description, which holds them in equipoise without the need to revert to strict topography (except in a literally marginal manner at the very edges of the map on the wall). It would be question-begging to ascribe this remarkable conjunction to the supposed fact that Holland in the seventeenth century belonged to "the Age of Observation."[16] It is more helpful to ponder, as does Alpers, the meaning of "description" as what draws mapping and painting together—not only at this historical moment but at others as well (e.g., in the American landscape lithographs of the 1850s that mapped the Western landscape with an eye to the laying down of a transcontinental railroad); for it is precisely as a "hybrid term that [description] brings together things basically dissimilar in nature."[17]

Consider more closely Vermeer's *The Art of Painting* (Plate 11A). In its framed space are brought together several modes of describing. The model is being described by the artist, who is seen just beginning to paint her head. The artist himself, viewed from the back, is in turn described in his activity of describing: here is description to the second power! At the same time, the interior of the room in which artist and model are together situated is elaborately described—as is dramatically illustrated by the swept-back drapery at the left, looking like a stage curtain opening onto the scene. Accoutrements of acting are described on a table in front of the model, who dresses herself (and thus implicitly describes herself) as Clio, Muse of History. And the map—depicting an actual map manufactured by Nicolaas Visscher whose sole remaining copy is now preserved at the Bibliothèque Nationale in Paris—is shown hanging on the wall. Because the map is itself a description of the Netherlands and its main cities, another instance of double description is here presented by Vermeer. To this double description is added the inscription of words, including the very word *Descriptio.*

In these various ways, the richly descriptive character of the painting returns to the very root of description in *scribo* (and thus to its Greek equivalent *graphō*).[18] As de/scriptive, *The Art of Painting* is about writing; or, more exactly, it is about the representation of pictorial space conceived on the model of writing, that is to say, on the model of a flat surface on which are inscribed images as well as words—just as occurs in the case of maps themselves, which offer two-dimensional surfaces filled with depictive as well as verbal elements.

Both mapmakers and painters conceived of their respective forms of representation as involving a surface on which "views" of various sorts are described. In both instances, we can agree that "images are drawn or inscribed like something written."[19] As the *de-* of *descriptio* implies, images created under the aegis of description are both like something written *about*—the identifiable subject matter of the image—and something written *down*: down on the surface of the map or painting. Thanks to this double aspect of description, "maps and pictures are reconciled,"[20] both being modes of describing the world. In this light, it is not surprising to find them coexisting so elegantly and equably in *The Art of Painting.* The art of painting, like the science of mapmaking, is to describe in both senses.

This complex descriptive undertaking present in maps and paintings alike would not be possible without a common conception of pictorial space. In this conception, pictorial space is *the space of survey. Survey,* a word we shall meet again in chapter 11, is a term that significantly applies both to maps and to landscape paintings. In fact, the Dutch word *landschap,* from which the English words *landskip* and *landscape* both ultimately derive, refers to the land regarded as an object of description for professional surveyors as well as painters.[21] For both classes of landscapists, the graphic description of the land was premised on a sense of space as surveyable.

By "surveyable" is meant three things. First, such space is *unframed.* Unlike Albertian space, which projects the world onto a separate picture plane with its own frame (i.e., a frame that functioned as a window through which the world is seen), survey space is not projected onto a framed surface. (A corollary of this is that devices of framing, such as Lorraine's *repoussoir,* are of less interest to Dutch painters than to many of their European contemporaries.) Second, the viewer is not placed in a fixed position outside of pictorial space—as occurs on the Albertian model—but is himself *situated in the same space that he surveys.*[22] Third, as the word "*sur*-vey" implies, the view is typically from *above,* from an elevated point of view. But the point of view is not itself a "point," that is, a fixed position in space. The viewer is suspended in an indefinite viewing area that lacks determinate location. Nevertheless, the viewing itself is felt to take place at a considerable distance, though a distance that represents a privileged access to the scene surveyed below.[23] The privileged access stems from the basic sense that the viewer, however distantly situated, nevertheless belongs to the surveyed space.

Thanks to sharing this threefold notion of surveyable space, the descriptions that occur in maps and landscape paintings are remarkably confluent—so much so that maps can be depicted within painted pictures, and pictures within maps. Indeed, the commonality is such that on certain occasions we cannot discern whether a given representation is a map or a painting—not just in the already mentioned case of *The Siege of Haarlem* but even more strikingly in Jan Christaenszoon Micker's astonishing *View of Amsterdam* (see Plate 12), in which the beclouded city is seen from a great (and indefinite) height yet with a decidedly cartographic clarity that is reinforced by the unfurling of a cartouche on the lower right as if to reclaim the painting as a map at the last minute. In this instance and others like it, we encounter a deeply ambiguous entity that can only be designated as a "map-painting" or a "painting-map."

From the inaugural drawings of Hendrick Goltzius in the first years of the seventeenth century to Van Goyen's and Rembrandt's landscape sketches, and from Koninck's *Landscape with a Hawking Party* to Jacob van Ruisdael's *View of Haarlem*—which is to say, over a fifty-year period—Dutch landscape painting was at least quasi-cartographic. The flatness of the actual landscape conspired in this development; but it was the way in which the land was *represented* that made it pictorially maplike. Landmarks were picked out in paintings as if they were identificatory icons on a map. The land itself was treated as a planiform pictorial surface tilted upwards and outwards, while also being divided into an informal grid of receding planes of space. Instead of thinking of the painting's pellicle—its surface-place—as a window *through which* to view the actual landscape (as again in much Renaissance painting), the working surface was regarded as itself like the space to be depicted: flat, two-dimensional, and opaque. Just as a mapmaker deals directly and exclusively with the surface of the plate he is engraving, so the Dutch landscape painter worked over the surface of the canvas as if it were the surface of a map he was inscribing.

One concrete effect of this circumstance is that for the landscape painter "surface and extent are emphasized at the expense of volume and solidity."[24] It was against this emphasis that Cézanne attempted to restore the density of natural things to landscape painting and that Cubism attempted much the same thing in a more abstract modality. Another effect was the favoring of panoramic landscape views: views that themselves constitute surveys of the depicted world. Here we find a legitimate sense of panorama that, unlike the ill-fated works that were so popular in nineteenth-century American culture, followed logically from a coherent conception of pictorial space. If such space is viewed as laid out on an unframed surface that the artist views from above, it is only to be expected that he or she will regard this space as panoramic. Thus the resulting painting can be construed as a pictorial description of a panoramic view.

No wonder this stretched-out, panoramic notion of surveyable space—the painterly equivalent of cartographic space—had such a powerful appeal to Constable, who set out to survey the Stour Valley two hundred years after Goltzius first sketched *Dune Landscape near Haarlem*. Dutch engineers had laid out the irrigation system in East Anglia, which shared with Holland an insistent flatness.[25] No wonder, either, that American Luminists such as Lane and Heade were so taken by their Dutch predecessors: they wished to create a legitimate form of delimited panoramic painting that invited the viewer into its precinct without the special framing effects of Lorraine, Poussin, or Rosa.[26] In their case as well, the viewer senses that she stands—as the spectator of the painting's purely *pictorial* space—in the same continuous space as that in which the represented landscape itself is set. In the United States as in England, the effect of the Dutch mode of mapped landscape was profound. It was as decisive as the influence of the Claudian schema of a framed and projected pictorial space, a sense of space that stemmed ultimately from Alberti.

One important reason why Cole's *The Oxbow* can be regarded as emblematic of American painting in the nineteenth century is that it succeeds in bringing together two different features of modernist space, that of the Albertian "vanishing point" (in the foreground) and the Dutch "distance points" (at the horizon) within the framework of a single painting.[27] The foreground containing the vanishing point frames a scene that is panoramically presented in the farness established by distance points. But in between the Claudian foreground (which the viewer does not share) and the Dutch background (which is maplike in its precision) lurks an abyss that is unique to American art. This abyss, as we know, reflects a feeling of distinct danger in the wilderness that made up so much of American space in the early nineteenth century. Nothing like this natural menace is present in Dutch art, which is composed and self-reassured in comparison. The panoramic tendency of this art is at one with the serene sense that the natural world *has already been traversed* and is now fully inhabited. Thus, little remains to be done except to survey the scene and to represent it accordingly. Whereas the American landscape is there *to be explored*—with all that this entails of peril and risk—the Dutch landscape is there *to be described,* whether in a map or in a painting (ideally, in a single work that combines the cartographic precision of the one with the pictographic evocativeness of the other).

The allure of description is virtually irresistible in the settled scene of seventeenth-century Holland, a scene in which action in the form of global exploration had given way to a passion for observation shared by mapmakers and painters (not to mention scientists in their "cabinets"). Vermeer emblazoned the word *Descriptio* on the edge of the map in *The Art of Painting* for good cause. Description stands between mapmaking and landscape painting as their common boundary. As such, it replaces topography as their *tertium quid.*

It replaces something else as well: the tendency to assume that narration is the raison d'être of landscape painting. From the allegorical landscapes of Roman and medieval art to the dramatic landscapes of the late Renaissance—in all of which landscape is the backdrop to an essentially theatrical action—it had been presumed that landscape painting exists to dramatize nature or, at the very least, to set the stage for human drama. This is still the ruling thought in Dürer and the Brueghels, Lorraine and Poussin, Rosa and Rubens: for all these artists, a landscape painting should suggest, if not explicitly depict, a dramatic action of some significance. Lorraine's use of the *coulisse* and the *repoussoir*— both of which serve to create a quasi-narrative sense of imminent action and suspense— is only an exemplary instance of a pervasive practice. When Cole adopts similar conventions in his early Catskill paintings, he is only following suit. But in *The Oxbow* he adds to these dramatic and dramatizing devices—narrativized in the blasted stumps of the foreground and the minatory storm breaking over them as well as in the precariously perched painter himself—the decidedly undramatic and nonnarrative space of a rural paradise that is beyond danger and even perhaps beyond time. In doing this, he is in effect repeating the quasi-cartographic commitment of Dutch painters to the depiction of an already known and traversed realm that is surveyable in the timeless moment of a single glance. Concerning this part of his painting, a part that maps the flatlands of pastoral promise, we may say what Alpers says of the seventeenth-century painters of the Netherlands: their "works are descriptions, but not in the rhetorical sense, for description in these cases is not a rhetorical but a graphic thing. It is description, not narration."[28]

PICTOGRAPHIC PROPENSITIES

To be descriptive rather than narrative is to engage primarily in space rather than in time. This is just what we might expect of an art—indeed, an entire culture—that has become newly reliant on images, finding in imagistic or pictorial representation a truth that formerly resided in texts. Texts tend to be narrative in nature, if only because of their endemic linearism: when they do not actually tell a story, they give an account (e.g., an allegory or a fable, an explanation or an argument) that is storylike in its sheer successiveness. As I have suggested, painting before the Dutch era modeled itself on textuality and was explicitly or implicitly narrative in intent and structure. Landscape, if it is allowed any part at all, formed a literal pre-text: it was "scenery," a mere setting for the historical, religious, or fabulous events that received the primary attention of the viewer. The action was in the action itself, the *when* of its succession, rather than in *where* the action takes place. Even in the remarkable precursory steps taken by the early Flemish masters such as the van Eycks, Robert Campin and Hugo van der Goes, Joachim Patinir and Hieronymus Bosch, landscape, though brilliantly depicted, remains at best a context for a central story. Only in Dürer's series of watercolors done on his return to Germany from Italy in the spring of 1495 do we find anything like a purely descriptive approach to landscape for its own sake; but these remarkable works remain sketches and not completed paintings.[29]

It took the Dutch painters of the seventeenth century to appreciate landscape as something more than pretext or context, as the focal subject of full-dress paintings. They accomplished this by removing painting from the mystique of history and time and by showing that, *regarded as pure image,* it had an interest of its own. This interest is an interest in description—in the pure pictoriality of the picture, its sheer show, its traced and painted surface—rather than in any implicit or explicit story it might have to tell.

This brave step, unprecedented in Western art, would not have been possible without the inspiration of maps, which give us images of the larger world; they provide "pictures of all parts of the known world," in Ptolemy's phrase cited in the epigraph to this chapter. They bring the unknown world, the *terrae incognitae* of unexplored domains, into the known image—not unlike Freud's exploration of the unconscious.[30] Although maps rarely tell a story as such (implicitly, however, they convey a tale of exploration and all too often, exploitation), they embody genuine geographical knowledge, and were so taken by the Dutch. If a map is "an image that functions as a kind of description,"[31] then it was only natural for it to be something of a paradigm for art as well as science. What could be represented on a map was fit knowledge for the culture at large. At once highly condensed and economical in format, and yet eminently readable as a surface of inscription, maps became the cognitive cynosure of an entire culture; for on a map much can pivot: "astronomy, world history, city views, costumes, flora, and fauna came to be clustered in images and words around the center offered by the map."[32] History itself is subsumed into cartography—and thus time into space.

Maps represent measurable space and thereby describe it. This is the key to their emblematic power. A formula of Kant's, though not intended for maps as such, here finds effective application: "measurement of a space (as apprehension) is at the same time a description of it."[33] The mensurational properties of maps are at once epistemological (i.e., as aiding in the depiction of world-space) and orientational (e.g., as guides to navigation). These properties assure the value of maps for science and human conduct, respectively. At the same time, as pictorial in status, maps can become allied with art, so much so that (as we have just seen) they are even indistinguishable from it at certain moments. Speaking of the supposedly strict distinction between maps and works of art—a distinction that is based on the false premise that all art is narrational in essence—Alpers observes that "at a time when maps were considered to be a kind of picture and when pictures challenged texts as a central way of understanding the world, the distinction was not firm."[34] Among several reasons, the distinction was not firm because in seventeenth-century Holland maps and landscape paintings both arose from a survey space they shared in common.[35]

Mensurational and spatial features of these maps and paintings are so immanent to them—so deeply buried from explicit view—that we must look elsewhere for a more vivid emblem of the descriptive *nisus* of the age. This emblem is found in the blatant and brilliant pictoriality that characterizes so many of the cartographic and artistic productions of the seventeenth century. By "pictoriality" I here mean that aspect of a visual image which *shows itself as an image*—which, as it were, announces and advertises itself as such. This pictoriality is even more evident in maps of the period than in the paintings. To gaze at the map of Africa in Willem Janszoon Blaeu's *World Atlas* (1630) is to look at a pictographic extravaganza: a map featuring images of elephants and various exotic animals on the land, ships and sea dragons at sea, pairs of native figures at its right and left borders, and detailed topographic studies of cities and islands at its top edge

(Figure 8.1). All of these images, along with appropriate words of identification, are co-present on the same surface plane, jostling each other amicably within that space. The oval and rectangular frames of the insets around the margin coexist with the gently rounded parallels and meridians, as well as with the rugged contour of the African continent itself. A veritable society of images is brought before us in rich and multifarious display, satisfying not only a cognitive curiosity but a distinctly visual appetite for the purely pictorial.

When an image of such a map is in turn set within another, painterly image—as happens so strikingly in *The Art of Painting*—we are treated to a double feast of vision. Then map images and painterly images collaborate, each enhancing the pictoriality of the other. The painting becomes in effect the image of an image or, if you will, the description of a description. We begin to appreciate the rich ramifications of an approach in which spatiality is favored over temporality; in that approach not only can we condense information in an efficiently communicative manner—such is the virtue of the first-order cartographic image as such—but we can combine images in ways that reveal new aspects of depth and meaning. Thus, to see a map *depicted* in Vermeer's painting is to sense its pictographic potential with special vivacity; it is to grasp, even more lucidly than when we are looking at the map by itself alone, the potentialities of the alliance

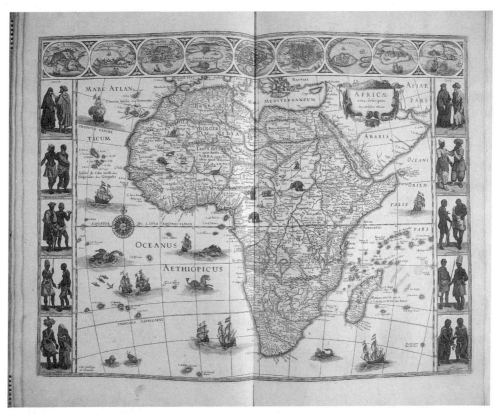

Figure 8.1. Map of Africa in Willem Janszoon Blaeu, *World Atlas,* 1630. Edward E. Ayer Collection, The Newberry Library, Chicago.

between the cartographic and the painterly. And it is to understand how, in a view of the world as itself a picture (for such is Vermeer's sense of the world), maps become paradigmatic presences. In "the age of the world-view" (Heidegger's epithet for all of early modern Western culture, including seventeeth-century Dutch culture as a paradigm case), *the world itself is put on view*; and it is put on view by maps and paintings—by maps that have the allure and look of paintings and by paintings that are cartographic in their very conception.

Reinforcing the pictographic propensity of this age is the prominent use of the mirror. If both maps and paintings alike attempt to represent "the vivid appearance of things seen,"[36] their sheer pictoriality, then the mirror is an altogether appropriate ally. For a mirror give us images—nothing but images!—of the world that confronts and surrounds it. Furthermore, in the early modern Dutch outlook the mirror enacts the world-as-image, the world-as-viewed, the world-as-represented; for the most direct way to represent the world is to present a mirror image of that world.

Thus geography itself was often thought of in terms of its mirroring powers. A sea atlas would be titled *Spieghel der Zervaerdt* (Mariner's mirror), and Petrus Apianus, in his *Cosmographia* (1545), claims that cosmography (i.e., universal geography) "mirrors the image and appearance of the universal world as a mirror does one's face."[37] In looking at the map of Africa in Blaeu's *World Atlas,* then, one is in effect looking at a mirror of this part of the known world: the page purports to furnish us with a direct reflection of Africa. In the introduction to his *Grand Atlas* of 1663, Joan Blaeu speaks of geography as "the eye and the light of history" because of the fact that "maps enable us to contemplate at home and right before our eyes things that are farthest away."[38] They can do all of this insofar as they are comparable to mirrors that reflect back to us, from the pictorial surface of the pages on which they are printed, the surface of the actual world they capture in the form of an image: hence the high priority of depicting relief. Given the idea of the map as mirror, it is not surprising that earlier mapmakers' preference for the language of *graphikos* ("drawn," "traced") gives way to the Dutch preference for *schilderij* ("picture"), that is, the language of purely pictorial representation.

The language of mirrors—and of eyes, glasses, and sight generally—was extended to landscape painting as well. Edward Norgate, in his *Miniatura* (c. 1648–50), wrote that "Landscape is nothing but Deceptive visions, a kind of cousning *[sic]* or cheating your owne Eyes, by our owne consent and assistance, and by a plot of your owne contriving."[39] Because the "Deceptive visions" of landscape are realized by one's own contriving, they are not to be regretted but celebrated; for landscape paintings are as much mirrors of the world as are maps: they, too, give us back a reflection of the world on view; they put the visible world on display for our delectation and knowledge. They reflect the surface of the known world on their own surface, entertaining and instructing us by this painted surface, which is as mirrorlike as that of maps.

CARTOGRAPHY/CHOROGRAPHY

Despite a common commitment to description and spatiality (rather than to narration and temporality), to imagistic representation and pure pictoriality, to vivid appearance and mirrorlike reflection, there remains one decisive difference between maps and paintings. This is that maps represent places for the most part incidentally—for example, in the literally topographical images on their margins—whereas landscape paintings represent places centrally and essentially. In Ptolemaean nomenclature, maps mainly belong to geography, which is largely indifferent to place, while paintings belong to chorography, which is devoted to the description of place. How is this so? I shall conclude this chapter with some reflections on the role of place in seventeenth-century Dutch pictography.[40]

We need to start with the primal fact that whereas a Dutch map of the time typically represents a sizable segment of then-known world-space—a continent or a country or, at the very least, a city—a landscape painting from the same period represents a particular place or set of places. Even when a map zeroes in on a comparatively limited subject (e.g., "the Siege of Breda"), the subject itself is viewed as merely part of a much more inclusive geospatiality: hence the virtually irrepressible construction of atlases, which put together side by side (literally, page by page) closely concatenated sections or parts of a single comprehensive geographical unit, be it a nation or the earth itself. No such ideal of all-inclusiveness is presumed or sought in landscape painting, which is content with the particularity of place: even a panoramic painting presents a group of coherently connected places. What Heidegger says of the spatiality of the ready-to-hand world holds true for the spatiality of landscape painting: here "space has been split up into places."[41]

Much the same contrast can be seen in the basic difference between geography and chorography. Whereas geography puts its representations of the earth within the embrace of a unitary space that reduces places to points or positions—that is, regards them as *sites*—chorography takes any such space as *already diversified into concrete places*. For the chorographer, even a region is viewed through its constituent places and not as a simple block of space; regions themselves, as Heidegger also avers, "always are ready-to-hand already in individual places."[42] This is why Constable, that consummate chorographic painter of East Anglian localities, paints the region of the Stour Valley by the pictorial rendering of specific places in that valley; rather than saying that these places are merely *located in this region* (that would be to speak as a geographer), we should say that the region resides *in its own places*. To paint a region is to paint its constituent places—a feasible project. But there can be no landscape painting of the earth as a whole, which exceeds even the vastest panorama. Yet it is precisely the earth in its totality that Blaeu's *World Atlas* claims to represent. In other words, no painting can give us what Ptolemy assumed to be the proper object of geography: "picturae *totius* partis terrae cognitae"—a picture of *all* parts of the known earth.

What, then, does a landscape painting give to us if it cannot provide us with the image of the totality of the earth? It gives us the images of particular places that constitute particular regions: it *chorographs* them. We do not doubt that these places are integral parts of a larger geographical unit such as a state or a continent, but these latter (much less the earth to which they belong) cannot be given effective pictorial representation in a painting. The effect is something like that of seeing the larger unit through a microscope (to allude to a favorite scientific instrument employed in the Netherlands of the period). In microscopic vision, we see a very small part of a given space, but we see it in much more detail than were we to stay with an unassisted view of that space. The loss of the macroscopic view is a gain in the discriminating look. So too with chorographic landscape painting: we lose the region as a geographic entity, but we grasp its material essence in terms of the discrete localities that are its integral parts.

What is unique about the representation of places by a master landscape painter such as Jacob van Ruisdael is that this representation gives us the *shape of the land* in which these places are located: the *schap-* of the *landschap*. The Dutch suffix *-schap*, like the affiliated English suffixes *-ship, -skip,* and *-scape* (as well as the German *-schaft*), connotes "shape." This suggests that the chorography of place as it is realized in landscape painting is concerned with the morphology of these places, their shapefulness. The chorographically sensitive landscape painter picks out the shapes of places (including the shapes of things-in-places)—shapes that are not to be confused with the formal shapes of geometry or even of geography. Where geometric and geographic shapes must perforce be precisely drawn—drawn either to a preestablished pattern (as in plane geometry) or according to the mensurational requirements of cartographic representation (i.e., in geography)—chorographic shapes need not be precisely traced, nor are they prefixed or measurable as such. In ordinary perception, we continually discern such shapes in the natural landscape itself. Only in extreme Cubist reductionism will the morphism of shape be constrained to the geometric; otherwise, it enjoys the latitude of the vaguely described: what appears in the clearing of landscape need not be—indeed, typically is not—altogether lucid, much less rigid, in form. The informal morphology of landscape (in contrast with the formal geomorphology of earth) resists, even if it does not altogether refuse, the rigors of geometry.

What this means in practice is that painters of landscape such as van Ruisdael and his many successors in Europe and the United States are engaged in a common descriptive enterprise that seeks to re-present the shapes that define and saturate places. To re-present is not to replicate such shapes but to re-implace them: it is to create in the place-of-representation what has its place-of-origin in the natural world: creation is here re-creation. What is thereby re-created is the morphic pattern of the landscape, its implicit eidetic or formal structure. *The painter aims at an eidetics of the landscape, its inherent form or material essence: he describes landscape as land-shape.*

It follows that the morphology of a given landscape may be as fully realized in drawings, that is, literal de-lineations, as in full-fledged paintings. Indeed, Rembrandt conveys more complete and sensitive images of the landscape in his pencil sketches and etchings than in his finished paintings. As Clark observes:

> In his landscape drawings of the 1650's, every dot and scribble contributes to an effect of space and light; problems which had baffled earlier landscape painters, for example, the difficulty of the middle distance, of getting into a picture smoothly from a low point of view, did not exist for him . . . Yet when he came to paint he felt that all these observations were no more than the raw material of his art . . . Looking at his drawings and etchings we may regret this high ambition which has deprived us of masterpieces.[43]

A case in point is an etching of 1651 (Figure 8.2). Notice the difference between the case of Rembrandt and that of Cole or Lane. In none of these latter instances do we regret the fact that they regarded their line drawings as mere raw material; their paintings, even their oil sketches, carry forward their image of landscape in ways that we admire and

Figure 8.2. Rembrandt van Rijn, *The Goldweigher's Field,* 1651. Etching and drypoint. Photograph courtesy of The Art Institute of Chicago. Reprinted with permission.

savor. In Rembrandt's case, however, there was no follow-through at the level of completed paintings.[44] We are left to appreciate unabashedly preliminary works, whose sheer linearity nevertheless realizes a vision of landscape as a scene of the shapes of places. For all their spare economy, these works remain descriptive of these places: *descriptio* in its original and literal scope means "drawing."

Does this mean that the seventeenth-century Dutch artist was a cartographer after all? This might seem to be suggested by the emphasis I have just been placing on landscape forms and on the virtues of drawing and tracing these forms—that is, on the factor of *graphikos* that Ptolemy, in his original Greek text, makes essential to geography: "Hē geōgraphia mimēsis esti dia *graphēs* tou kateilēmmenu tēs gēs merous holou." (Geography is the mimesis, *by graphic means,* of part of the earth taken as a whole; my italics). As if to confirm this line of thought, Clark adds that Rembrandt "could immediately find a graphic equivalent for everything he saw."[45] As an exclusively graphic landscape artist, was not Rembrandt a graphologist of the earth and thus a literal geo-grapher?

Tempting as this interpretation may be, I prefer to think of Rembrandt as a *quasi*-cartographic or *quasi*-geographic artist. In his landscape drawings, he gives us not a "mapped landscape" (in Alpers's preferred term) but a landscape that is presented as *mappable* but not yet *mapped.* It is mappable by dint of Rembrandt's astute exploitation of a survey space that is shared with fully realized maps. This is not, as Alpers claims, a matter of "the transformation of a mapping mode *into* landscape representation, where such representation remains a form of mapping."[46] On the contrary, mapping is never attained by Rembrandt, much less sought by him. Instead, his etchings offer us a landscape representation that could subsequently, in a different context and with different means and intentions, be *given* cartographic representation by someone else—as happens when maps are actually made of an artist's familiar haunts (e.g., "Constable's England," "Cézanne's Aix"). The pictorial space presented to us by Rembrandt is only secondarily cartographic, just as it is only secondarily a matter of mensuration. Even if it is conveyed to us by graphic means, and even if it shares with maps an origin in surveyable space, Rembrandt's pictorial space is primarily chorographic, for it conveys the inherent shapefulness of things and places in the perceived and felt landscape.

Far from it being the case, then, that "maps give us the measure of a *place* and the relationship between *places*,"[47] we ought rather to say that maps give us the measure of *sited spaces,* while landscape paintings give us the *shapes of places.* In fact, it is fundamentally mistaken to believe that a map represents places as such—that, in Alpers's claim, "places, not actions or events, are its basis."[48] Places are truly *represented*—that is to say, given a new place to be, re-implaced—only in drawings or etchings or paintings of the wild or inhabited world. Landscape art in this coherent sense conveys the morphic structures of places, not their geometric or geographic forms. It is for this reason that in his etched and sketched landscapes we have to say that Rembrandt was practicing chorography and not cartography; only on *this* basis did he give us "an enriched description of place rather than the drama of human events."[49]

We would have to look to faraway places to find anything quite like the achievement of Dutch landscape painting in the seventeenth century. In the West, its closest analogue is Luminism. In the East, it is anticipated by Chinese painting of the Northern Sung era. In this latter case, we find a comparable pursuit of the morphology of landscape, that is to say, an eidetics of place. Despite the fact that the Chinese were also great mapmakers (as we shall soon see), their landscape painting is no more cartographic than is Rembrandt's work. Certainly, we can say of Northern Sung painting that it, like a seventeenth-century Dutch map, presents "a surface on which is laid out an assemblage of the world."[50] Moreover, Chinese art presents this assemblage panoramically and in such a way that we view it from a surveyor's position: the panoramic landscape of many Northern Sung paintings places the spectator in an elevated but nonprecisely identified viewing area. From this vantage point, we experience much the same paradoxical combination of "distance preserved and access gained"[51] that obtains in Dutch mapping of the early modern era. But the affinity with maps breaks off here. For *what we see* in a Chinese landscape painting is resolutely chorographic and not cartographic.

What we see in landscape art are places and not sites. Or, more exactly, we see the configurations of places rather than the formations of sites—the inlines of the former instead of the outlines of the latter. To grasp the morphic pattern of a place is to grasp its *ch'i* energy, its internal dynamics, what makes it just this place and not another. Sites, in contrast, lack *ch'i* and thus an animating inner form. This is why we can exchange sites for one another indifferently in cartographic space: unable to convey the morphic singularity of places, maps are confined to the representation of sites.

Landscape painting in the Northern Sung dynasty and in seventeenth-century Holland is a painting of places in their characteristic (in)formal patterns. Such painting represents the places of given regions of the earth in their intrinsic shapes, and it is in this sense a practice of geo-choro-morphology. This art seeks not just a lucid representation of landscape but a clarified depiction of land-shape. To paint (or to draw: Northern Sung scrolls, like Rembrandt's etchings, are situated on the very borderline between painting and drawing) is to give shape to the land, to draw out its inherent *morphē,* its inborn *ch'i.* Through line and color, form and image, the shapes of places come to be described in their pure pictoriality. Cartography and geography cede place to chorography, landskip reverts to its origin in *landschap,* and landscape comes into its own.

CHAPTER 9

Discursive and Presentational Symbolism in Maps

The Revealing Case of Portolan Charts

Yet I am the necessary angel of earth,
Since, in my sight, you see the earth again.

—Wallace Stevens, "Angel Surrounded by Paysans"

MAPS AS SYMBOLIC PRESENTATIONS

Depth leads to projection: the representation of the former is intimately linked with the deployment of the latter. It is as if the depths of the earth have given rise to the projection of the earth's image—a projection that is at once infinite and incomplete. The very idea of "pro-jection" means to hurl or throw *out* or *before*. Maps can be considered to be what the earth (aided by the ingenuity of the mapmaker) throws out before itself as a major means of its own unconcealment; taken in this spirit, they are the self-representations of Gaea.

To grasp the necessity of projection in the making of maps is thus to be apprised of the way in which geometry arises out of geography. Projection is a creation of geometry as it makes the critical transition from solid to plane figures. The figures of solid geometry are figurations of the earth; they regularize the ruggedness of landscape by compressing its primary features into well-formed three-dimensional shapes. From landform comes form itself: out of geo-morphology (i.e., the forms of earth) there issues geo-metry (i.e., the measurement of earth), a process that Husserl has traced in "The Origin of Geometry."[1] And the converse holds as well: once geometry has become established in ancient Greece (precisely during the time of the first full-fledged maps of the *oikumēnē*, the known world), it can be taken back onto the rough earth, shaping it and limiting it by superimposition of form (as we are reminded vividly by Blake's etching of Newton laying down geometric shapes on the barren soil).[2]

The decisive difference between prehistoric and early Greek maps is that the former have no system of pure geometry, that is, a science of solid or plane figures that are abstracted from the particulars they inform: if circles and other regular figures are used (as they were by the Babylonians, Egyptians, and other ancient peoples), they are not regarded as separable from the objects they denote.[3] On the Bagnolo Stone, a cosmic map discovered in the Valcamonica region, a circle and a rectangle are juxtaposed not as two pure shapes but as signifying "the relationship of the earth to either the cosmos as a whole or the sun in particular."[4] The earth is represented by a rectangle and the sun by a circle, but rectangularity and circularity are not at stake *as such* as they are in plane geometry. Shape is here wholly in the service of indicating cosmic bodies. It has not yet come into its own. When it does, it can be set within a grid, thereby making projection possible. As Strabo writes in his retrospective consideration of the origins of geography in Greece, "using these lines as 'elements,' so to speak, we can correlate the regions that are parallel, and the other positions, both geographical and astronomical, of inhabited places."[5] By "elements" *(stoicheia)* Strabo means the lines of latitude and longitude as they intersect to form axes of coordinates, much as in Cartesian analytic geometry.

The positions of inhabited places are therefore located, given "true positions," on the cartographic grid that allows different regions, otherwise widely separated in space, to be regarded as parallel to each other on a map because they are at the same latitude or longitude. Places and regions are housed in this grid or "graticule."[6] The grid itself, constructed from rectangles or squares, represents the projective geometrization of Gaea. It was the enduring contribution of the ancient Greeks to make geometry "the foundation of geodesy and cartography,"[7] that is to say, the basis both for measuring the earth and for graphing its shapes—for projecting it into the ideal (if limiting and limited) space of the map.

A projection is a symbolic construct. As Susanne Langer makes clear, projection is a primary means of symbolization, whether it occurs in geometry or in language, in music or in painting. In general, "the transformation of 'facts' or experience into symbols is accomplished by a system of projection."[8] Map projections are a leading instance of such transformation, even if, vis-à-vis the solid geometry of the earth, they can give rise to false inferences: "Geometric projection is the best instance of a perfectly faithful representation which, without knowledge of some logical rule, appears to be a misrepresentation. A child looking at a map of the world in Mercator projection cannot help believing that Greenland is larger than Australia: he simply *finds* it larger."[9] To see Greenland as larger than Australia is to experience a Mercator map as a spatial *presentation* that offers compelling (if literally misleading) visual evidence of comparative size. If a map is a representation in its projective power—a power of which the child is unaware—it is also a presentation in its immediacy. In Langer's language, it is a "presentational symbol" rather than a "discursive" one.

To say that a map is a presentational symbol is to stress the way in which it organizes space. Concerning such organization, we may pick out four primary traits. First, a map presents itself as a visual configuration that makes use of lines but is not read line by line; unlike a discursive symbol, which is apprehended in a strictly linear manner such that one must read it in a certain sequence, a map may be entered at any point or (as frequently happens in large-scale maps) all at once. This is why the attaching of four cardinal directions to maps comes comparatively late in the history of cartography: in the Middle Ages, "East" was presumed to be at the top of a map because this was the position of honor and Jerusalem was known to lie in this direction (hence the idea of "orient-ation" by a map); it was only in the last few hundred years that "North" took the place of "East" at the top of a map, for a scientific reason: this is where the primary magnetic pole is to be found on earth. Second, the spatiality of a map operates by a process of self-inclusion whereby every part of it is what it is only as an integral part of the whole space that is presented. As Langer puts it, the spatiality of a presentational symbol is such that "the meanings of all

[the] symbolic elements that compose a larger, articulate symbol [such as a map] are understood only through the meaning of the whole, through their relations within the total structure: their very functioning as symbols depends on the fact that they are involved in a simultaneous, integral presentation."[10] This is a crucial claim, for it points to an important fact about the parts of a given map: namely, their capacity to coinhere with each other and with the whole they jointly compose. It is this coinherence that accounts for the simultaneity of presentation that is essential to the perception of maps as configurative formations that bring disparate elements together in a common cartographic space.

A third feature of maps as presentational symbols is their capacity to *represent*. It is this feature that underlies projection proper. As Robinson and Petchenik observe, "the cartographer intends to communicate some knowledge of the arrangement of things in real space by using a system of arraying graphic marks on a smaller, representative space."[11] These marks indicate the parts of the earth they represent. In this respect, they are indexical signs whose own perceived form signals various features of the earth: the perceived existence of the former induces belief in the presumptive existence of the latter.[12] For this reason, maps can be said to "present an image of the same [ultimate] object, the earth milieu."[13] As spatialized images, maps are presentational symbols that, despite their self-enclosed and self-totalizing character, refer to those portions of the earth's surface that have been projected in the map.

A fourth trait is the *manner* of this very reference. Unlike discursive symbols, which tend to stand in for general ideas and relations, maps as presentational symbols refer in a discrete and particularizing way. Even if it is not true that "a map provides information about particular places,"[14] it does furnish information about various portions of the earth's surface: it conveys sites, if not places. This is true whatever the scale of the map: at the limit, the earth itself can be mapped as a totality of sites. Thanks to its projective status, however, every map presents a view that is particular and not general: a view *from somewhere*. Indeed, many maps present views that no one may ever have seen before—for example, mountains perceived from a practically unassumable point of view, as in Raisz's physiographic map of China (Figure 7.13)—and yet they are perfectly valid as representations of that which they present. The relationship between the viewpoint and the site viewed is precisely a relationship of scale; and it is the factor of scale that distinguishes maps from other kinds of presentational symbol: "the attribute that seems to set maps apart is the way in which they represent reality with respect to a particular scale of spatial relations."[15]

Thus, specificity of scale joins with specialness of viewpoint to create a cartographic vision of the earth and its milieu. The paradox is that this vision—which aims to be definitive of the geographic object, as regular as it is reliable—is achieved by the device of projections that, along with rectilinear grids and geometric shapes, are abstractive in relation to the uneven surface of the earth. The offspring of Gaea represent her well, but they do so by means that no longer share her raw irregularities. These means are at once spatial, presentational, and symbolic; nonnarrative yet eminently readable; self-inclusive but indexically referential; simultaneously perceived as a single whole and yet highly specific; regular and smooth yet standing in for what is intrinsically irregular and rough; two-dimensional and flat but representing, by projection, what is three-dimensional and solid.

THE COLLUSION OF IMAGES AND WORDS IN MAPS

Let us grant these presentational features of maps. The issue then becomes: do these features, indispensable as they are, exhaust their symbolic status? I think not. Maps are presentational symbols; but is this the full extent of their signifying power? What is their relation to discursive symbolism, that is, explicitly verbal language? Robinson and Petchenik comment that maps "do not fit clearly into the categories of either discursive or presentational symbolism."[16] I would put it differently: maps are *at once* discursive and presentational symbols. The brunt of the discussion in the preceding chapter was that both modes of symbolism, verbal and imagistic, are active in the construction of maps—and have been from the very beginning. Words have been attached to maps from the first Babylonian clay-table representations of the known world, and they continue to be essential to contemporary road atlases. If this is so, it cannot be the case, as Robinson and Petchenik maintain, that language and maps are "incompatible."[17] As we have seen on a number of occasions, they are not only compatible, they call for each other, whether this be in the form of cuneiform signs identifying Babylon and surrounding places or an "Explanation of Symbols" attached to modern maps. We do not have to adhere to the belief that maps always contain narrated stories—that they are forms of history or literature—to acknowledge the importance of words in their formation and understanding.[18]

In fact, what makes maps so effective as means of representing landscape is precisely their capacity to employ images *and* words—pictographs *and* logographs—in their own special blend of semiosis. This potent combination not only makes maps special in their mode of symbolism; it also makes maps unusually informative, much more informative than if they were restricted to words alone or images alone.[19] Powerfully presentational as a map is as a "unique spatial complex"[20]—exhibiting as it does all four of the properties we have just traced out—it is even more effective as a composition of simultaneously perceived images and successively grasped words. The two basic elements, in their very hybridization, potentiate each other, much as in the parallel cases of rebuses and calligrams. Just as Freud declared that dreams can be considered forms of "picture writing" *(Bilderschrift),* so maps are modes of pictural discourse: they are "picto-ideograms," in Conley's suggestive term.[21]

As composite symbols, maps contrast with other ways of representing landscape. One such way is to give a straightforward verbal account. This was the preference of John Muir and other naturalists who chose to write of their experiences of wilderness. Muir also sketched, but he kept the resulting images clandestinely concealed between the covers of his private journals, wishing to make known to the world only his lively verbal descriptions of Glacier Bay and the Sierra Nevada mountains. Another way to represent landscape is by painting it, in which case landscape is articulated by image rather than by word. Here, as well, however, there is a strong temptation to supplement a monovalent medium, this time by adding a verbal commentary that is kept just as much out of sight as were Muir's drawings: Constable's precise metereological observations were usually written on the reverse side of his cloud studies. For him and many other painters, what counts most is a convincing image, not a persuasive account in words.

In maps, by contrast, what matters most is the very conjunction of images and words, their coordination and mutual validation. Maps are mixed modes of representing landscape (and seascape)—in short, "earthscape"—as constituted by the places and regions of the earth's milieu. Far from being a mere compromise of means, the mixture is mutually enhancing. Rather than a reduction of the earth—as the move to two dimensions might suggest and as the conversion of place to site often effects—this doubly representational process, putting together images with words, enlarges our understanding of the world in which we live; it carries forward in a new, and newly illuminating way, our perception of this world. Again, like Antaeus, maps gain strength from their continuing contact with the earth they represent by symbolic projection in two essentially different yet literally cooperative media; but they are Epimethean as well, giving back to their creators an incomparable insight into the character of the earth's peculiar layout not otherwise available. The complex configurations of the earth, its roiling seas and highly variegated terrain, call for the collaboration of word and image in the symbolic constructions that chart them.

ART AND ARCHIVE

If maps are indeed unique as redoubled representations of earthscapes—combining as they do presentational with discursive symbolism—this uniqueness is especially evident in *portolan charts,* a distinctive map form that was prevalent in the Mediterranean world between the late thirteenth and early sixteenth centuries. Situated as they are between medieval *mappae mundi*—notable for their purely fanciful and theocentric symbolism—and early modern Dutch maps constructed in the wake of Gerhardus Mercator (1512–94), portolan charts embody a spontaneous mapmaking activity that was alert to the particularities of land and sea and that was unusually resourceful in its representation. The part of the earth to which they were most sensitive was that of harbors and coastal regions, which, for purposes of navigation, they mapped so reliably; but they also set forth compelling images of certain major features of interior lands. Despite their limited focus and their origin as practical aids to mariners (they were "usually made for [and sometimes by] sailors . . . and to satisfy a commercial demand"),[22] they are of undisputed significance in the history of mapmaking in the West. Not only were the makers of portolan charts in Majorca, Genoa, and Venice the first to institutionalize cartography as a specialized enterprise (carried out in workshops created for this very purpose), but they realized a major revolution in the very conception of maps: "the advent of the portolan chart [is] one of the most important turning points in the whole history of cartography."[23] Why is this so?

To begin with, these maps were exceedingly accurate. Portolan charts were "the most geographically realistic maps of their time,"[24] providing reliable shapes of the Mediterranean Sea, Africa, and parts of North and South America and of the Far East. In particular, they gave detailed representations of coastlines and their features—more detailed even than the verbal descriptions found in *portolani,* the ancient sailors' manuals from which these maps derived their own name. With their dual focus on region (e.g., the Mediterranean)[25] and on place (i.e., the string of places along a given coast), they banished to the periphery the mythical regions and places that were of primary concern in medieval maps—the land of "Prester John" and of "Gog and Magog," the four rivers of Paradise.[26] They refused to accord to a spiritually significant place such as Jerusalem, often the center of the paradigmatic medieval "T-O" *(orbis terrarum)* map, anything other than its actual location vis-à-vis other locations. For mythical and mystical presences, they substituted images of actual kings and known animals.[27]

Portolan charts were also remarkable for their artistic merit. Whether accomplished in the austere "Italian" style or the more flamboyant "Catalan" manner, they were a distinct pleasure to behold. Painters were often commissioned to work closely with mapmakers: in portolan workshops, "the labor would often be divided between a scribe, a rubricator, and one or more painters."[28] Many voyages of exploration to the New World took with them a painter whose task was to record the virgin landscape for eventual inclusion on a portolan chart.[29] The inclusion could be striking indeed: the aesthetic quality of many of the landscape views or "vignettes" compares favorably with the landscape paintings of early Flemish masters such as Jan van Eyck or Hugo van der Goes.[30]

Accuracy and artistry here collude in a characteristically premodern way. Just as the painterly representations in portolan charts are decidedly premodern in their delimited but elegant miniaturism—no sweeping vistas à la Ruisdael or Koninck here!—so the cartographic elements are notable for their preoccupation with the minutiae of coastal outlines at the expense of larger landmasses, as is evident in comparing the quaintly picturesque and Eurocentric Catalan Atlas of circa 1375 with Ortelius's encompassing atlas *Theatrum Orbis Terrarum* (Theater of the world) of 1570. In the former, the premodern spirit shows itself in a passion for detail and for the emblematic view (e.g., of idealized Indians of the New World), in striking contrast with the synoptic and all-embracing representations of Ortelius's masterwork, which includes no fewer than fifty-three maps of various parts of the earth and claims to be "a portrait of the world."[31] Where Ortelius's atlas was a complex venture whose later editions included contributions from numerous other mapmakers, portolan charts were created, and intended for use, by a few hands only. The publicity-mindedness of the modern age—symbolized by the fact that twenty-eight editions of Ortelius's *Theatrum Orbis Terrarum* had been published at the time of his death in 1598—stands starkly contrasted with the extremely limited circulation of portolan charts, which were rarely copied out of fear of disclosing navigational secrets.[32]

Portolan charts also combine archival documentation with sheer utility. A close scrutiny of the place-names on these charts reveals detailed geographic and demographic data not elsewhere available.[33] Thanks to a process of continual refinement in the toponyms ascribed to a given coast—a refinement in terms both of sheer number and of comparative importance—one historian of the subject concludes that "the wealth of place names [on these maps] constitutes a major historical source."[34] Portolan charts constitute virtually the only cartographic representations of the explorations of Africa, Asia, and the New World during the "Age of Discovery." Christopher Columbus doubtless used portolan charts in his epochal voyage, and the same is true of Magellan, Drake, Vasco da Gama, Cortés, Verrazano, and numerous others, some of whom (e.g., Sebastian Cabot and Juan de la Cosa) were themselves talented portolan mapmakers. At the same time, portolan charts were enormously useful, as Columbus knew well from his own experience. They were invaluable practical guides not only with respect to coastal configurations but also on the high seas—thanks to their elaborate system of projective "rhumb lines" radiating out from centers ("wind roses"), lines that enabled navigators to sail an invariant course across great distances by charting both the distance and the direction of a given voyage.[35]

The archival aspect of portolan charts continues the Roman and medieval passion for surveying and recording plots of land.[36] The densely packed place-names of portolan charts, each name inscribed perpendicular to the coastline, offer an impressive survey of the cities and towns of a given country. The voluminous notation of toponyms answers to Roger Bacon's call for an inductively based empirical science of the factual—one leading form of which was geography, strongly recommended by Bacon for intensive study. Such a science is a far cry from Descartes's paradigm of a deductive science modeled on geometry. Moreover, the sheerly utilitarian aspect of portolan charts contrasts with the Cartesian ideal of a purely rational or "analytic" geometry, whose usefulness is a merely secondary feature. Where Greek maps were highly theoretical in inspiration (they were as much illustrations of theories as actual geographical guides) and where medieval maps were theocentric ("the cosmographies of thinking landsmen"),[37] portolan charts reflect the daily challenges of mariners and thus "preserve the Mediterranean sailors' firsthand experience of their own sea."[38]

I do not mean to imply that portolan charts, in their premodernity, are simply antigeometric in spirit. Both ancient Greek and early medieval maps employed geometry (primarily plane geometry) as a matter of course. So too portolan charts are not without *l'esprit géométrique*. The spirit, of course, is not that of analytic geometry or even of standard Euclidean geometry. It is a spontaneous geometry that proceeds by the schematization of shape rather than by the abstraction of form: indeed, gauged by the standard of geographic exactitude, these maps fall far short of modern criteria of cartographic merit. Instead of the actual configuration of a shoreline, a portolan chart will often trace a set of graceful arabesques that connect promontories on a given coast. Distances between the promontories, which are what mainly concern the sailor, are quite accurate; but the precise shape of the shore in between (a shape that does not affect the sailing course as such) is replaced by a spontaneously drawn geometrical figure. As Tony Campbell describes it, "in some sections the stretch between headlands has been formalized into regular arcs, owing more to geometry and aesthetics than to hydrographic reality."[39] Other landscape features are equally schematic: mountains are represented by repetitive triangular shapes (an ancient motif!), rivers by elaborate corkscrews that emanate from oval-shaped lakes, a cross stands for dangerous rocks, and red dots indicate sandy shallows.[40]

Just as portolan charts are charmingly artistic in their playful inventiveness of decorative color and form, so their geometry is captivating in its very capriciousness. Art and geometry not only coexist here, they augment each other's presence; and their creative collaboration is in turn compatible with, and perhaps even facilitative of, their remarkable reliability. Instead of searching for the perfectly regular form by a procedure of abstraction followed by superimposition, portolan charts exhibit a flowing geometry in which there are abstract but not literally abstract*ed* shapes, as we see in the Miller Atlas's dazzling depiction of islands near Sumatra (see Plate 13).[41] Here we witness an elegant economy of flowing lines and brilliant colors that complements the straight red and green rhumb lines that subtly structure the chart. No attempt is made to educe ideal forms (ideal as judged by Euclidean standards) from the irregular natural shapes of the Sumatran islands. Instead, their very irregularity is celebrated, thanks to the free play of the cartographer's imagination. Exhibiting an *esprit de finesse,* this imagination transmogrifies the actual shapes of the landmasses, their inherent geomorphology, into a schematized set of fanciful forms. The result, a unique mélange of art, geometry, and empirical knowledge, is at once a work of art and a contribution to the precise representation of one part of the known world.[42]

FROM DISORIENTATION TO ORIENTATION

Another sign of the premodernity of portolan charts is the experience of radical disorientation that we experience on first confronting them. This disorientation is such that we may not even realize that we are perceiving a map to begin with, much less a map of a particular part of the world. When I look at one of the early charts in Petrus Vesconte's atlas of circa 1321 (see Plate 14)—an atlas that consists of nine parchment leaves pasted on boards (the standard Greek word for map is *pinax,* literally "board")—I find a lozenge-shaped format encircled by sixteen points from each of which nine rhumb lines, in alternating colors of red and green, extend to points on the opposite side. Gazing onto the same complex presentation are two saints ensconced in corner panels, one of which also contains what appears to be a scale for the map. But is this a map? The saints seem as perplexed as I am. All three of us peer between the intersecting rhumb lines in order to discern an emerging configuration. The configuration, however, is not yet recognizable as a geographic entity; it consists of graceful arabesques on which (at an angle of ninety degrees) are written barely discernible place-names, only a few of which can be made out: "Marbella," "Mallios," "Tura." Even to read these names, I have to keep turning the map around and upside down, since they are oriented entirely to the twisting coast. Finally, I recognize "Valencia" and "Barcelona" and, with a start, I realize that this must be a map of Spain!

One reason why I did not recognize the subject matter of the map—or even recognize that it *was* a map—is that Spain is presented with north at the bottom of the chart and west to the right, exactly the converse of current conventions. In contrast with both medieval *mappae mundi* and modern maps, portolan charts have liberated themselves from fixed conventions by freely locating any of the four cardinal directions in the uppermost position, with no preference given to any single direction. Only toward the end of the evolution of these charts—for example, in Jorge de Aguiar's chart of 1492—is north clearly indicated at the top of the map by a special fleur-de-lis symbol affixed to a compass rose.[43] But even after the establishment of this familiar convention, considerable complications abound on portolan charts. The inscription of coastal toponyms still runs perpendicular to the coast; and the place-names of inland cities and even of entire continents are oriented not in relation to a cardinal direction but only in relation to the nearest coast. What this means in effect—and what helps to explain why the perception of such maps can be so confusing—is that the map reader's own position is not taken into account as a stable vantage point from which to view a map. Whereas the reading of modern maps presumes a fixed position on the part of the viewer—just beneath its lower (and typically "southern") edge—no such cartographic Archimedean point is provided by portolan charts. The result is a disconcerting lack of proper viewing point, which accords nevertheless with the use of these charts in the map rooms of ships (where, being already at sea, the arbitrariness of directionality is altogether fitting): "intended to be rotated, portolan charts have no top or bottom . . . there is no way of telling which, if any, of the four main directions they were primarily intended to be viewed from."[44] One exiguous orientational clue is provided in the map of Spain in Plate 14 because here we can align ourselves with the two guardian angels, whose upright posture proceeds from south (their heads) to north (their feet). Otherwise, and in the absence of such orienting figures on the margin, we would find ourselves at sea indeed—lost in the maze of the map, not knowing which way to turn next.

This is a strange situation: a map, whose primary purpose is to orient us more securely in the world, instead bewilders us. Or at least it does so until we learn to focus on the one factor that is always genuinely orientational in a portolan chart. This is the coastline itself. However schematically it may be drawn (and it is very schematic in the chart of Spain), it is quite reliable as a cartographic guide. *It is this line that orients our looking.*[45] Once we take it into account, once we make it our primary "plumb line," our confusion begins to dissipate. No longer are we caught in a bewildering mesh of rhumb lines, whose formal-geometric orthogonality only serves to confuse the modern viewer (rhumb lines, determined by a fixed compass direction, help to orient the navigator on the open sea, not the sailor in sight of shore or ourselves in sight of the map). No longer does the map seem entirely topsy-turvy: it now starts to "make sense" (especially if we hear in this last phrase the echo of the French *sens*, "direction"). Not only is our reading of the strictly coastal toponyms facilitated, but we suddenly understand why other locatory names, as well as images of major cities (Venice, Genoa, Jerusalem), are presented upside down or sideways: they, too, are oriented in relation to the coastline. The same is true of landscape features: where before we were befuddled at perceiving the Atlas Mountains upside down in relation to the presumptive bottom of the map, now we realize that they are *right side up in relation to the north African coast.* We have been oriented—or rather, reoriented—thanks to the tenuous and often only dimly discernible coastline. It is around this line, and this line alone, that the map (and thus our perception of it) turns; everything else, including the placement of the cardinal directions and our own bodily relation to the map, is dependent on this primary delineation.

In a portolan chart, then, one special landscape feature, the coast, guides and orients the making, reading, and actual using of a map. In maps made after the beginning of the seventeenth century—just when portolan charts became extinct—cardinal directionality (and above all the cartographically determinative direction of north) is fixed in place and the viewer's position is settled. This position, itself as definite as the direction of north (the viewer always looks upward into north at the top of the map), is one wherein the map as a representation presents itself to the map reader in a consistently readable manner.

The modern map, like the modern landscape painting it mimics in this regard, is conceived as a stable and stabilizing entity situated literally *before* its viewer. It is a *Vorstellung* in the strict sense of this most characteristically modern of epistemological terms:

"placing before." Such a map exists for the knowledge or satisfaction or use of its viewer as a mirror of the known world that casts its cartographic image *in front of* itself and into the sight and mind of the observer. As mirrors offer a consistent vertical orientation to those who look into them (their celebrated reversal affects only left-right ordering), so post-portolan maps furnish a steady north-south orientation that orients the position of their viewers in turn. And as mirrors exist for the delectation of their onlookers—this way lies the origin of narcissism and of the self-absorbed ego[46]—so modern maps exist for the sake of their omnivorous and self-regarding readers. In this sense, they are, in Conley's apt term, "self-made."[47]

GETTING AROUND THE COAST

But let us return to portolan charts, which are of considerable significance for a study of how landscape is represented in maps. Because the coastal line that magnetizes a portolan chart and makes it decipherable is itself a direct representation of a landscape feature, we may say that this line is what renders such a map *topographical*. A topographical map is definable as "a map showing detailed features of landscape; in its origin it would be a map of a limited area known to the draughtsman through his own observation."[48] Such a map is indeed a map of a *place* (as *topos* literally signifies), and not just of a *site* (the proper object of nontopographical maps that concern themselves with positions in cartographic space). Examples of topographical maps construed in this sense include the ancient wall map from Çatal Hüyük in Anatolia and the Mesopotamian clay tablet from Nuzi (Figures 7.3 and 7.7, respectively), both of which depict rivers and mountains that would have been known to their creators. Local maps of the later Middle Ages—those that are not full-scale *mappae mundi*—are also topographic in character, showing as they do particular landscape features: for example, the picture maps of the Lower Scheldt made in 1468.[49] These latter are contemporary with portolan charts and differ from them in not being fixated on coastlines; they accord at least equal attention to inland features.

To be fixated on coastlines is to focus on the *line* as a means of representation. At once the simplest graphic means of representation and yet arguably the most dynamic, the line has the peculiar property of representing landscape in terms of dyadic pairs of exclusive alternatives: inside/outside, this side/that side, above/below, and so on.[50] As employed on portolan charts, the line defines *land's end*: on one side of the line lies the sea, on the other side, land. Between is the line limit—the *limites*, as the Romans called the boundary line of the centuriated land they surveyed so systematically[51]—a line that is at once real (as a drawn or printed line standing in for the literal termination of a landmass) and imaginary (as a highly schematized representation of this very termination). It was perhaps the either/or exclusivism of the line as limit that, by the very austerity of its representation, led portolan chart makers of the Catalan style to embellish their charts so lavishly: the insistent emphasis on the coast in its linear simplicity elicited supplementation by ingeniously painted images of what lies inland from the shoreline. Even so, the line of the coast remained definitive, because it served to frame (and thus to limit) these colorful images.[52] The very word *coast* points to this framing action: it derives (via Middle English *coste*) from Latin *costa*, "rib, flank, or side," that is to say, a definitive structure of a body by which it contains the viscera within it.[53]

Plate 1. Hans Memling, *Virgin and Child with Saints Catherine of Alexandria and Barbara,* c. 1479. The Metropolitan Museum of Art, Bequest of Benjamin Altman, 1913 (14.40.634). Photograph copyright 1981 The Metropolitan Museum of Art.

Plate 2. John Smibert, *Vew* [sic] *of Boston*, 1738. Oil on canvas, 30 x 50 inches. Childs Gallery, Boston.

The Falls

Above, below, where'er the astonished eye
Turns to behold, new opening wonders lie,

of Niagara

With uproar hideous first the *Falls* appear,
The stunning tumult thundering on the ear.

This great o'erwhelming work of awful Time
In all its dread magnificence sublime,

Rises on our view, amid a crashing roar
That bids us kneel, and Time's great God adore.

18

25

Plate 3. Edward Hicks, *The Falls of Niagara,* 1825. Oil on wood panel, 31 ½ x 38 inches.
The Metropolitan Museum of Art, Gift of Edgar William and Bernice Chrysler Garbisch,
1962 (26.256.3). Photograph copyright 1998 The Metropolitan Museum of Art.
Reproduced with permission.

Plate 4. Anonymous, *The Plantation*, c. 1825. Oil on wood. The Metropolitan Museum of Art, Gift of Edgar William and Bernice Chrysler Garbisch, 1963 (63.201.3). Photograph copyright 1984 The Metropolitan Museum of Art. Reproduced with permission.

Plate 5. Albert Bierstadt, *Valley of the Yosemite*, 1864. Oil on canvas, 29.7 x 48.9 cm. Gift of Martha C. Karolik for the M. and M. Karolik Collection of American Paintings, 1815–1865. Courtesy Museum of Fine Arts, Boston. Reproduced with permission.

Plate 6. Frederic Edwin Church, *Heart of the Andes*, 1859. Oil on canvas, 66 ⅛ x 119 ¼ inches. The Metropolitan Museum of Art, Bequest of Margaret E. Dows, 1909 (09.95). Photograph copyright 1979 The Metropolitan Museum of Art. Reproduced with permission.

Plate 7. Frederic Edwin Church, *Cotopaxi*, 1862. Oil on canvas, 48 x 85 inches. Founders Society Purchase, Robert H. Tannahill Foundation Fund, Gibbs-Williams Fund, Dexter M. Ferry Jr. Fund, Merrill Fund, Beatrice W. Rogers Fund, and Richard A. Manoogian Fund. Photograph copyright 1985 The Detroit Institute of Arts. Reproduced with permission.

Plate 8. Frederic Edwin Church (American, 1826–1900), *Twilight in the Wilderness*, 1860s. Oil on canvas, 40 x 64 inches. Copyright The Cleveland Museum of Art, 2000, Mr. and Mrs. William H. Marlatt Fund, 1965.233. Reproduced with permission.

Plate 9. Thomas Cole, *View from Mount Holyoke, Northampton, Massachusetts, after Thunderstorm (The Oxbow)*, 1836. Oil on canvas, 51 ½ x 76 inches. The Metropolitan Museum of Art, Gift of Mrs. Russell Sage, 1908 (08.228). Photograph copyright 1995 The Metropolitan Museum of Art. Reproduced with permission.

Plate 10. John Constable, *Dedham Vale, Morning*, 1811. Oil on canvas, 31 x 51 inches. Courtesy of Meredyth Proby. Collection Richard Proby Barr. Reprinted with permission.

Plate 11A. Jan Vermeer, *The Art of Painting*, mid-seventeenth century (exact date unknown). Oil on canvas. Kunsthistorisches Museum, Vienna. Reprinted with permission.

Plate 11B.
Detail of Jan Vermeer,
The Art of Painting.

Plate 12. Jan Christaenszoon Micker, *View of Amsterdam,* seventeenth century (exact date unknown). Oil on canvas. Amsterdams Historisch Museum. Reprinted with permission.

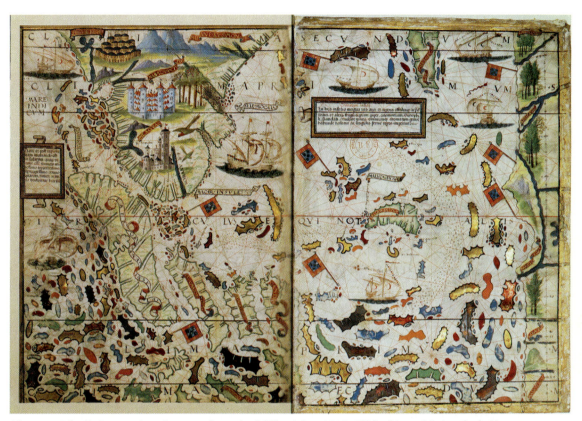

Plate 13. *Islands Surrounding Sumatra,* from the Miller Atlas, 1519. Bibliothèque Nationale de France. Reprinted with permission.

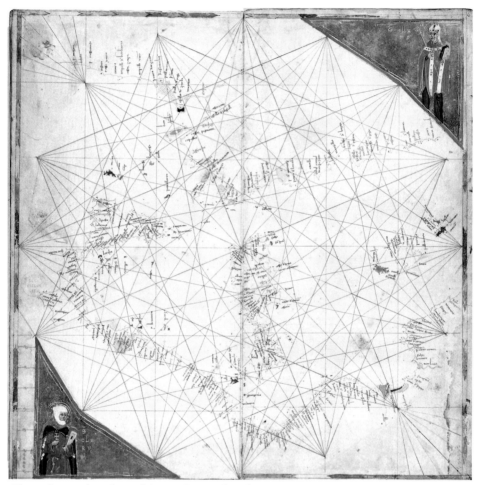

Plate 14. Portolan chart of Spain, from *Atlas of Petrus Vesconte,* c. 1321. Courtesy of Bibliothèque Municipale de Lyon.

Plate 15. Jan Huygen van Linschoten, *View of the City of Angra on the Island of Terceira*, Amsterdam, 1595. Maritiem Museum, Rotterdam. Reprinted with permission.

Plate 16. Kuwagata Keisai, *Nihon Meisho No E*, panoramic map of Japan, before 1824. From Takejiro Akioka, *Collection of Old Maps in Japan* (Kajima Institute Publishing Company, Ltd., 1971). Owned by National Museum of Japanese History, all rights reserved. Reproduced with permission.

Portolan charts are topographic maps by virtue of the fact that they depict a particular delimited area that was known to their makers by perception and memory: namely, the coast of the Mediterranean. These charts constitute, in Campbell's phrase, "a living record of Mediterranean self-knowledge."[54] In this regard, they are genuinely *regional* in status, and are thus also "chorographic" in Ptolemy's original sense of the term: "chorography, selecting certain places from the whole [of the known world], treats more fully the particulars of each by themselves, even dealing with the smallest localities such as harbors, farms, villages, river courses, and such like."[55] In the case of the Mediterranean, the most definitive locality is the coast. As Mollat du Jourdin and La Roncière remark, it was indeed "entirely logical that the portolan charts should originate in the Mediterranean area";[56] for in this internally diverse region—whose overall extent Ptolemy not surprisingly overestimated by a factor of 20 percent—what stands out are the coasts. In fact, the ancient Mediterranean world consisted largely of coasts along which and between which to navigate. It was a world of coastal circumnavigation. In Fernand Braudel's astute observation, "If Mediterranean sailors persisted in using the old coastal routes . . . it was because the old ways fulfilled their needs and suited the complexities of the coastline; it was impossible to sail far in the Mediterranean without touching land. And a coastline in sight is the navigator's best aid and surest compass."[57] If coastal life is indeed "the underlying reality"[58] of the early Mediterranean world, to map the coastline of that world is to map its lifeline. It is to trace the line of the shore to which early sailors clung from fear of the mostly uncharted high sea *(pontos)* and the radically unchartable ocean *(oceanus).*[59] No wonder a special word for such coast-hugging was coined by early Italian mariners: *"costeggiare,"* signifying to avoid the open sea by holding to the coast, in contrast with the reckless method of *camin francese,* to go straight ahead across the water.[60] An even earlier word, the Greek *periplous,* means literally the circumnavigation or "sailing around" of coasts; early sailing manuals were called *"periploi"* and consisted in written directions for proceeding along the coast from one point to the next, points separated by no more than a day's journey. Given that the coast almost always remained in view in this cautious circumnavigation, such *periploi* were all that a sailor required for successful voyages.[61] Later termed *portolani,* they are the written equivalent, and almost certainly the literal predecessor, of portolan charts.[62]

It has been speculated that Homer's *Odyssey* can be construed as a sailing manual for considerable parts of the Mediterranean. This is not altogether unlikely in view of the ancients' opinion of Homer as "the father of geography."[63] His celebrated description of the shield of Achilles in the *Iliad* appears to be a description of a world map *(periodos gēs),* with the earth, sea, sun, moon, and stars at the center of the shield, ringed around by a city at peace and a city at war, by agricultural and pastoral activity, and finally by the ocean on the outer fringe.[64] The ocean itself, "that vast and mighty river" *(Iliad,* book 18, ll. 607–8), was regarded by the early Greeks as ringing around the *oikumēnē,* the known habitable world whose central region was precisely the Mediterranean sea.[65] Almost all of the extant *periploi* from the ancient world bear on the Mediterranean region, either exclusively or as a point of departure and return for voyages beyond the Pillars of Hercules (i.e., the Strait of Gibraltar).[66] It is very probable that these circumnavigational guidebooks were in part replaced by portolan charts and in part survived on their own, often supplemented by illustrations of coastal profiles.[67] Indeed, comparable navigational handbooks are still in use for coastal sailing in many countries.

Classical *periploi,* early medieval *portolani,* and late-medieval and Renaissance portolan charts all focus on the Mediterranean as their primary region of representation. Or, more exactly, we should say that they take the Mediterranean as their *focus geographicus* to the extent that it is itself a collocation of discrete regions, namely, the various subseas and corresponding coasts into which the Mediterranean is physically subdivided. As Braudel says, "the Mediterranean is not a single sea but a succession of small seas that communicate by means of wider or narrower entrances."[68] Taking this geographical observation as a clue, David Woodward has proposed a theory of how portolan charts could have become so accurate without the use of modern navigational methods. Perhaps, he suggests, they evolved from separate self-enclosed maps of each of the primary basins of the Mediterranean. In a given basin, for example, the Adriatic Sea and the contiguous Dalmation and Italian coasts, sailors could have established a self-monitoring tradition of continual circumnavigaton until they had finally mastered its local geography:

> The average distances derived from these sailings between pairs of ports—both along the coast and across the sea—could then have been used in the construction of a series of separate charts of the individual basins. If these routes were plotted to form networks in each of the basins . . . each network might have assumed the form of a self-correcting closed traverse approximating the shape of each basin . . . These discrete compilations could then have been amalgamated into charts of the entire Mediterranean.[69]

I cite this intriguing theory as yet another manifestation of the profoundly chorographic intent behind the construction of portolan charts—and before them, written guides in the form of *periploi* and *portolani.* In all three cases, though most dramatically in the instance of portolan charts, a deep-seated regionalism encouraged a remarkably complete and accurate description of the entire extent of the Mediterranean— *"mare nostrum"* or *"mare internum,"* as it was called in ancient times. More exactly, intimate regional knowledge made possible a comprehensive account of one outstanding feature of the landscape of that fateful part of the world: its coastal configurations; for it was a painstaking description of these configurations that all three kinds of chart offered to sailors of the region, to geographers of the time, and to ourselves looking back in wonder at the way in which "our sea," the "internal sea," was once chorographed so effectively in images and words.

IMAGE VERSUS WORD IN PORTOLAN CHARTS

"In images and words": here we are invited to return to the opening issue of the modes of symbolism at stake in maps. This issue becomes particularly acute in the case of portolan charts. Their very focus on a delimited set of salient features in an equally delimited geographical region—both of these in conjunction with a demand for practical utility—places a special premium on their descriptive power. Where more ambitious world maps can afford to be deficient in detailed descriptions, this cannot be true of sailing charts, on which livelihoods and actual lives depended. In this light, it is not surprising that descriptions conveyed by portolan charts are given both in image and in word, as if to reinforce their reliability. Our concern will now be with the respective merits of these two mapping media and their relationship to each other.

In his widely read book *The Story of Maps,* Lloyd A. Brown tells what could be considered the "official account" of the relationship between image and word in the evolution of portolan charts from *portolani* and *periploi*:

> The coast pilot [i.e., the written guide] was a book designed to aid mariners in negotiating stretches of coastline and intricate harbor approaches; it described the location of reefs and shoals and prominent points along the shore from which a mariner could get a bearing. *The portolan chart was a supplement that evolved from the need of more graphic descriptions of navigational hazards and from the inadequacy of words to describe the various situations a pilot might encounter in the course of routine coastwise sailing.*[70]

This seemingly straightforward and unobjectionable statement brings with it a host of assumptions: the "inadequacy of words" to convey the specificities of coastal sailing, a corresponding adequacy of graphic images to accomplish this same task, the supplementary status of the portolan chart as a set of such images. Each of these assumptions is open to question. Consider, for example, the contrary view of E. L. Stevenson: "more useful to [sailors] than a seafarer's chart, which might be employed in navigating from port to port across a trackless and unknown sea, would be a written description of the seas over which they were prepared to travel and the coasts they had to visit."[71] Who is right here—Brown, who maintains the superiority of image over word, or Stevenson, who holds the opposite?

Let me begin by citing from the oldest known written description of coastal sailing in the Mediterranean, the *periplous* of Scylax of Caryanda, an admiral of the Persian fleet under Darius I between the years 519 and 512 B.C. A later copy of his text, which dates from 361 B.C., offers an account of a complete circuit of the Mediterranean, with distances given in terms of stadia and sailing days. The text opens as follows:

> I shall begin at the Pillars of Hercules in Europe, and shall continue to the Pillars of Hercules in Libya, and to the land of the great Ethiopians. The Pillars of Hercules stand opposite to each other, and the distance between them is one day's sail. Not far distant lie two islands by the name of Gadeira. On one of these is a city which is distant one day's sail. Beyond the Pillars of Hercules which are in Europe, there are many trading stations of the Carthaginians, also mud, and tides, and open seas.[72]

Had Scylax's description remained entirely at this level of generality, Brown's charge as to the inadequacy of written language "to describe the various situations a pilot might encounter" would be to the point: so far, no sense of the particularity of sailing situations has been conveyed. But Scylax becomes more specific as his text proceeds—for instance, in this passage concerning Libya:

> From Thonis the journey to Pharos, which is a desert island, is 150 stadia. In Pharos, there are many harbors, but the ships get drinkable water at Marian. From Pharos to this port is a short sail. Here is also a peninsula and a harbor. To this point is 200 stadia.

> Beyond lies the Bay of Plinthine. From the mouth of the Bay of Plinthine to Leuce
> Acte requires a sail of one day and one night, but if you should sail around the head
> of the bay twice as much time would be required.[73]

The attention to harbors is significant: portolan charts are sometimes designated as "harbor-finding charts." *Portus,* the Latin word from which "portolan" as well as *portolani* derive, signifies harbor or haven, and in fact most of the place-names on such charts designate harbors, that is to say, ports with navigable entrances. As Thomas Blundeville wrote in 1622, "navigation is an Art which teacheth by true and infallible rules, how to governe and direct a Ship from one Port to another, safely, rightly and in shortest time."[74] Even if we cannot invoke "infallible rules" in the case of Scylax's *periplous,* we can concede that this ancient piloting guide does attempt to direct ships from one port to another in a safe and expeditious manner. In doing so, it gives estimates of sailing times—something difficult to convey by image alone.

Consider also the only other *periplous* to survive from ancient times, the *Stadiasmus Maris Magni* of circa A.D. 250–300:

> It is 20 stadia from Hermaea to Leuce Acte; nearby is a low island which is distant
> two stadia from the land. Boats carrying merchandise can anchor here, entering by
> the west wind, but near the shore below the promontory there is a wide roadstead for
> vessels of all kinds. Here is a temple of Apollo, a famous oracle. Near the temple there
> is water . . . From Pedalium to the islands is 80 stadia. Here is a deserted town called
> Ammochostus; it has a harbor, and may be approached by all winds, but there are
> low rocks at the entrance. Enter with care![75]

In this fuller description, we find mention not just of distances between harbors but of the particular winds, rocks, and currents that characterize the harbors themselves; noteworthy architectural structures are cited as well: these, too, are "landmarks," especially as viewed from the sea. By the time we reach Marcian's *Periplous of the Outer Sea* of circa A.D. 400, still more precise measurements of harbors are given, while at the same time the author is able to warn the reader frankly of the fallibilities of such measurements:

> I maintain that it is not easy to establish with accuracy the number of stades over
> every part of the sea. If the shore is straight and has no indentations or promontories,
> there is no problem in reckoning its measurement; but if it is full of bays and projec-
> tions it is impossible to be accurate. One does not sail in a fixed route in the same
> way as one travels on a military road. Let us take as an example a bay which round
> the shore measures 100 stades. Anyone sailing close to the shore will find he covers
> fewer stades than the man walking along the beach.[76]

Marcian here not only points to the difficulties of dead reckoning and to the special problems raised by the method of triangulation as applied to rugged coasts; he also indicates the important difference between two means of traversing a coastline: sailing compared with walking. The difference does not consist merely in the total number of stades measured out when sailing versus those measured out in walking: between these, he remarks, "there will be not be very much difference," unless the arc traversed is that "of a smaller circle."[77] Here the invocation of geometry is not in the interest of exactitude of representation (much less of the schematized representation that characterizes portolan charts proper) but is employed merely as an easily graspable way to help the reader realize that traversal of a bay by boat will differ significantly from traversal by foot in terms of time as well as of space.

I underline this apparently innocuous observation of Marcian's because it is an instance of a description in writing that would not be given representation in the imagery of a map, which rarely, if ever, concerns itself with such detailed differences in modes of traversal. Concerning these differences, written description is manifestly superior, and it is also superior as a reflection of the thoughts of the navigator when he traverses certain intervals at sea. Where writing down such thoughts is an integral part of composing a *periplous,* any such thoughts must be superadded in script to a portolan chart alongside, or within, the image of a given geographic entity—as when verbal inscriptions are attached to certain continents (e.g., Brazil in the Miller Atlas)[78] or included in a formal "Address to the Reader":

> It was several times reported to me . . . by many owners, skippers and sailors proficient in the navigational art, that the island of Sardinia which is in the Sea, was not placed on the charts in its proper place . . . Therefore, in Christ's name, having listened to the aforesaid persons, I placed the said island in the present chart in its proper place where it might be.[79]

Even if writing is the appropriate and justified medium of ancient *periploi,* their gradual evolution into portolan charts suggests that imagery possesses certain distinct advantages. Among these are the following:

1. Imagistic displays can be read all at once and in such a way that every part is immediately perceived as belonging to a larger (geographic) whole.

2. Such displays, when suitably pictographic in character, are readily recognizable as standing in for actual features of the landscape. They can do this without any requirement of strict isomorphism: rocks and shoals are both standardly designated by highly schematic yet still recognizable images.

3. The use of imagery encourages creative combinations of iconography that would require considerable circumlocution to describe in words. I am thinking here of the miniature landscape tableaux that, inserted into certain regions of a map, convey to us what life in these regions is like. Such tableaux are in effect *representations within a representation,* static vignettes or scenic images set within a map qua total presentation.[80]

4. The employment of cartographic images allows for a remarkable elasticity of *scope.* We realize this when we trace the history of portolan charts from their beginnings as maps of delimited locales such as harbors, islands, or short stretches of particular coastlines to the charting of whole basins of the Mediterranean and to the eventual mapping of other discovered parts of the world.[81] This is in effect a movement from topography to geography—from "the smallest localities" to the

largest units of the globe (i.e., non-European continents and "external seas")—rather than the reverse, as was attempted in ancient Greece and earlier in the Middle Ages, when the totality of the known world was the first and most important object of representation.

5. The imagistic component of portolan charts allows for an easily distendable *scale*: from that appropriate to the most minute, close-up consideration of a single harbor (i.e., a large-scale map) to that of the view from afar (as on a small-scale map of an entire continent or hemisphere).[82]

KEEPING IMAGE AND WORD TOGETHER

Despite the powers of visual imagery to which I have just pointed (and despite the awkwardness of adding written inscriptions to maps), it remains mistaken to hold that words merely exhibit cartographic "inadequacy" (Brown's word). Not only are words often quite adequate, they may even be indispensable, as we see most clearly in the role of place-names. If the proper name "Venice" is compared with the various icons of that city that were devised on portolan charts—some of which resemble the city and some of which do not in the least—it is difficult to deny that the mere inscription of this venerable city's name at the proper place on a chart constitutes a much less ambiguous designation than is given even by a quite detailed set of images. Place-names on maps possess a strict singularity of bound reference that no image can equal. These "rigid designators" (to adapt a term of Saul Kripke's for a very different purpose) are invaluable elements not only of portolan charts, but of maps of every imaginable sort. Their absence cannot be compensated for by any combination of other elements, including images of pellucid pictographic quality.[83]

This is not, however, to say that words are *in general* superior to images as vehicles of cartographic representation, as in Stevenson's view. Whatever their specific referential value, they cannot by themselves compose a map. The very idea of a map made up of nothing but words, a "verbal map," is an oxymoron: some factor of purely spatial representation, however attenuated or minimal, is indispensable to all maps.[84] As we have just seen, images have distinctive and irreplaceable virtues in realizing such representation. This is especially the case when it is a matter of topographically discrete landscape features, each of which is represented by its own image: grasslands, rivers, mountains, savannas.

What, then, is the right way to regard the relationship of words and images, particularly in the exemplary case of portolan charts? Rather than adopt either of the extreme positions proposed by Brown and Stevenson, I suggest that we regard the two *media mappae* as coessential, thus as enjoying a "practical complementarity."[85] This last is the phrase of Mollat du Jourdin and La Roncière, who base their judgment on a scrutiny of the coastal toponymy of "la carte pisane," regarded as the oldest surviving portolan chart (it is dated c. 1290), and of *Lo Compasso da navigare*, a *portolano* known to date from 1248 to 1256. The place-names are remarkably similar in the two cases, differing only in their total number (the *portolano* has 1,177 names, the chart 927 names). Moreover, other late-medieval manuals such as the Parma-Magliabecchi *portolano* of the early fifteenth century reveal an exactitude of observation that begins to seem maplike, or at least suggestively topographic, as in these characteristic descriptions:

> From Carminar to Cartagena is 20 miles—northeast by east. Cartagena is a good port at all seasons, before which port there are islands a mile distant. You may pass between any of these islands and the mainland which forms a point. As you enter the port, beware of the shoals. Sail close to the middle of the channel, but towards the northeastern shore, where you may anchor. Beware of sailing too close to a shoal recently discovered on the east side . . . The landmark of Cartagena is a high bald mountain on the east. On the west lies another mountain.[86]

This description is so direct and vivid that we could almost say of it that *it might as well be in images.* It is not difficult upon reading it to form a mental image of the harbor of Cartagena filled with small islands and flanked by two mountains. More tellingly than many ancient *periploi* (which often consist merely of lists of place-names), it is an instance of what we have called a "picture-story."[87] It is therefore not surprising to learn that in a carefully controlled experiment Jonathan Lanman was able to use this passage and others from the Parma-Magliabecchi text as the sole basis for drawing a map of the Mediterranean that is remarkably similar to a portolan chart of the same area, complete with schematic shorelines and sets of attached place-names. Lanman accomplished the same thing for *Lo Compasso da Navigare,* concluding that

> 1) The sailing directions in *Lo Compasso da navigare* and the Parma-Magliabecchi [manual] could be used, independently of any ancillary sources of information, to produce clearly recognizable charts of the Mediterranean and its adjacent seas, and 2) the two charts were skewed in the same direction (counterclockwise) and to approximately the same degree as were contemporary portolan charts.[88]

The passage cited earlier from the Parma-Magliabecchi *portolano* ("From Carminar to Cartagena") may be said to stand midway between an inventory-like listing of place-names (such as we find in a classical *periplous*) and an image-rich portolan chart in the Catalan style. Its very existence (and its role in Lanman's experiment) attests to the intercommunicability of image and word in mapping more generally—indeed, to their deep affinity, their constructive cohabitation. This cohabitation is vividly exhibited in portolan charts. Despite their indisputable uniqueness, these charts exemplify a complementarity of word and image that is as theoretically legitimate as it is practically useful: otherwise, the coexistence of these two media would be a merely contingent matter and could be easily dispensed with. On the contrary: such charts (and many other less striking kinds of maps) consist essentially and not by accidental arrangement in the conjunction of images and words.[89] This is what makes them such distinctive forms of symbolism; they are at once discursive and presentational as well as representational; and they accomplish this feat of being diversely symbolic in a remarkable variety of creative combinations.

MAPPING THE DEPTHS

Kenneth Clark has remarked that "landscape heightens our sense of well-being by enlarging the range of our physical perceptions."[90] One fundamental way in which any such enlarging of our perceptions occurs is through *depth,* a dimension to which I alluded in the first sentence of this chapter. Just to step onto the back porch and to look onto the lawn receding into the near distance is to experience the augmenting power of depth: it is to experience one's body extended backward and outward beyond the confinements of domestic space, which is suddenly expanded and supplemented, given a new dimension, as it were. This new dimension is that of depth, and it is continually provided by landscape vistas, ordinary as well as extraordinary. No wonder that painters have made such concerted efforts to bring a vivid sense of depth into their paintings from time immemorial: *to bring the land alive in painting is to represent it in its depth.*

At first glance, it might seem that maps have nothing to do with the depth of landscape. We might even wonder if they concern landscape at all: Do they not attempt to represent what is known of the "earth," ideally, the entire earth—which is always abstract with regard to ordinary perception? Does not the very word *geo-graphy* literally mean the tracing of the whole earth? Do not maps characteristically represent the earth in its sheer surface, its landforms if not its landscape, thus in its depthlessness? Isn't their overriding concern distance and direction, area and form, all of which belong to surface rather than to depth? It would be tempting to answer these questions affirmatively, but there are good reasons for hesitating before doing so. To start with, we know that mapmakers are not all equally ambitious; not all of them need go to the ends of the earth to accomplish their task. From the very beginning, local maps of particular parts of the land sufficed—for example, the prehistoric maps we have seen from the Mont Bégo area, from Valcamonica, or from Çatal Hüyük. Moreover, sometimes, as in ancient Babylonia, detailed "cadastral" maps used for regional land surveys existed side by side with far-reaching cosmographic maps of the universe.[91] As we have seen, Ptolemy sanctioned the difference between the two kinds of map with his distinction between *geography* and *chorography,* the latter referring specifically to the mapping of local regions and thus the particular configurations of land in these regions: hence their felt and perceived landscape.

If there is no doubt as to the legitimacy of mapping landscape—a legitimacy reaffirmed by the Dutch tradition of *landschap,* a word that we know to apply to the surveying of land as well as to its representation in landscape painting—can we be as confident with regard to the question of mapping its depth? Ptolemy himself was exercised with this question, as we know from the fact that he wrote a treatise with the intriguing title *Planisphaerium.* Although the treatise has not survived, the idea has. The word *planisphere* itself is defined in a lexicon of cartographic terminology as "a map or chart formed by the projection of a sphere, or part of one, on a plane."[92] A planisphere undertakes the representation of three dimensions—of land or sea or earth *in depth*—on the two-dimensional surface of a map. It is a matter, in short, of representing depth by means of projection.

Such representation need not be strictly stereographic (i.e., itself three-dimensional), as it is in the case of a globe. A globe is a literal plani/sphere in that it manages to represent the three-dimensionality of the earth's mass by a combination of three dimensions (in its own physical mass as a solid sphere) and two dimensions (in terms of the map featured on its surface). Short of a globe is a properly *plani*spheric representation such as is found in the "Cantino" planisphere, one of the earliest and most complete of extant por-

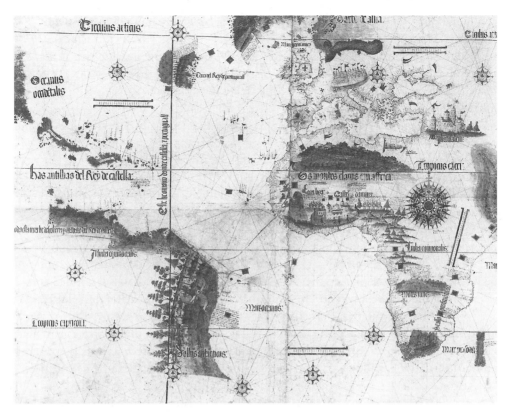

Figure 9.1. "Cantino" planisphere. Anonymous map of Africa, Europe, and South America, 1502. Biblioteca Estense Universitaria. Reprinted with permission.

tolan charts (Figure 9.1).[93] In this sumptuous map of the world as it was known at the beginning of the sixteenth century, we appear to have a cylindrical projection from a spherical paradigm.[94] I say that we "appear" to have this, because it is questionable whether we can speak of projection in any strict sense in the case of portolan charts: an issue to which I shall return in a moment. What is certain is that in a planisphere the depth of a globe—inherent in its very sphericity—is only abstractly represented. In the Cantino case, little trace of the earth-*as-sphere* is retained; and even in a full-fledged cylindrical projection such as that devised by Mercator in the late sixteenth century— which exaggerates the size of landmasses toward the North and South Poles—we are not given a fully convincing representation.

For a convincing representation of depth, we need to turn to different aspects of the Cantino planisphere and to other comparable portolan charts. Even if the earth's special form of spherical depth ("geographical depth" in a Ptolemian sense) is inadequately represented, other forms of depth fare better. These are the depths of (and in) the earth's surface, the depths of its landed character, its "landfall," as we might call it; for it is a matter of representing how the land surface falls out—or more exactly, *how it falls into shape.* Shape here means land-shape—the disposition of the land, its articulations and configurations, "the lay of the land." Integral to the lay of any finite portion of land, any region, is its depth, that is to say, how it recedes outwards or downwards or upwards from a given viewing place. Depth in this delimited sense—the depth of landscape—is indeed representable on maps, as we have already observed in the case of those modern maps that show landscape features drawn in relief by contour lines, hachuring, and still other techniques.

But the depths of landscape were already represented, fully and vividly so, in the premodern era. The Cantino planisphere itself is a case in point. It contains several landscape vistas in the form of miniaturized paintings in a Flemish mode. Each of these tableaux offers a prospect of landscape as seen from an imaginary vantage point at the base of the view itself. In one case, a comprehensive view of the Atlas Mountains is embedded in the otherwise empty space of north Africa: cliffs in the foreground give way to a lush middle ground of green hills and finally to deep blue mountains in the far distance. In addition to other views of precipitous mountain peaks (each of which presents the prospect of a depth *upwards,* i.e., "height"), we are treated to a vision of an exotic city before which stand natives drawn in scale and behind whom are tiny distant huts, as well as a scene of the Brazilian wilderness, positioned almost perpendicularly to the directionality of the African vignettes. In the Brazilian case, a remarkable representation of depth is achieved by subtly demarcated receding areas of the landscape: three parrots gambol in a foreground studded with tall trees whose trunks begin to recede into the middle ground; the latter is subdivided into at least three horizontal planes by lines of underbrush; and a distant horizon beckons, punctuated by tiny trees seen from afar. We are in the presence of a surprisingly sophisticated grid of recession in depth in this premodern and supposedly backward representation of a landscape.

The Cantino planisphere is a projection of (reported or imagined) depths of particular landscapes into the (perceived and felt) flatness of the actual map surface. This flat surface is complicated by the pictorial depth, much as if a curtain had been drawn aside to reveal landscape scenes depicting the interior splendor of the newly discovered continents of Africa and South America. There is something distinctly theatrical about these scenes, reminding us of the early English sense of "landscape" as theatrical backdrop and of the fact that the word *map* itself derives from the Latin *mappa,* literally, "cloth." We have to do here with cartographic vignettes: *tableaux vivants* frozen in time. It is a matter of something *scenographic,* something at once ornamental and theatrical, drawn and presented.

It may be objected that it is precisely this scenographic element, this factor of decorative drama, that falls short of cartographic reality and that is reflected in the literal marginalization of these landscape tableaux on portolan charts. It cannot be denied that such vignettes lack outright cartographic utility, given that many of the scenes depicted in the Cantino manner represent the artist's fantasy of what was to be found in the depths of mostly unknown continents. But is there not another way to regard the situation—a third way, one that steers between the Scylla of a purely dramatic or ornamental depiction of depth and the Charybdis of the purely practical employment of depth measurements?

Such a way exists, and is to be found on portolan charts long before it was to be rediscovered in the twentieth century and dignified with the term *physiographic.* The depth in question is often most evident in the mapping of an *island,* that is to say, an isolated and circumscribed geographic entity whose bounded terrain lends itself to pictographic representation more easily than would a larger and unbounded landmass.[95] The attempt to represent the landscape of islands appears to have begun with Cristofero Buondelmonte's atlas of Mediterranean islands first published in 1420 under the title *Liber Insularum Archipelagi*. In his introduction to this work, which became a best-seller in its day, Buondelmonte announces proudly that "You will see [in this atlas] verdant mountains white with snow, springs, pastureland . . . plains . . . ports, with neighboring capes and shoals, fortified towns and stretches of sea."[96] And indeed, we do see just this in the remarkable pictorial display characterizing insets in maps of islands such as Chios and Corfu. Although the detail of these insets is painted in bold and colorful brushstrokes, they are not merely miniaturized paintings of imaginary landscapes but meticulous representations of natural and man-made landscape features, including trees and towers, marshlands and lighthouses, all represented in convincing depth. The same is true of André Thevet's *Grand Insulaire* of 1586, which shows, for example, the Falkland Islands (then called "Les Isles de Sansom ou des Geantz") in lucid, quasi-physiographic relief, with carefully delineated groups of trees placed amid hills and a circumambient set of realistically rendered shoals. By the time we reach Jan Huygen van Linschoten's *View of the City of Angra on the Island of Terceira* (see Plate 15) in his *Itenerario, Voyage ofte Scheepvaert* of 1595, we are in the presence of a fully sculpted landmass whose features—agricultural, mountainous, urban—are forcefully and graphically depthful.

Given the curvaceous and well-rounded character of a landmass such as this, the two-dimensional image here appears to be returning to its origin in the planisphere, recapturing *per impossibile* the three-dimensionality projected onto the latter. Other instances remind us, however, that no map can attain a complete representation of spherical geography. Alvise Gramolin's portolan chart of 1624 presents two islands (Crete and Euboea) of the Aegean Sea in full relief, several others in semirelief, and the surrounding coastal areas of Greece and Anatolia in utterly flat projection.[97] The Archipelago—that is, the original archi-pelagos or "first sea" of Minoan and early Greek civilization— is shown in dramatic depth, a depth all the more effective because of its contrast with surrounding flat lands: here is an instance of a literal plani/sphere, a map that thematizes its own bivalent origin.

In the various cases just cited, we witness the spectacle of depth representation that is neither merely ornamental nor altogether utilitarian. It is the spectacle of a projection of three-dimensional landscape space into a two-dimensional format—a projection that, unlike the projection of the whole earth onto a world map, is local in scope as well as pictorially persuasive. Thus, even if portolan charts are only imperfectly projective in a strictly geometrical sense,[98] they are vividly and successfully projective in a topographical and chorographical respect. They are maps of places and regions rather than of sites.

FROM CHART TO ATLAS

Soon after portolan charts reached the height of their representational power, the center of professional mapmaking began to move from the Mediterranean up the western coast of Europe, passing by Lisbon and Dieppe on the way to Antwerp and Rotterdam. Indeed, one of the four portolan charts I have just been discussing was made by a Dutch mapmaker (van Linschoten) and another was almost certainly influenced by Dutch cartography of the seventeenth century (Gramolin).[99] This change in cartographic authorship reflected a shift in supremacy at sea—a transition from the earlier centers of sea power in or near Italy (Majorca, Genoa, Venice) to Portugal and Spain, and then to Holland and England. (Some of the very last portolan charts were made by members of the "Thames School" in London.)

It was in Holland, above all, that the premodern world of portolan charts ceded place to the resolutely modern world of Descartes, Rembrandt, and Jacob van Ruisdael—and their counterparts in cartography. We have already observed the close connivance between seventeenth-century painters and cartographers in their commonly pursued *ars descriptio*.[100] Just as Dutch artists turned to an increasingly naturalistic approach to painting in their pursuit of the "landscape of fact" (in Clark's phrase),[101] and just as Descartes put nature to the test of objectivity and its accurate representation in the knowing subject, so in this age of the world picture cartographers made their distinctive contribution by depicting the shapes of the known world in ever more concise and precise images and words. Thanks to the recent explorations of Africa and the New World (including South America and the Pacific)—explorations that would not have been possible without portolan charts—the *oikumēnē* now began to reach out to the entire earth: "geography" at last lived up to its own name. Once more, there were efforts to represent a cartographic cosmos, a *cosmographie universelle,* after the earlier flawed efforts of Hecataeus and Ptolemy (and, even before them, of anonymous Babylonian and Egyptian cosmographers).[102] The age of the world picture was at the same time the age of the world atlas.

In this new world, the portolan chart did not fare well. It survived, but only in limited circumstances and as the source of selected features of the new cartography. Thus, for example, a system of portolan rhumb lines is still evident in the maps of Willem Janszoon Blaeu's 1608 *Het Licht der Zeevaart* (The light of navigation).[103] But there is no effort at strict planispheric portrayal, and Blaeu graduates his maps so as to indicate latitude. His map of the West Indies is based explicitly on a Mercator projection; and the reader is advised to take full advantage of the arts of astronomical navigation such as the night dial, the astrolabe, and the cross staff. These latter made possible not only nighttime sailing but a decisive independence from the visibility of coasts—a visibility on which the premodern method of dead reckoning had relied. The practice of *costeggiare* was giving way to a much-improved and technologically assisted version of *camin francese.* As a direct consequence, coastal maps, whose supreme exemplars in Western mapping are portolan charts, were no longer in demand. Nor were maps that could not easily be reproduced by copperplate engraving, a printing process that a master mapmaker such as Blaeu brought to perfection in the early years of the seventeenth century.

The emerging era of the reproducible image was also the time of mirrorlike representation; the abstraction of the Mercator projection conspired with a comparably abstract means of reproduction to bring about a characteristically modern mapping of the world in which the painted imagery of portolan charts, realized colorfully in planispheric projections and scenographic depictions, was no longer highly prized. Pictographic images of local or regional landscapes in their singular depths held much less interest than abstract and depthless representations of the world-whole that had only recently been drawn together as one transversible globe. The circumspect circumnavigation of coasts, their literal *peri-plous,* was replaced by the bold circumnavigation of the earth as a single totality of surveyable sites. This was an earth unified, not by a circumfluent mythical Oceanus, but by actual oceans and seas external to the Mediterranean world— a world whose coasts had been the privileged subject matter of ancient *periploi* from the time of the Phoenicians and the Carthaginians, as well as of portolan charts, those idiosyncratic, resplendent representations of portions of the earth as it was known from late-medieval to early modern times.

CHAPTER 10

Far-out Mapping

Both in East and West there seem to have been two separate traditions, one which we may call "scientific or quantitative: cartography," and one which we may call "religious or symbolic: cosmography."

—Joseph Needham, *Science and Civilization in China,* vol. 3

GOING EAST

Nowhere is Eurocentrism—that most insidious and long-lived form of ethnocentrism—more manifest than in the case of Western cartography. Did not the Phoenicians' skillful sailing and early colonization justify the belief that the only civilizations that mattered were to be found in the Mediterranean basin, whether Egyptian or Minoan, Mycenean or Athenian, Carthaginian or Roman? Were these not therefore the only civilizations worth mapping? In view of affirmative answers to such questions—answers that were difficult to resist offering, given a Eurocentric bias—it is not at all surprising to find that the specific geography of the Mediterranean was the basis for that proto-*periplous* titled *The Odyssey,* a tale of circumnavigation that is still paradigmatic for the European mind, as we see so eloquently in Joyce's *Ulysses*—modeled so closely on the Homeric epic. Nor are we amazed to learn that ancient coastal guides for Mediterranean mariners were the models for portolan charts of the rest of the world as it became known by the time of the Renaissance. And we should not wonder either at the fact that during most of the Dark and early Middle Ages the primary forms of map placed the Mediterranean at the center of the world: on the emblematic T-O maps, it was the definitive vertical stroke that separated Europe from Africa and above which Jerusalem defiantly occupied a privileged orbocentric position (see Figure 10.1). Whether in the versions of Isidore of Seville (A.D. 570–636) or Beatus Libaniensis (who died in A.D. 798), Asia was placed at the top or eastern direction—in which region was also to be found Paradise, the Garden of Eden, enormous mountains (especially the Taurus range), and several rivers (typically, the Tigris, Euphrates, and Indus). If Asia was accorded the most cartographic space, it was also acknowledged to be the most completely unknown: here was the Land of Gog and Magog, the mythical barbarians of East Asia fenced off by a wall supposedly built by Alexander the Great (but in fact probably reflecting reports of the Great Wall of China).

In the face of the massive and virtually irrepressible political and state power of the Mediterranean—a power in place from at least the second millennium B.C. until the age

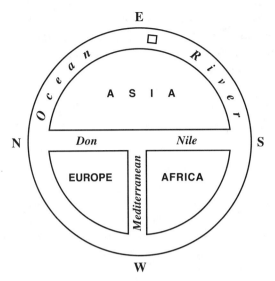

Figure 10.1. Early Western Christian *mappa mundi*: "map of the world" as divided by the three major waterways symbolizing the cross, with east at the top and Jerusalem at the center.

of Phillip II of Spain at the end of the sixteenth century—it is all too natural to neglect developments, including cartographic developments, elsewhere on earth. This is so despite the fact that the very Mediterranean hegemony documented so movingly in Braudel's *The Mediterranean and the Mediterranean World in the Age of Phillip II* was flanked on the east by two civilizations that possessed their own indigenous and ingenious methods of mapping. The first of these was Babylonian and gave rise, as we have seen, to extremely early cosmographic and cadastral maps. The other was that of China and Japan, which we must now take into consideration. For in the Far East remarkable contributions to mapping were made—contributions, moreover, that decisively influenced mapmaking in the Mediterranean world itself. These contributions fall into three quite distinct groups: multiplicity of map media; scientific cartography; and the merging of landscape painting and mapping. In each instance, Chinese and Japanese mapmaking preceded or outstripped the creation of maps in Europe, including those made in the Mediterranean area; and in each case as well, important lessons are to be learned regarding the processes and possibilities inherent in the making of maps, lessons that are obscured when one focuses exclusively on the history of European cartography.

MAPPING IN RELIEF

A veritable explosion of map media—that is, vehicles for mapping—greets us as we look into early Chinese mapmaking. The array of alternative means for the creation of maps, their material base of presentation, is astonishing and unequaled in the West. The first Chinese maps appear to have been made on wood, as seems also to have been true in ancient Greece (*pinax* or map, once again, meant "board"), but after this in China we find bronze maps, ceramic tile maps, maps on stone steles and on diviners' boards, silk maps, maps on paper, woodcut maps, maps on fans and screens, and relief maps on bowls. (To these the Japanese added maps on sword guards *[tsuba]* and on enamel medicine cases, Imari-ware plate maps, maps on cufflinks and on sake bowls, maps on the back of metallic mirrors, and netsuke maps carved on ivory.) Take, for example, the celebrated map of the Nine Cauldrons. A document of 605 B.C. describes this map:

> Formerly, when the Hsia had attained the height of its greatness, the distant regions made pictures of the [material and spiritual] things natural to them, and the governors of the nine provinces made tribute offerings of metal. With this [Yü the Great] caused cauldrons to be cast on which these pictures were represented. In this way the people were instructed so that they could recognize all things and spirits both good and evil. And thus when they traveled over the rivers and marshes, and through the mountains and forests, they did not meet with any adversities. They did not go in fear of the weird spirits and genii of mountains and waters.[1]

The mention of "weird spirits and genii"—distant cousins of what the Romans were to call *genius loci*—indicates that the map inscribed on the Nine Cauldrons must have included considerable local topography.[2] It was evidently not an attempt at general cosmography, which inevitably leads to the predominance of (world) geography over (regional) chorography. Instead, it reflected the Chinese preoccupation with the rigors of representing localities and regions—in this case, the nine regions or provinces into which ancient China was traditionally divided. Hence the inclusion from the earliest times of a specifically geographical map-laden section of the official history of each region.[3] The importance of mapping in classical Chinese society is evident in the fact that, beginning in A.D. 39, there was an annual ceremony during which the Minister of Works presented a map to the Emperor—a map that varied considerably in its specific medium of presentation.[4]

What is instructive about the multiplicity of map media is not merely the range of choice it provides but the fundamental insight that difference in medium means a difference in the entire function and status of mapping, including its social and political effects. Thus, the use of nine solid and quasi-permanent bronze cauldrons as the material ground for regional maps reinforced the *stabilitas loci* that provinces possessed in early Chinese history. In contrast, the stability of China as a whole—as one nation—was embodied in the hard stone stellae on which are sculpted two extant maps of the first half of the twelfth century, that is, from the early Sung dynasty.[5] In still further contrast, the use of silk in early maps brought with it a conducive means of indicating exact geographical location, thanks to the intersection of threads of the warp and the woof at determinate points; it has even been suggested that the cartographic grid system, which the Chinese were the first to employ systematically, was an effect of this feature of the silk textile on which maps, from the most ancient times, were meticulously painted.[6]

Perhaps most significant of all was the creation of a variety of relief maps unique to China. Produced from the eleventh century A.D. onward (and thus constituting the earliest relief maps of which we have any knowledge),[7] such maps were often modeled in clay or wood, and in one notable case a map was constructed of eight pieces of wood that could be folded up and made portable.[8] The very first forms of relief maps are found in the "hill-censers" or "Vast Mountain Stoves" of the Han dynasty or earlier. Made of bronze or pottery, these incense burners (and sometimes mortuary jars) featured a cover or lid shaped with surprising realism like a group of closely concatenated mountains, with holes through which the incense escaped (Figure 10.2). The mountains on a censer jar such as this represent the sacred island-mountains of the Eastern Sea.[9] Notable here is an effort to map landscape features not by reducing three dimensions to two—as happens in virtually every other form of mapping in the West, most conspicuously the planisphere— but by *maintaining actual tri-dimensionality in the map itself.* Further, unlike the globe

Figure 10.2. Bronze incense burner; relief model of sacred mountains, Han dynasty. Height 18.4 cm, diameter of dish 20.3 cm. V & A Picture Library; copyright The Board of Trustees of the Victoria and Albert Museum. Reprinted with permission.

(about which we know of only one example in China before the time of Matteo Ricci),[10] the Chinese relief maps represented specific items in delimited landscapes: instead of the earth as a whole, the precise contours of particular parts of the earth's surface are displayed. Once more, the temptation to the generalities of geography is suspended in favor of the localities of chorography, but in this case a chorography in three dimensions!

Expressed otherwise: if planispheres and all other maps on flat surfaces attempt to represent the three dimensions of world-space within the confinement of two dimensions (i.e., by projections of various sorts), and if certain painters such as Caravaggio and Picasso try conversely to extract three dimensions out of the two to which the surface of their canvas constricts them,[11] Chinese relief maps represent the three dimensions of a local landscape in and by the three dimensions of the relief itself. These maps therefore represent landscape *by dimensional parity*. Hence the crucial importance of the material medium, which allows for, indeed, actively encourages, such parity. In a case like this, the mapping medium is not an indifferent vehicle of cartographic information; *it informs the information itself* by furnishing for it a "bearer"[12] that determines the concrete perceptual character of a given representation. Even if they are distinguishable, map and medium are here inseparable—as inseparable as we have seen discursive and presentational symbolism to be in Western maps. The material medium, whose polyform possibilities were explored so daringly by the ancient Chinese, is for them the very basis of map presentation, and thus of map representation as well.

For the ancient Chinese, the making of maps also enjoys a close proximity to the making of landscape paintings. Emperor Sun Chüan "searched for an expert *painter* to draw a *map* with mountains, rivers, and all physical features."[13] In the case of mapmaker and painter alike, the precise nature of the material means of representation is deeply influential upon the outcome—as we have observed in the case of American and British landscape painters. When Cole or Constable "transfers" a drawing of a particular landscape to a painting of that same landscape, a decisive difference in the resulting representation is immediately perceptible. Even in the fastidious effort of Lane to ensure accuracy of transfer by means of a set of grid lines laid down over his drawing—lines that may be compared to early Chinese grid maps in their delicate insistence—we cannot help but notice the disparity when Lane paints the same scene, even though the painting faithfully replicates the pictorial content of each square of the original drawing. Not only in terms of color, but thanks to texture, consistency, luminosity, modular gesture, and so on, we are confronted by a work so critically different from the drawing that, between them, there is only a loosely knit sameness of subject matter. So too in the case of mapmaking: a topographic drawing of the sacred island-mountains of the Eastern Sea, however detailed it may be, cannot be subsumed without remainder into a landscape painting, much less into the bronze or plaster modeling in depth of these "same" mountains. One kind of map is not simply assimilable to the other, even if they are intimately related by virtue of sharing a common scale of representation and a common conception of their subject matter.

I just spoke of "modeling *in depth*," for this is precisely where the exemplary case of Chinese relief maps takes us—into the depths of the landscape. These depths are indicated when the philosopher Chao Shih-shu said that the cartographer Chu Hsei had made "a wooden Map of the Countries of the Chinese and Barbarians, upon which the convexities and concavities of mountains and rivers were carved."[14] A three-dimensional relief map gives direct representation to such "convexities and concavities," and in this respect is superior to the most refined contour map drawn out on flat paper. This becomes apparent upon inspecting a seventeenth-century Chinese contour map of one of the same sacred mountains of the Eastern Sea that was sculpted on incense and mortuary jars. Skillful as are the graphic elements in this contour map—which uncannily resembles contour mapping of the late nineteenth century in the West[15]—it cannot compete for convincingness or vivacity, let alone accuracy of depiction, with what is found in the fully modeled relief maps atop censers of the Han period. In the early Chinese relief map, then, *depth comes into its own*; depth is here represented in a medium whose own dimensionality is continuous, and not conflictual, with the dimensionality of the landscape from which it takes its rise.

GRIDDED MAPPING

After art, science! The ancient Chinese are remarkable for having been at once the first fully accomplished landscape painters and the first scientific cartographers. Just as they were painting full-fledged landscape vistas many centuries before these were first created in the West—by some estimates, a thousand years earlier!—so they were mapping with precision long before the Western science of cartography became solidly established in the late Renaissance and early modern era. By this I do not mean to imply that the Chinese did not engage in highly speculative mapping of the sort that was also prominent in Babylonia and, above all, in the Middle Ages. In fact they produced a number of cosmographic charts, but these remained a minor genre compared to mapping that was cartographically precise.[16] The practical motive for such exact mapping doubtless derived from the Chinese obsession with land surveys, whether at the regional or national level, and then often for military purposes: during the period of the Warring States (403–221 B.C.), a special government office was set up specifically for the care of maps.[17]

Whatever the ultimate motive, it cannot be doubted that the Chinese developed cartography as a serious scientific discipline at the very same time that Western cartography, having gotten off to a promising start with Erastothenes, Hipparchus, Marinus, and especially Ptolemy, was entering into what Needham calls "the Great Interruption." Marinus was an almost exact contemporary of the founding father of Chinese cartography, Chang Hêng (A.D. 79–139), and yet in the Dark Ages of the post-Ptolemaean period an interest in scientific aspects of cartography was virtually extinguished in the West, whereas this interest (really, a passion) flourished in China for an entire millennium and more. Just when scientific cartography was being superseded by a tradition of religious cosmography in the West, it came to be pursued at a high level of sophistication in the East.[18] For example, interest in the problem of coordinates vanished from Western purview (with one or two exceptions to which we shall return), while it was a focal point of discussion in China during the entire fourteen hundred years that separated Chang Hêng from the arrival of Jesuit missionaries in the late sixteenth century, at which point a Ptolemaic model of cartographic space was introduced to the Chinese. On the other hand, it is arguable that Western portolan charts of the late-medieval and Renaissance periods would not have been possible without the use of the magnetic compass, itself a Chinese invention of many centuries earlier. When we observe Chinese portolan charts dating from the seventeenth century and after, supposedly the result of Westernization, we may very well be perceiving a form of mapping that, despite its Westernized look, could not have arisen in the first place without a critical discovery made by the Chinese themselves in the fifth to eighth centuries B.C.![19]

But the construction of the magnetic compass—with its important implications for determining "true" versus magnetic north, and for converting rhumb lines of dead reckoning into loxodromes, and wind roses into compass roses[20]—was overshadowed by yet another Chinese contribution to scientific cartography: the use of the "square grid" in the making of maps. Despite the absence of explicit techniques of map projection, the Chinese succeeded in constructing maps on the basis of grids as early as the twelfth century A.D. The idea of a grid—definable as "a referencing system on a map in which points are defined by their distances from two perpendicular axes, usually with the aid of marginal graduations"[21]—has multiple roots in early Chinese culture.[22] I have cited the use of silk in mapmaking, but there is also the fact of land allotment in regular shapes; the model provided by extremely early maps consisting of concentric rectangles; and even the paradigm provided by the abacus.[23] Whatever the exact origin, there can be no doubt as to the utility of a grid system for mapping. Unlike a nongridded reference system— which only allows for locating places positioned within a given space of representation by means of identifying numbers or letters in the margin, as in the case of many conventional contemporary maps—the system of the square grid invented by the Chinese allows for the determination of location by simple position within a given unit or between units. It also permits the accurate determination of distance, area, and direction on a map.[24] Figure 10.3 shows an example of a square-gridded map from a gazetteer of the Tang Dynasty (A.D. 618–907).

Eurocentrism must make way for the fact that the seventeenth century in the West, the "century of genius" (Whitehead) in science and philosophy and of "the art of describing" (Alpers) in maps and painting, was anticipated with regard to a quite fundamental conception by the Chinese of the Han and Chin dynasties. Chang Hêng is usually credited with the discovery of the grid system; he is said to have "cast a network of co-ordinates about heaven and earth, and reckoned on the basis of it."[25] Chang Hêng's successor, Phei Hsiu (A.D. 223–71), often considered "the Ptolemy of Chinese cartography" (perhaps we should say rather that Ptolemy himself was "the Phei Hsiu of Greek cartography"), was the first to describe explicitly how to construct the grid system. In the Preface to the presentation of his maps in the *Chin Shu,* "the earliest surviving statement in China about the application of mensurational technique to cartography,"[26] he begins by lamenting the loss of the older maps. None of the new maps of his era, as he goes on to say, "employs a graduated scale *(fên lü)* and none of them is arranged on a rectangular grid *(chun wang).* Moreover, *none of them gives anything like a complete representation of the celebrated mountains and the great rivers*; their arrangement is very rough and imperfect, and one cannot rely on them."[27] This single brief statement throws down a crucial challenge: how may a cartographer be at once exact and exacting (i.e., by means of *fên lü*

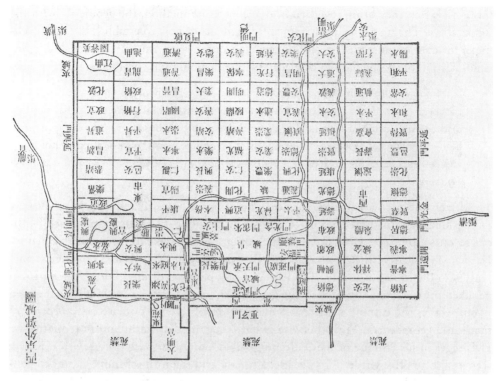

Figure 10.3. Gridded map from the Tang Liang Jing Chengfang Kao. Xu Song, compiler. Reprinted by permission of the Widener Library of the Harvard College Library.

and *chun wang*) and yet also convey "a complete representation" of such intrinsically irregular landscape features as mountains and rivers? Is it possible to achieve both at once?

Phei Hsiu thinks that it is indeed possible, and he proceeds to demonstrate this by describing six principles of mapmaking, whose abbreviated expression is as follows:

1. The graduated divisions *(fên lü)*, that is, the means of determining the scale to which the map is to be drawn, must be determined, thus allowing for the distinction of the far from the near.

2. The rectangular grid *(chun wang)* must be laid down so as to be able to depict the correct relations between the various parts of the map.

3. The sides of right-angled triangles *(tao li)* must be paced out so as to fix the lengths of derived distances.

4. One must measure the high and the low.

5. One must measure right angles and acute angles.

6. One must measure curves and straight lines.[28]

Formal and remote as these principles may seem at first sight, it has been said that "they indicate concisely the major problems in present-day cartography."[29] One reason why this is so is that they bring together, in one composite method, the geometric with the topographic. The first three principles are statements of how to establish the right technique for determining scale, position, area, and distance on a map. The last three principles bear on the particularities of place. As Phei Hsiu himself comments:

> These [latter] three principles are used according to the nature of the terrain, and are the means by which one reduces what are really plains and hills [literally, cliffs] to distances on a plane surface . . . Thus even if there are great obstacles in the shape of high mountains or vast lakes, huge distances or strange places, necessitating climbs and descents, retracing of steps or detours—*everything can be taken into account and determined.*[30]

This statement is remarkable for at once conceding the necessity for reducing the three-dimensionality of natural objects to the two-dimensionality of the map surface (without recourse to projective techniques, including those at stake in three-dimensional relief maps) while nevertheless proclaiming the adequacy of that surface's representational powers to the in-depth reality of "what are really plains and hills"; for "everything can be taken into account" not *despite* the geometrism of the grid (as we might be inclined to suspect in the case of a Cartesian conception of space), but precisely *because of* that same geometrism. Put otherwise, what is regarded as perniciously reductive by Heidegger in his critique of the "enframing" that Cartesian space brings with it is here seen as a distinctive virtue.[31] To be sure, a square grid is a form of frame meant to contain and to relate various features of the landscape. But it is a frame that, far from closing off the reality of what is represented therein, makes its cartographic expression possible. Not only does a properly scaled grid system represent "the reality of the relative positions"[32] of the mapped items in a given landscape, it also allows the perceived character of these items to be taken into account by means of appropriate calculative methods, for example, leveling heights, determining diagonal distances, and straightening curves.[33]

Early Chinese mapmakers, like their later Western counterparts, employed pictographic symbols for landscape features such as settlements, roads, rivers, mountains, and trees.[34] Especially striking is the fact that Chinese ideograms for such topographic features often incorporate these same features: the character for "river" is a graph for flowing water, that for "mountain" derives from a drawing of a three-peaked mountain, and that for "map" itself contains a schematic map.[35] Indeed, the relationship between maps and writing, as that between maps and landscape art, runs deep, as we can see from this vignette of education in China in the early twentieth century:

> Some memories of the past are still fresh to me even today. As a young pupil, each week I would receive from my teacher two sheets of chequered paper with red squares forming lines. On one sheet I was required to do handwriting exercises by writing from a calligraphic copy and for each square only one Chinese character was required. On the other sheet I had to give the enlarged version of a given map and add place names along with my own name. I did not realize then that this was the very method traditionally employed for millennia in China in making maps.[36]

Calligraphy and cartography here go hand in hand—literally so!—and their educational alliance, an alliance that shows up in the extensive use of legends on Chinese maps from the earliest times, prefigures the alliance to which Phei Hsui points so tellingly. In this alliance, purely verbal descriptions are of equal importance with geometric and topographic imagery. It was traditional, and is still an extant custom today, to consult, in addition to maps, various written sources such as gazetteers and travelers' reports as a way of gaining intimate knowledge of a local region. Phei Hsui himself, in compiling his treatise on mapping, carefully examined ancient textual accounts of particular regions. Indeed, he was known to his contemporaries more for his literary than for his cartographic prowess. Instead of the tension between text and image that we have witnessed in the contrast between written *periploi* and highly imagistic portolan charts in the West, in China the collaboration between image and word was very close from the beginning, and it stays close later on. In a typical atlas, such as the *Guang Yutu* of Luo Hongxian (printed circa 1555), much of the work consists in verbal commentary on maps. Overall, we can speak of an abiding "complementarity of text and map."[37]

There is also a continuing complementarity between chorography and cartography. The localism and particularism of the former—which calls for discrete idiographic representations "according to the nature of the terrain"—is accommodated by the generalism and geometrism of the latter. What at first blush may seem to be diminution or reduction—as when we attend mainly to the grid system—ends by being enhancement and exfoliation. As Phei Hsui himself says at the conclusion of his Preface, "when the principle of the rectangular grid is properly applied, then the straight and the curved, the near and the far, can conceal nothing of their form from us."[38] The straight and the curved, the near and the far, as primary geomorphic shapes and modes, collude and come to "complete representation" in the unlikely place of the square grid, as we can see in a striking map of the nine frontiers from a Ming encyclopedia, a map that combines grid, pictorial elements, and writing (Figure 10.4).

None of Phei Hsui's own maps have survived, not even his celebrated atlas of China (titled *Yü Kung Tin Yü Tu* [Regional atlas of Yü Kung]) or the enormous map he made from eighty rolls of fine silk. But many maps that appear to be constructed on his six basic principles are extant, including the *Yujitu* and the *Guang Yutu*.[39] It has been speculated that the Chinese square grid system may have influenced portolan charts in the West: thus, for example, a fifteenth-century Venetian nautical atlas sports a measured grid in place of rhumb lines.[40] Still more intriguing is the survival of two maps from the period of the Great Interruption in the West that show the grid system fully in place: a Byzantine map of Ceylon of circa A.D. 1250 and a map of Palestine by Marino Sanuto from about A.D. 1306 (Figures 10.5 and 10.6). These extraordinary maps suggest an early, subterranean presence of Chinese scientific cartography in the West—a presence that, as in the case of the magnetic compass, predates the arrival of Jesuit missionaries such as Matteo Ricci in China during the sixteenth century (though perhaps not that of Marco Polo in the thirteenth century).

These two maps are striking for their combination of a strict grid system with lively landscape features: the first shows Ceylon's contoured landforms from a bird's-eye view and the second situates specific landscape features of the Palestine landscape (mountains,

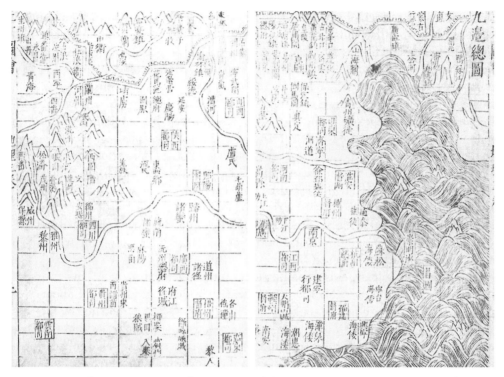

Figure 10.4. *Juibian Zongtu,* comprehensive map of the Nine Frontiers. Ming; two pages from the *Sancai tuhui* (printed 1609); compiled by Wang Zi. Reprinted by permission of the Widener Library of the Harvard College Library.

rivers, gulfs) in relation to each other. In the latter case, the features in question would be chaotically disparate from each other, unreadable fragments, if the grid system they co-inhabit did not bring them together in one coherent cartographic space. Here, indeed, in these anomalous maps of the Middle Ages in the West, the presence of the Chinese grid system is palpable; for despite "great obstacles in the shape of high mountains or vast lakes, huge distances or strange places," both maps manifest considerable cartographic clarification and unification: thanks to the square grid system, "everything can be taken into account and determined." Just as maps in general—Western as well as Eastern—can be said to be compositions of discursive and presentational symbols, so maps such as these, whether their origin be Chinese or Mediterranean, exhibit a convincing combination of geometry and landscape—two factors that, like the words and images with which they are allied and on which they depend, are no longer competitive but coinherent.

NEW DIRECTIONS: JAPAN

But let us return from science to art, specifically to landscape art in (or, more exactly, *as*) mapping. This we shall do by turning from Chinese to Japanese maps—indeed, to a relatively late chapter in the history of the latter.[41] In its early history, Japanese cartography was greatly influenced by Chinese cartography, albeit in a deferred fashion. This is not to say that there was not in Japan a native tradition of mapping as well as a lively interest in the very process of mapping itself. As Nauba Matsutaro writes, the "Japanese had a fondness for maps that was unusually deep and strong; it was almost a feeling of reverence. It is no accident that they were prolific map makers even long ago."[42] As long ago as we know about is the eighth century A.D., when *kondenzu* maps, in effect land surveys, were created following the Taika Reforms of A.D. 646, which placed all land in the control of a central government.[43] Usually inscribed on hemp, these earliest surviving maps are notable for their use of a loosely applied grid scheme that corresponded to the *jori* system of land allocation. If this grid resembles that which was to be developed much more systematically in China, it was by no means universally employed by the Japanese, who at

Figure 10.5. Map of Ceylon on a Byzantine grid, in a manuscript from Mount Athos, c. 1250. As reproduced in Joseph Needham, *Science and Civilization in China* (Cambridge: Cambridge University Press, 1959), vol. 3, Figure 221. Reprinted with the permission of Cambridge University Press.

Figure 10.6. Marino Sanuto, Map of Palestine on the Chinese grid system, from his *Liber Secretorum Fidelium Crucis,* c. 1306. As reproduced in Joseph Needham, *Science and Civilization in China* (Cambridge: Cambridge University Press, 1959), vol. 3, Figure 241. Reprinted with the permission of Cambridge University Press.

about the same time evolved an indigenous mode of mapping called *gyoki*. Reputedly devised by a Buddhist monk during the Nara period (A.D. 697–780), *gyoki* maps consist in egg-shaped rings (each of which designates a province) aligned in terms of major roads. Only the region in which the capital (Kyoto) is located is fully circumscribed by its own boundary line, the others being partly overlaid by the outlines of neighboring provinces. As a result, the coastline is "distorted beyond recognition,"[44] reminding us of the schematic coastal configurations on portolan charts. Except for increasingly bright coloring by hand, and despite sometimes being drawn surrounded by a protective dragon,[45] *gyoki* maps remained largely unchanged in character for nearly eight hundred years, from the eighth to the sixteenth centuries A.D. An example from a codex published in the middle of the sixteenth century A.D. is shown in Figure 10.7.

The prolonged history of *gyoki* maps was an era, if not of Great Interruption, at least of Long Stasis. As Muroga Nobuo comments, "no significant developments were made during [this] early period."[46] It was not until the Edo period (A.D. 1603–1868) that dramatic things began to happen in the world of Japanese cartography. Strikingly, this occurred under the inspiration of the Dutch, who in the course of establishing a vigorous

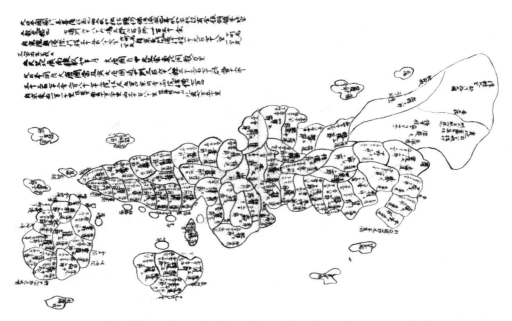

Figure 10.7. *Dainihonkoku Zu* from the Shūgaishō, codex of 1548. Tenri Central Library, Nara Prefecture.

sea trade with Japan founded a map factory at Nagasaki. This map factory presented Joan Blaeu's world map of 1648, grandiosely titled *Nova totius terrarum orbis tabula,* to the Tokugawa government, and the Japanese in response coined a word, *Rangaku,* to designate a Dutch approach to learning in general and to maps in particular. The sudden interest in world maps, for example, was almost certainly a result of the Dutch presence, as we can observe in colorful and elaborate exemplars painted on screens.[47] Before long, the Japanese and the Dutch were exchanging maps of each other's countries: Mercator's fumbling efforts of 1575 in his *Indiae orientalis insularumque adjacientium typus* being replaced by Nicolaas Visscher's much more accurate map of "Iaponia Regnum" of 1680, while the Japanese gradually perfected their maps of Holland (as we see in Takami Senseki's "Newly Translated Map of Holland" of 1849).[48]

Nicolaas Visscher, it will be recalled, was the cartographer who made the map of the Netherlands that was so meticulously copied by Vermeer in *The Art of Painting.* In Visscher's map, various "topographic views" of cities filled the borders of his otherwise straightforwardly chorographic work. In this respect, it formally resembled Japanese city maps also created in the seventeenth century, except that in these latter a rigorous grid plan of the city proper occupies the center of the cartographic space, while around the edges are placed various architectural and landscape views.[49] In both kinds of map, however, a bird's-eye view of the central subject (whether nation or city) is surrounded by profile views of buildings or mountains, thus curiously replicating the ancient pattern that combines "in-plan" (bird's-eye view) with "perspective" (sidewise) standpoints— a pattern we first encountered in the wall painting at Çatal Hüyük and in petroglyphs from the Valcamonica region of Italy. Such hybrid cases stand in contrast with the strictly in-plan maps favored by the grid system of the Chinese and Romans, allied in this regard as in their common imperialism.

The Japanese pursued another significant variation by creating maps in which an entire region is seen from an elevated oblique angle. These maps are to be met with in China, in certain portolan charts, and in American cartography after 1930 (when they were reintroduced by Erwin Raisz). The Japanese case is distinctive, however, insofar as in maps of this kind *the oblique view merges altogether with the pictorial representation of landscape*. Unlike those maps that keep the topography of the land on the literal margin, here concrete landforms are brought centrally into view and shown as such, thereby achieving a full fusion of landscape representation with cartographic fidelity. In creating such maps, the Japanese surpass their Dutch map masters even as they remain indebted to them, achieving a new plateau in the art of mapping.[50] If Vermeer's *The Art of Painting* confined landscape to the *parergon*—the literal "by-work" of the depicted map's edge—and thus doubly displaced its presence (first to the wall of the room in which the artist is painting and then to the margins of the map hanging ceremoniously on this wall), the Japanese made landscape integral to the central cartographic image. In this way, they made the map into the functional equivalent of a landscape painting. Thanks to this bold step, the Dutch were outdone at their own game: what had been for them a matter of ingenious interplay and instructive parallelism between maps and works of art, a redoubled art of description, became in the hands of their Japanese inheritors a revolution in the very conception of what a map can be. It can be a work of art in its own right.

By this last claim, I mean something more than the easily assumed opinion that Japanese maps of the Edo period are "picturesque" and thus "artistic." I also mean something other than is maintained by Matsutaro when he writes that "most old maps made in Japan, hand-drawn and printed, were rendered in bright colors and possess an artless kind of beauty."[51] Just as spontaneous beauty is not the essence of the revolution in question, so decorative beauty (such as we cannot help but notice in folding screen maps) is also not mainly at stake: "an ornate style accentuated by gorgeous colors."[52] And just as we have seen that the peculiar power of landscape painting lies in something other than ornamentation and practical use, so Edo mapmakers encourage us to discover something in mapping beyond pleasingness or utility—something having to do with the irrepressible presence of landscape itself. In the new form of Japanese cartography, landscape is as fully at one with maps as the perceived and remembered landscape is with the paintings of Constable or Cézanne, Cole or Lane.[53]

This revolution did not happen in a vacuum. Beyond the crucial Dutch connection, there were other historically decisive events that made it possible: above all, the re-unification of Japan in the late sixteenth century, followed by a long period of comparative tranquillity. Strip maps for travelers were made since at least the middle of the seventeenth century, responding both to a royal edict requiring an annual visit to Edo and to a newfound leisure and prosperity.[54] As in the case of the Peutinger Table of the late twelfth century (Figure 7.8), these "picture-scrolls" brought together direct representations of the landscape with information of special value to the traveler. By the end of the eighteenth century, *ukiyoe* ("pictures of the floating world"), that is, multicolored wood-block print maps, had come into widespread circulation at the very time when a popular demand for maps reached a new height. The result was the dramatic development to which I am pointing: the creation of maps that were at once cartographic and painterly, equally and fully both. In this new vision of what maps can be, the geographic merges with the chorographic, three dimensions with two, depth with surface. We could also say that the factor of *ch'i,* that vital natural force which had so thoroughly infused early Chinese landscape painting, now comes to reside in Japanese maps.[55]

I do not want to imply that the only use of *ukiyoe* woodcuts lay in the creation of map masterpieces.[56] Nevertheless, something extraordinary happened when *ukiyoe* art teamed up with cartography in such intimate companionship that they were virtually indistinguishable from each other. In what does this companionship consist, and how exactly does it occur in Japanese maps of the late Edo period?

Let us take as an exemplary case a map of Japan made by Kuwagata Keisai in the early nineteenth century (see Plate 16). As we gaze at this map—or is it a painting?—we are at first struck by its brilliant pictoriality: its detailed realism, clarity of delineation, brightness of color, immediate readability, and sumptuous layout. These are qualities that we find as well in many nineteenth-century Japanese woodcuts, including those by Hiroshige (who in fact made his own map of Japan at about the same time as did Keisai).[57] Indeed, we would find many of the same attractive pictorial qualities in the landscape vignettes that so gracefully bedeck the Cantino planisphere examined in the preceding chapter. Something else is at work in Keisai's map, however, and it behooves us to determine what this is. We shall do so in terms of four features: depth, encompassability, partial exaggeration, and density of composition.

FOUR FACTORS IN *UKIYOE* MAPS

Depth. Our gaze is at once swept around and away from the putative center established by the dramatic presence of Mount Fuji (almost entirely in white). We see Japan receding in two "wings," upward and to the right toward Edo and leftward and outward toward an archipelago of disparate islands. Hovering at the horizon are the mountains of Korea. Much more effectively and subtly than in the Cantino planisphere—which established depth by a series of superimposed planes (e.g., in the tableau of Brazil)—here we are treated to a virtual *celebration of depth,* thanks to two closely concatenated depth-bestowing factors. First, a complex *intra-insular* topography generates the perception of depth within the depicted landmasses. Our look is led gracefully but forcefully across and through the island(s) by the sinuous progress of roads that wind around mountains and occasional temples. Stacked-up objects (primarily mountains) suggest depth by their very occlusion of each other.[58] Placards with affixed place-names grow smaller with increasing distance from our implicit viewing point; and distances, written in Japanese numerals, reinforce the perceptual experience of depth. All of this creates recession within what Husserl calls the "internal horizon" of the landscape. Second, and in direct contrast with this first factor of depth, is the equally efficacious sense of recession constituted by the "external horizon"—by the *extra-insular* aspects of the map. I refer both to the coastline of Japan as it outlines the country against the surrounding sea and to the distant horizon, limned by a single horizontal line and broken only by the profile of the remote

Korean mountains. While the coastline represents the place where Japan vanishes into the sea (thus giving to the map as a whole a depth *downwards*), the horizon represents the vanishing of the entire island complex in a depth *outwards* toward the unseen and unknown (heightened by the presence of the moon's rim hanging just over the sea's horizon). Moreover, both kinds of depth complement and enhance each other: the intricate and involved depth of the landforms rejoins the more open and sweeping depth of the coastline and horizon. Their interplay is at once aesthetically satisfying and cartographically significant.

Encompassability. The Japanese are not alone in their focus on islands as the preferred subject of mapping. Early Western practitioners of topographic mapping—for example, Gramolin and van Linschoten (who, incidentally, made the first fully accurate map of China's coastline)[59]—also gave to islands a privileged position. What is it about islands that lends itself so readily to in-depth representations in maps that emphasize the wholesale inclusion of landscape features? The straightforward answer is: their easily encompassed extent. When the eye of the map beholder is able to take in, at one continuous glance, the entire layout of a bounded layout of land, he or she is encouraged to linger on particular topographical features that populate this land—to concentrate on these features as contained within the internal horizons that are determined by the land's perimeter. In contrast, to look at the map of a region of a large landmass whose boundaries are nowhere indicated is to be led *away from* particular features of this mass—thus to wonder *what else* is to be encountered in such a partially represented landscape.

It is not surprising, then, that the Chinese, for all of their interest in local landscape features, did not develop the cartographic representation of these features to the point of perfection attained by the Japanese; in particular, they did not succeed in that full merging of landscape scene and mapped reality that is so prominent in the *ukiyoe* school. Can it be accidental that Japan itself is just small enough to be imagined as seen in its entirety from a sufficiently high point of view and thus to be encompassed in one sweeping look, as occurs so effectively in Keisai's panoramic map—and yet, at the same time, to be seen as sufficiently close to be representable in terms of its most prominent landscape features?

Partial exaggeration. These landscape features are not, however, represented according to actual scale. If they were, their prominence would be much reduced, and the landmass of Japan as a whole would be presented as much duller and flatter than in Keisai's exemplary rendering. His shrewd strategy and that of other *ukoyoe* map artists is to select only the most conspicuous such features—the "famous places," as the Japanese call them, by which term they mean to include architectural sites as well as places of nature—and to give to them a deliberately exaggerated stature within cartographic space that aids in their identifiability and recognizability. This strategy also has the advantage of suppressing less interesting secondary features as well as empty spaces. In all of these respects, the exaggeration in question can be said to be both harmless (no deeply misleading distortion is involved)[60] and useful (insofar as cartographic clarification and simplification are thereby achieved).

Density of composition. Closely related to the employment of exaggeration is the pictorial density that is so characteristic of Keisai's map and others like it. The cartographic space depicting the Japanese landmass is closely packed; it is visually active and busy, giving us the distinct impression that there is *much to do* and *much to see* in the country therein displayed. A special effect of intimacy draws us as map viewers into the landscape by arousing our curiosity (what is on the other side of Mount Fuji?) and often satisfying it (i.e., by the wealth of detail with which any given part of the landscape is represented). We feel that this is a land and a landscape that we would like to explore and to inhabit; it is the very opposite of a scene of desolation such as a trackless desert might present. The threatening foreignness of uninhabited wilderness, in which we cannot imagine ourselves to be ever fully at home, is replaced by a scene of invitingly inhabitable places. Even if the comparative size and scale of these places are not exact in a strict cartographic sense—we are far away from Phei Hsiu's concern for precise measurement—the map exudes a humanly responsive ambience in which we feel ourselves at home by projecting ourselves into it. And project we do: not now in any technical sense of cartographic projection (*ukiyoe* maps are notable for their absence of any grid or graticule, much less of any planispheric pretention) but in the concrete sense of feeling ourselves to be moving in imaginative projection into the intimate corners of the landscape so persuasively portrayed in these extraordinary maps.

THE CONVERGENCE OF ART AND CARTOGRAPHY IN THE LANDSCAPE MAP

The overall effect of the foregoing four factors—factors that are characteristic of *ukiyoe* maps even if not unique to them (Linschoten's map of Angra [Plate 15] embodies at least two of them)—is to leave us feeling more actively engaged as map viewers than is the case with many other kinds of map, including even portolan charts. Thanks to the vivid representation of an encompassable depth and to the alluring aspects of exaggeration and density of composition, we experience a special sense of immersion and inclusion. Despite the lack of consistent scale and rigorous projection, we experience ourselves deeply and immediately *oriented* in the space of the map. This is so even though many *ukiyoe* maps are either nondirectional or directional only in terms of a particular landscape feature (frequently Mount Fuji or some other conspicuous landmark).[61] Indeed, rather than a scientifically determined magnetic north or a theocentrically imposed east, the felt and intuited directionality of the entire landmass of Japan in Keisai's map guides and structures our orientation, whatever its cardinal directionality. The important fact is not just that "the diagonal alignment of Japan from southwest to northeast favored a northwest orientation [in its mapped representation], with Japan shown horizontally, to conserve space and save paper,"[62] but also that our view of the land in its full sweep as seen from an oblique and elevated point of view draws us into a close relationship with it. This relationship, and not some externally imposed requirement of cartographic exactitude or proper direction, determines our sense of orientation. We regain strength as map viewers from contact with the earth, an earth at once limned and mapped. Part of this renewed strength is a renewed sense of concrete directedness (not to be confused with direction as such) that is determined by and sensitive to salient features of the landscape under representation.

The lesson the instructive case of *ukiyoe* maps offers is twofold. To begin with, such maps show that landscape can be the simultaneous subject of artistry and of cartography—

that it need not be represented *twice,* first abstractly in a neutral cartographic space and then, once more, in a separate set of concrete images. This was a lesson lost on the very next generation of mapmakers in Japan: post-Edo mapmakers reverted to the seventeenth-century Dutch practice of putting topographic views of special landscape features in insets at the edges of maps that were otherwise drawn entirely in plan.[63] Such bowdlerization of landscape representation contrasts poignantly with the ambitions of an artist-*cum*-mapmaker such as Keisai, who was able to fuse landscape shown in pleasing pictographic detail with the *same* landscape taken as cartographically informative. For him, the representation of landscape in the first sense is not distinct, or even distinguishable, from the representation of landscape in the second sense. The two have become one in the single representation that is the *ukiyoe* map itself. A pictorial, indeed, even a decidedly painterly, portrayal of the land is at one with its cartographical discernment.

A second thing to be learned is that in such an interfused representation the role of the human body is brought into play—if not by express actions, then by its imaginative projection into the landscape mapped and painted; for it is only by the map viewer's free movement of his or her virtual body in the represented space of the map, its place-of-representation, that a sense of intimate inhabitation in its recesses (and its in-depth recession) can be accomplished. Just as the body is indispensable in the full experience of human dwelling—that is, in built place—so it now demonstrates itself to be sine qua non in the full experience of mapped places. But this only becomes manifest when mapmakers are themselves bold enough to represent these places in forms that invite the active participation of the body in its perceptual and orientational powers.

One such form, and doubtless the most effective one, is that of landscape painting itself. In a breathtakingly daring gesture, *ukiyoe* cartographers brought the resources of such painting directly into their mapping. Ptolemy had said that "chorography needs an artist, and no one presents it rightly unless he is an artist."[64] In Kuwagata Keisai mapmaking found such an artist. He and his compatriots in early-nineteenth-century Japan created what we can only call "landscape maps": map-paintings that call on us to enter them as scenes shown in works of art. The categories of "landscape painting" and "mapping," which we normally regard as distinct and discontinuous modes of representation, are here creatively confounded.

It is one thing to speak of mapping the landscape—we have been exploring this for several chapters now—but it is quite another thing to talk of *landscaping the map.* Yet it is precisely the idea of transforming a planiform map into a veritable landscape that *ukiyoe* cartographer-artists present to us as a prospect both disturbing and delightful: disturbing to our conventional notions of mapping and delightful to our discerning senses.

CHAPTER 11

Rectangularity and Truth

Among practitioners of siting, straight lines were regarded as signs of malign influences, of a lack of qi, *and traditional Chinese painters tended to hold straight lines in low esteem for the same lack of vitality . . . Even on grid maps, one frequently finds that the uniformity imposed by the squares is broken up by pictorial elements.*

—Cordell D. K. Yee, "Chinese Cartography among the Arts"

IN FAVOR OF THE FLAT

Having just recognized one extreme in the range of mapping—the *ukiyoe* map, which makes landscape the central content of maps—we also need to recognize another, equally fateful extreme. This is found in maps that deliberately ignore landscape features and even suppress them altogether; for we cannot afford to assume with P. D. A. Harvey that there is a natural, much less an ineluctable, progression from "symbolic" to "picture" maps—where by *picture* is meant a representation that is sensitive to the topographic detail specific to given landscapes. (Nor can we presume that relief maps, or for that matter "picture-stories," simply form an intermediate stage between these extremes: we have just witnessed the sui generis status of Chinese relief maps atop censer jars, a status that also obtains for storylike itinerary maps of Japan and Tibet.) There is an entire class of maps that favor flatness or "planitude" of representation rather than the plenitude of landscape presences, and yet are considered fully valid maps. The prominent features and famous places that were given emphatic representation by a mapmaker such as Keisai are here disregarded in view of other concerns. The maps I have in mind are not in the least embarrassed by being merely two-dimensional and make no claim to being planispheric projections of the round earth onto flat paper. Considerations of projection are minimized or overlooked in favor of whatever aids in establishing a consistent and reliable representation on the engraved or printed page.

Such maps are *survey maps*—"land maps," as they are sometimes called. (When they are used expressly for determining ownership and real-estate value, they are termed "cadasters" or "cadastral maps.") Despite the special interest we have taken in the pictorialization of landscape, we cannot in good conscience bypass these maps. Not only do they form a distinctive genre, but they offer a valuable corrective to a primary focus on the detailed visual treatment of landscape, its pictographic representation. Moreover, land or survey maps are historically the main form of regional map and thus of chorography in Ptolemy's strict sense: "the end of chorography is to deal separately with a part of the whole."[1] But where Ptolemy called on the chorographer to "describe the smallest details of places" and to "paint a true likeness"[2] of these details, survey maps have traditionally been unresponsive to issues of detail and likeness, being concerned instead with the mathematical determination of "exact position and size"[3]—a task that Ptolemy assigned to the geographer proper. Hence a survey map represents the planification of chorography: its flattening out or leveling down. (*Planus* means level, flat, smooth; we encounter its offspring in terms such as *planigraphic, planimetric,* and *planiform.*)

The appropriate medium of land maps is what I have called "survey space," and its proper module is a "site." A site represents the radical simplification of a place: its regularization by a formal pattern that sheers off the irregularities that landscape always brings with it. In the East as in the West, the regularization occurs most characteristically by the imposition of geometric structures, that is, plane figures such as squares, circles, and rectangles. The concatenation of such figures produces a system of sites, whether this be a group of rectilinear shapes (e.g., on the earliest Japanese *kondenzu* maps of reclaimed rice land or in ancient Chinese maps of the *yujitu* type that exhibit nested rectangles) or of circular shapes (as in the Babylonian world map of circa 600 B.C., Japanese *gyoki* maps, and in many medieval *mappae mundi*).[4] Regular polygons are already evident in the Sumerian plan map of Nippur (1000 B.C.) to which I have made previous reference.[5] Indeed, after 1000 B.C. in Greece an increasing geometrization of mapping—as of all spatial entities and their representations—is to be found. Whoever "the first geometer" was (if not Thales, then an unknown predecessor), he may well have been a contemporary of Homer. Although Homer himself did not geometrize the Mediterranean world in the *Odyssey,* this masterpiece of circumnavigation—providing as it does an accurate account of islands and coasts, albeit entirely in words—made possible the geometrizing of Greek cartography that was undertaken in the sixth century B.C.; for by this time land surveys had been institutionalized, beginning with those of Miletus, the birthplace of important pre-Socratic philosophers (including Anaximander, the first Greek known to have constructed a world map). Miletus itself was laid out in accordance with a system of squares reminiscent of the ancient Chinese grid plan, and in general "the Greeks surveyed within a system of squares or rectangles for towns and of rectangles for rural areas."[6] The variegated terrain of the eastern Mediterranean was already being represented as orthogonal—as indifferent to the presence of local landscape features such as hills and rivers, trees and houses.

The Romans, who were the great land surveyors of the Mediterranean world, learned much from the Greeks in this domain as in others. They invented a revealing name for surveyors: *"mensores,"* literally, "measurers." Records of their activity in the field (typically, newly conquered land) were kept in the *Corpus Agrimensorum.* True to their name, the *mensores* posited a rigorous system of squares into which the land was divided. These squares or *centuriae* consisted of absolutely regular squares of 2,400 × 2,400 Roman feet, each with two perpendicular axes called, respectively, the *cardo maximus* and the *decumanus maximus.* From the intersection of these two major lines—which resemble nothing more closely than the double axes of x and y coordinates in analytic geometry—subdivisions of smaller squares were determined.[7] The resulting "centuriation" of the land represents its radical geometrizing; it amounts to the planification of landscape, its effective conversion into sitescape. Not that such conversion was ever complete: the Roman surveyors had to recognize the existence of the irregular pieces or *subseciva* that remained between centuriated land and the outermost boundary of the land under survey. They were also forced to acknowledge "land known as *arcifinius,* which is unmeasured, [and which] is bounded . . . by rivers, ditches, mountains, roads, trees."[8] Indeed, their maps of centuriated lands included pictographic representations of natural features such as rivers and mountains as well as architectural entities such as castles, forts, and even statues.[9] Even the most assiduous efforts of the *agrimensores* to convert recently acquired land into fully surveyed sites failed to eradicate all traces of what was incommensurate with these efforts—in short, all traces of place in its configurational idiosyncrasies.

RECTANGULARIZING THE LAND

An even more instructive instance of the fate of survey mapping is found in the United States. As imperialistic as Rome in its attitude toward its uncharted territories, the United States turned in the late eighteenth century to the systematic surveying of these territories, which had been brought under common federal ownership to help pay war debts.[10] But it was the Dutch who were the most important precursors: the idea of applying a network of regular linear shapes to the land had come from Holland's efforts to irrigate its flatlands in the most efficient way. In his *Historical Account of Bouquet's Expedition against the Ohio Indians* (1764), Colonel Henry Bouquet advocated the Dutch system of using a rectangular (and, more strictly, equilateral) grid pattern in the surveying of public land. Thomas Jefferson was the author of a special congressional committee report issued in 1784 that recommended the adoption of such a system, which was passed by Congress on May 20, 1785, under the title "An Ordinance for Ascertaining the Mode of Disposing of Lands in the Western Territory." The disposal itself consisted in the division of these lands into six-mile-square units called "townships," each of which was subdivided into thirty-six mile-square tracts for the sake of settlement. One example among many others of such a proposed division is provided in Figure 11.1. Notice that the specification is so regular and precise that no ordinary place-names, not even that of the state, are mentioned in the official title.

The ordinance sparked intense debates in the beginning. The Southern states insisted on their right to honor already-existing irregular "sections" that were often drawn up along "natural lines," while the North considered such higgledy-piggledy units to be "indiscriminate locations."[11] In the end, the Northern states prevailed, and the "Rectangular Survey" (as it came to be called: the official title was "United States Public Land Survey") was put into action, including its injunction "to divide the [new] territory into townships of six miles square, by lines running due north and south, and others crossing them at right angles."[12] In this seemingly innocuous and apparently easily enactable directive, indeed, in the very words "rectangular survey," we witness the ascendancy of *l'esprit géométrique.* Not since Palladio's attempt during the height of the Italian Renaissance to design villas and other buildings in accordance with various regular geometrical figures had there been such an efflorescence of this spirit in the West. But where Palladio was acutely sensitive to the interplay between architecture and the surrounding landscape—

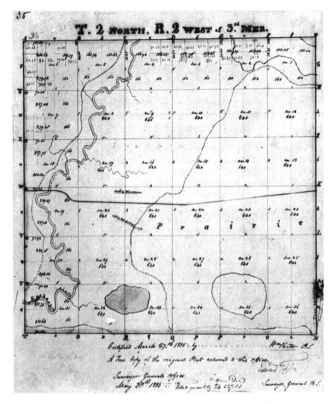

Figure 11.1. General Land Office, *Plat* [i.e., plot] *of Township No. V. North, Range No. IX West of the 3rd [Principal] Meridian,* November 18, 1815. Manuscript with watercolor wash, 39.5 × 32.5 cm. Records of the Bureau of Land Management, National Archives, Washington, D.C.

as was Thomas Jefferson himself in his building designs—the early American land surveyors assumed the Promethean posture of the pure Geometer who lays abstract patterns down upon the bare earth. It is therefore not surprising to learn that geometry played a prominent part in the early education of a public land surveyor such as the celebrated William Austin Burt: "he spent much time in the study of works on geometry, navigation, and surveying."[13] A geometrically inclined surveyor like Burt attempted to impose regular plane figures on the land he was surveying. As a *sur*veyor he looked *over* this land—and in this very capacity he was tempted to overlook its ideolocal features, which are inherently recalcitrant to geometrization. Much in the manner of Pythagorean town planners in southern Italy in the sixth and fifth centuries b.c., American land surveyors assumed that geometry could conquer landscape. The Rectangular Survey was even said to represent "the triumph of geometry over geography."[14] But was this in fact the case?

It is revealing that the very first application of the Rectangular Survey—to eastern

Ohio—had to contend with "natural lines." Ohio is bounded on the east by the Ohio River, and this meant that the six-mile-square units bordering the river were in every case truncated into quite irregular shapes. These ungeometrical units, reminiscent of Roman *subseciva*, had to be included in any survey that purported to be complete and yet they fell hopelessly short of the ideal of pure geomorphic geometry (see Figure 11.2).

Landscape irregularities were even more pronounced in states with swamps and bayous as well as twisting rivers.[15] Not to mention mountains, which had to be confronted

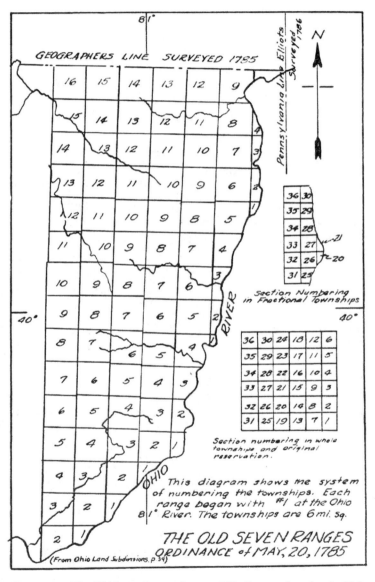

Figure 11.2. *The Old Seven Ranges Survey: Ohio.* Surveyed under the U.S. Public Land Survey, in accordance with the Ordinance of May 20, 1785.

as the Rectangular Survey headed further west! In the case of California, for instance, a special act of September 27, 1850, was passed by Congress to the effect that

> whenever, in the opinion of the Secretary of the Interior, a departure from the rectangular method of surveying and subdividing the public lands in California would promote the public interest, he may direct such change to be made in the mode of surveying and designating such lands as he deems proper, with reference to the existence of mountains, mineral deposits, and the advantages derived from timber and water.[16]

As Jefferson had put it in his original committee statement of 1784, consideration must be taken of all "remarkable and permanent things, over or near which such [survey] lines shall pass."[17] Still more generally, in a phrase added to the Ordinance of 1785, "the quality of the lands"[18] should not be entirely neglected.

Problems in the dogmatic pursuit of rectangularity arose not only from such natural obstacles to surveying as the meanders of rivers and the slopes of mountains.[19] They also reflected the divergence between true and magnetic north, as well as the effects of the curvature of the earth, which meant that even a slight convergence of north-south boundaries (drawn to cohere with the principal meridians) had to be figured in.[20] Moreover, the methods of marking these boundaries were far from strictly geometric in their execution: "it was quite customary in surveying [land] to run, more or less crudely, two lines upon the ground, witnessing the three corners of these two lines to *the nearest trees at hand.*"[21] Early surveyors used not only trees but simple mounds of dirt or shallow pits to mark the intersection of survey lines; but, as the Iowa Surveyor General said in a report of 1850, "when the mounds thaw out they flatten, and the pits tend to fill up."[22] There was also the fact that surveying regions inhabited by hostile Native Americans often proceeded by guesswork and had to take into account the vicissitudes of tribal history: "the public domain was surveyed in pieces whose shapes and sizes depended upon Indian territories."[23]

We may discern three stages in the history of survey mapping in America:

1. In the first blush of a passion for ordering an undomesticated landscape, a massive superimposition of a rectangular grid system on virgin territory is contemplated. The presumption is that this system can apply anywhere and everywhere—that it has no effective limits. Such limits as present themselves are regarded as contingent and to be overcome or suspended. Thus, for example, we read in instructions from the Surveyor General (this very title, established in 1796, invokes the fervor of a military campaign) sent to the states of Ohio, Indiana, and the territory of Michigan that "in all measurements, the level of horizontal length is to be taken, and not that [which] arises from measuring along the surface of the ground *where it happens to be uneven, rolling, or hilly.*"[24] In other words, when the terrain gets rough—when what Husserl calls its natural "obtrusions" begin to loom large—one should ignore the irregularity and stick to the (literally) straight and narrow: to "the level of horizontal length," that is, to what is in effect a straight line laid onto the landscape as if the latter were a leveled-down plane surface, a piece of flat paper, as it were. (The use of the term *plat*—meaning literally "flat"—for "plot [of land]" is here revealing: as in the title of Figure 11.1 above.) We have just seen reasons why this geoplanar approach, tantalizing as it may have been in theory, did not work in practice.

2. As if by way of partial concession, in a second stage of surveying the nongeometric is allowed to invade the geometric—but only in highly selected ways. For instance, rivers are recognized (and thus represented on maps) *to the extent that survey lines traverse them.* Thus the *Niles Weekly Register* of August 23, 1817, in an account of local surveys, noted that "continued west from the same point on the meridian the base line crosses the Wabash River about three miles above its junction with the White River."[25] Special instructions were supplied for marking such points of intersection between perfectly straight lines and imperfect natural phenomena.[26] Further, slight anomalies of measurement were now accepted and even anticipated, and a practice of carrying over any excess or deficiency to the north and west ends of survey lines was adopted (evidently on the assumption that any excess land properly lies in these directions). But it remained the case that any landscape features that did not directly impinge on the measurement were omitted from survey maps even though they were carefully noted in the "field books" kept by surveyors—notebooks that are the analogues of ancient Chinese gazetteers. The regularities were regarded as cartographically suitable material, while the irregularities were relegated to the written word, as we can see from instructions to enter into the field book "the face of the country, whether level, rolling, broken, hilly, or mountainous."[27] What was unfit for the survey map, the unruly face of the earth, was confined to the survey book.[28] This repression of the nongeometric undulations of the landscape stands in stark contrast with the ancient Chinese attitude toward straight lines as expressed in the epigraph to this chapter. Rather than consigning irregularities to gazetteers, the aim was to *represent them in images,* even in the very midst of a grid plan. In one indicative incident, the first emperor of the Northern Sung dynasty (tenth century A.D.) found himself disgusted with the sheer rectangularity of a map that had included an extension of the city wall: "He personally took up a brush to smudge [the straight lines] out and ordered that a large circle be made on a sheet of paper. It wound [around] and turned, went up and down and slanted. [The emperor's] side note said: Build the wall according to this [image]."[29]

3. A third (and still extant) phase of survey mapping entails a much more complete recognition of detailed topographic features as rightfully belonging to a map and not just to an informal journal. In 1845, John Charles Frémont could write in his "Report of the Exploring Expedition to the Rocky Mountains":

> We were not yet at work on the map . . . Indeed, the making of such a map is an interesting process. It must be exact. First, the foundations must be laid in observations made in the field; then the reduction of these observations to latitude and longitude; afterward the projection of the map, and the laying down upon it of positions fixed by the observations; then the tracing from the sketch-books of the lines of the rivers, the forms of the lakes, the contours of the hills. Specially is it interesting to those who have laid in the field these various foundations to see them all brought into final shape—fixing on a small sheet the results of laborious travel over waste regions, and giving to them an enduring place on the world's surface.[30]

Frémont has certainly not given up on the ideal of accuracy; but instead of advocating the imposition of preestablished geometric forms on the earth, he recommends that the actual forms of landmasses themselves ("the lines of the rivers," "the contours of the hills") receive explicit representation on the map. Figure 11.3 is an example of a map created under his supervision. An otherwise wholly planimetric survey map is here supplemented by the inclusion of precisely rendered topographic features, much as the Romans were led to do two millennia before.

One sign of the changing attitude toward mapping exemplified by Frémont was the establishment of a special body of "topographical engineers" who undertook land surveys in the Far West—surveys that resulted in maps that were much more informative as to the specificity of local and regional landscapes. Some of these same figures were in fact artists of considerable talent, for example, John Mix Stanley, John R. Bartlett, and John E. Weyss. (Indeed, Albert Bierstadt accompanied an expedition of such surveyors on one of his first trips to California.) Not only were techniques of hachuring and contour lines (and, later, shaded relief) assiduously employed, but certain creations of the Corps of Topographic Engineers were difficult to distinguish from detailed pictorial landscape views: for example, William H. Holmes's *Panorama from Point Sublime* of 1882.[31] In this handsome lithograph of the Grand Canyon, we witness survey mapping not merely making peace with topography but ceding place to it—indeed, *becoming itself topographic.* Far from landscape features being passed over altogether or else displaced in verbal reports in field books, they now make up the entire map. We also see this happening in John E. Weyss's graphic pen and ink map of Yosemite Valley (Figure 11.4). Bierstadt's melodramatic paintings of Yosemite Valley (e.g., Plate 5)—not to mention Carleton Watkins's photographs from the floor of the valley (Figures 2.1, 2.2)—are here complemented by a less dramatic but far more comprehensive representation of this valley as viewed from an elevated position several miles above it: a representation that does not seek to impose geometric shapes but to bring out the inherent forms of the land itself.

ASSUMING THE RIGHT VIEW FROM ABOVE

At this comparatively late moment in the history of cartography we rejoin not just our starting point in this book but the situation we encountered in Japan at the end of the preceding chapter. By the close of the nineteenth century in the United States, survey maps, at first averse to the depiction of qualitative aspects of landscape, began to feature just such aspects—on which American landscape painters had consistently focused throughout the century—much as Keisai's maps of Japan made earlier in the same century are simultaneously cartographic and painterly. What had begun by being antithetical to the Eastern instance had, by a gradually increasing inclusion of the "remarkable and permanent things" of the landscape, come into convergence.

Once more, too, we see the map landscaped in the course of its own evolution—and, in the American case, despite an explicit intention to exclude the irregularities of natural topography in favor of the regularities of geometry. The intention itself is symptomatic of a site-obsessed and characteristically Western conversion of place into space. The very idea of survey, the view from on top—the *Überblick,* as the Germans would say, or *le point de vue de Sirius,* as the French put it—conveys the essence of the matter: if only one can take sufficient remove from the messy protuberances of the earth, one can imagine the earth's surface to be perfectly flat and to call for the inscription of plane

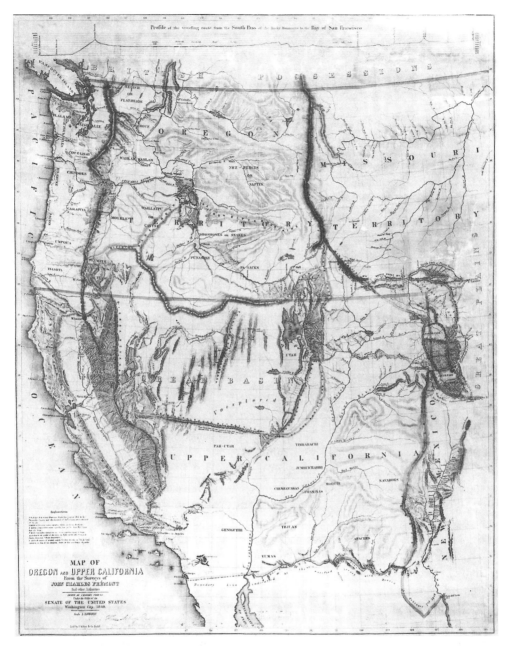

Figure 11.3. Charles Preuss, *Map of Oregon and Upper California from the Surveys of John Charles Frémont and Other Authorities,* 1848. Lithograph, 69 × 87 cm. Records of the U.S. Senate, National Archives, Washington, D.C.

Figure 11.4. John E. Weyss, *Topographical Map of the Yosemite Valley and Vicinity,* 1883. Manuscript, pen, and ink; 59 × 74 cm. Records of the Office of the Chief of Engineers, National Archives, Washington, D.C.

geometric figures upon it. This is projection with a vengeance—a very different kind of projection than we have previously encountered!

It was in the seventeenth century in the West that the idea of site began to crowd out the notion of place, a notion that the ancient Greeks and many in the Middle Ages had still honored. Not only in Cartesian analytic geometry—whose coordinates so strikingly resemble the perpendiculars of the Rectangular Survey—but still more pervasively in the Newtonian worldview, place became lost in Absolute Space, the utterly homogeneous space of sites. Blake's celebrated engraving of Newton expresses the situation tellingly by showing Newton crouched *over* the earth's surface and tracing plane geometric figures *on* it with a compass. Being in exactly the right position to geometrize the earth, Newton is in effect surveying it. He is assuming "the point of view from nowhere."[32] From this literal sur-view, the earth as an oblate sphere that is highly convoluted on its outer surface has been converted into a planimetric projection of the scientist's febrile fantasy. His assumed position is projected as well: given that it is only from such an imagined height that one would feel free to impose geometric structures on the unruly world, much as the demi-urge in the *Timaeus* superimposes Forms on a chaotic *chōra* (including, in particular, geometric shapes on the constituent elements of the universe).[33] From this height, the earth's variegated landforms fade into a blur and are no longer discerned in their distinctive differences; topography gives way to geometry as the surfaces below become amenable to the superimposition of forms from above.

Contrast this circumstance with that obtaining in Valcamonica, where, on actual and not merely imagined heights above an alpine valley, some of the earliest known topographic maps were chiseled in stone. From these heights one looks into a uniquely configurated place that resists transfiguration into geometric structures. And from these same heights one can also detect more perspicuously than from the valley floor the structure of topographic features in the region—its rivers, hills, and plains (some of which are still recognizably the same as those depicted in the prehistoric maps made there 3,500 years ago). Far from being obstacles to be ignored or "drawn through" (as occurs when a survey line intersects a natural feature), these features of the landscape present themselves in their own right as worthy of pictographic commemoration. In the petroglyphs by which this commemoration occurred we certainly witness a survey of the scene: the view is as much in-plan as the rigorous survey maps sanctioned by the 1785 ordinance discussed earlier in this chapter. This shows that we need not resort to elevation or profile views—or, at the limit, to the oblique views of physiographic maps—to capture topographic details adequately. Merely as seen from above, from the height of a bordering hill, we can perceive the local landscape and trace out its characteristic traits. But we can do so only if we are truly surveying a *place* or *region* and not a mere site. Whereas a site could be *anywhere* and its features are a matter of indifference, a place or region can be only *just here* and its features are only *just these*. And it is this just-here, just these aspects of the landscape that the petroglyphs of Valcamonica (most notably, the Bedolina map of Figure 7.4) share with their many successors in the history of cartography, whether these be portolan charts or Japanese city maps, the *Ti Li Thu* of 1193 (which shows all of China topographically), or John H. Renshawe's *Panoramic View of the Yosemite National Park, California,* a photolithograph of 1914. The view from above, far from constraining the mapmaker to a strict survey of site, may induce a special sensitivity to the peculiarities of "the smallest conceivable localities" (again in Ptolemy's phrase).

The very idea of panorama—an alluring if problematic direction in landscape painting, as we saw in chapter 1—here comes into its own. A topographically responsive panorama does not eliminate the earth's convexities and concavities in favor of a purely planiform surface; it shows these irregularities in their full sweep. Or we could say that it shows them *in their full depth.* The Newtonian surveyor, in contrast, is not concerned with depth: given his superior posture, he is interested only in how many of his geometric projections a flat surface can bear. But the surveyor of place, the true topographer, concentrates on the depth of what he sees spread out below: he moves his body—not his instruments, much less his mind—down into the valley of the shadow of places. Like Zarathustra, his primary action is that of *going under*—or, more exactly, of going *into* the earth's surface and its potpourri of features. The maps that exemplify such *Untergehen* allow us as their readers to follow the movement downward into the depths.

The issue, then, becomes not how to measure these depths (as in geodesy), much less how to reshape them as regular (as in geometry). The aim is to pursue chorography as topography, that is, to *trace* the inherent *places* and *regions* of the earth. The task of the choro-topographer is that of representing (in Frémont's words) "the lines of the rivers, the forms of the lakes, the contours of the hills."

This is a task worth undertaking, for, if properly pursued, it gives to us a survey that is no longer rectangular but multiform in its shapes—as polymorphic as the earth it-self in its profuse local geography. Such a survey is no longer planimetrically flat and empty (as flat and empty as only a site can be), but convoluted and plenitudinous. The dense configurations of a three-dimensional earth are given graphic representation in the two dimensions of a map onto which the results of this very different, but also very ancient, kind of survey have been inscribed. In its most accomplished realizations in the East as in the West, such a map is not only a "land map," a map *of* landscape; it is itself landscaped; it is a landscape-map.

At the same time, such a map constitutes a *placescape*. It is itself a place—a topographically nuanced surface-place, a place-of-representation—for the framing and manifesting, the "scaping," of the land-places it maps: places on the land and of the land, places not to be confounded with sites in sheer space.

GOING BEYOND DICHOTOMIZATION

Through the many meanderings of the last several chapters, a certain pattern, a textual landform, as it were, has begun to emerge. In closing this part, let me trace out this pattern in terms of a series of choices that have been offered to us. In each case, an initial opposition or tension gives way to a less divisive scene of interaction.

1. Take, to begin with, the distinction between presentational and discursive symbolism as it applies to maps. We have witnessed at numerous points the danger, and certainly the desiccation, that ensues from keeping these two modes of symbolism rigidly apart. The danger and the desiccation alike are especially evident in the case of American survey maps in their second stage of development—when topographic details were confined to a field book and mapping proper was primarily geometric in tenor. This separation amounted to a displacement of the full figuration of topography, and it was not until the contents of the field notes found their way into an adequate pictography—*logos* linking up with *eikon*—that a satisfactory representation of the landscape of the United States from Ohio to California could be attained. In this case, the discursive symbolism of writing did not survive on the final maps except in the limited form of sparsely situated place-names and designations of latitude and longitude. Elsewhere, we found a more intricate dialectic between the written and the depicted, the word and the image, on the map's surface. I refer especially to portolan charts, which sported toponyms attached to coastlines in addition to "legends" that lived up to their name instead of being mere keys to cartographic features, as well as to the maps of *ukiyoe* artist-mapmakers, who did not hesitate to write descriptions (in the form of small emblems or placards) directly into the landscape. But much the same intimate commixture of the presentational and the discursive is found in numerous Chinese maps of the Han, Tang, and Sung dynasties: their surfaces are as filled with Chinese ideograms as they are with topographic signs. Indeed, the oldest *printed* map of any known culture is a map of west China from the *Liu Ching Thu* (c. A.D. 1155) that exhibits a complex dialectic of words and images across its entire surface.[34] Here one form of symbolism is unthinkable without the other. An exemplary case such as this resolves the contest between Smith and Harvey as to the value of pictoriality in maps. (See their debate as discussed in chapter 7.) Contra Smith, we must say

that pictoriality is of indubitable value in cartography; but contra Harvey, we must also say this same pictoriality calls for supplementation by discursive symbolism.

2. Another instance of an initially divisive dichotomy is found in a basic enigma of all cartography, Eastern as well as Western: how can a three-dimensional entity such as the earth be represented in two dimensions? The ancient Greeks approached (but did not resolve) the enigma by devising the notion of the planisphere—that is, the projection of what properly belongs to a sphere onto a flat surface—and a number of early portolan charts adopted this idea as if it absolved them of any further reflection. Mercator's famous projection, first announced in 1569, indicated that the matter was far from settled—that, in particular, any projection of the three dimensions of the earth-sphere into two, however scientifically well founded, brings with it the price of cartographic distortion in some significant part of the map it helps to create.[35] On the other hand, the strict absence of projection in most gridded maps creates its own problem, namely, the need to take steps that compensate for the lack of projection—notably in the case of American land surveys, whose strict rectangularity fails to take account of the convergence of meridians. The truth of the matter is that just as there is no perfect form of planispheric projection (only an actual globe gives an undistorted representation of the earth in its entirety), so there is no perfect form of projectionless map. What there *is* is a variety of ways of treating the problem of the two-dimensional depiction of the spherical earth. Indeed, every mapmaker and every map reader must accept the fact that "no map can show the objects in nature in their true proportion."[36] Some modes of representation are suited for certain aims—for instance, Mercator's for projecting the earth accurately in its middle regions between the poles—and some for other purposes: for example, portolan charts for showing the configuration of coastlines, whose delimited extent is relatively unaffected by the convergence of meridians. There is also a nonreducible plethora of techniques (some of which were discovered independently, e.g., contour lines in China and in Western Europe)[37] for showing how the three-dimensionality of a given topographic feature such as a mountain can be effectively illustrated in a two-dimensional drawing. The depth of particular parts of the earth's surface as well as its overall landmasses finds multiple representational means. Rather than indicating an impasse, the enigma of two-dimensional representation of the spherical (and the high or deep) points to the fecundity and ingenuity of mapping itself. Here the resolution of an initial dichotomy does not reside in the combination of opposed terms (as in case 1 above) but in the sheer proliferation of modes of representation.

3. Or take the seeming irreconcilability between overhead and profile views. At first, these would seem to be incompatible and even warring approaches to the representation of landscape. What could be more disparate than viewing parts of a landscape from directly above versus seeing them from the side? Also, is either of these extremes compatible with an oblique view of forty-five degrees? Yet time and again we have come across creative combinations of such apparently antithetical vantage points, combinations that are not only cartographically readable but more informative than would otherwise be the case. The town plan of 6200 B.C. from Çatal Hüyük (Figure 7.3) gives us both a remarkably detailed idea of the town as seen from above *and* a profile of the twin-peaked volcano nearby. This makes perfect sense, because the view from the sky is the most apt for

the structure of the city, while the side view is the most adequate representation of the volcano. Similarly, a Mesopotamian "landscape jar," also of very ancient vintage, shows a river in-plan gliding like a sinuous snake between two rows of mountains seen in profile.[38] Both views are appropriate and informative and, in this case, fit together in a highly decorative way. Such cases as these argue for the thesis that in-plan maps are best suited for conveying the overall layout of places—and are thus ideally suited for land surveys—while maps showing views in profile are especially effective in transmitting a sense of concrete topographic features. And both approaches are perfectly compossible on a single map, as we see in countless cases, including Japanese city maps that, like that of Çatal Hüyük, show a town from above and the surrounding landscape in profile. Each is also compatible with an oblique view, which incorporates the virtues of both plan and profile views, as in the case of *ukiyoe* maps and certain maps from the Han dynasty in China that explicitly combine all three perspectives.[39]

4. So, too, maps show themselves to be capable of considerable complexity in regard to their orientation, without this complexity disolving into contradiction or disarray. Not only are the cardinal directions variously located on different maps—North at the "top" of a map replacing East in this upper position only comparatively recently[40]—but some maps are omnidirectional in that they make sense in *any* orientation, while still others are nondirectional insofar as no indication of definite direction is given. What would seem to call for a clear-cut choice in the circumstance (don't we have to know precisely which way the represented region is oriented?) ends by allowing something much more indefinite and open than we might have thought possible at first. Just as we are not forced to choose between pictorial and abstract symbolism, so we need not insist on the necessity or the rightness of any particular system of directionality, cardinal or otherwise, in the making and reading of maps. As with modes of projection, multiplicity triumphs in the realm of orientation, suggesting that mapmaking is an ever-proliferating, nonexclusivist enterprise.

5. A recurring dichotomy that has run throughout the last several chapters is that of grid versus topography, the formal geometric figure versus the representation of the natural landscape feature. Ever since Eratosthenes, Hipparchus, and Marinus in the West and Chang Hêng and Phei Hsiu in China proposed a grid system for maps—in the first case, as a representation of latitude and longitude, in the second as a scalar and locatory index—mapmakers have been fascinated with the rich potential of gridding cartographic space.[41] It is quite tempting to hold such gridding apart from the representation of landscape items—a temptation to which both Roman and American land surveyors succumbed in their programmatic ambitions. But we have seen that even in the extreme case of the passion for sheer rectangularity so evident in ancient Romans and modern Americans alike, pictographic representations of the landscape have entered the cartographic scene eventually and as if with ineluctable force. We might have expected as much from recalling such parallel instances as the Byzantine grid map of Ceylon (Figure 10.5), the map of Palestine in Sanuto's *Liber Secretorium Fidelium Crucis* (Figure 10.6), and, in the East, various maps in the *Kuang Yü Thu*.[42] The striking proximity in the dates of these last three examples (ranging only from A.D. 1250 to 1315) serves to show that a common fascination with the possibility of combining a grid system with topographic

detail was felt by cartographers at about the same time in widely separated parts of the earth. This fascination reflects the fact that, far from fighting each other off, the grid and the topographic image, the rectangle and the recognizable figure, can join forces constructively on the same map—as constructively as can image and word, two and three dimensions, in-plan and profile and oblique views, and various directionalities.

BODY, CHOROGRAPHY, GEOGRAPHY

Not only in the case of the five foregoing instances, but everywhere we look in the realm of maps, we find the mark of creative synthesis. For such seemingly idiosyncratic and isolated representations, this is nothing short of extraordinary. Maps bring not just land and water masses together in their representational capacities; they also knit together the most disparate factors of knowledge, history, and power. The close clustering of dates cited in the preceding paragraph, for example, points to the way in which Eastern and Western mapmakers have entered into considerable convergence, direct and indirect, at critical moments of their respective histories. Direct influence between the two great world regions of mapmaking has occurred since at least the travels of Marco Polo in the thirteenth century (in fact, probably several centuries before, thanks to itinerant Arabs), and it continues down to the present day (contemporary Chinese topographic maps are regarded as setting a world-class standard). Even apart from cases of actual contact, a number of common themes have emerged around such issues as pictoriality and topography, surveying and geometry, the exemplary status of coastal and island maps, the proximity between cartography and geography, the importance of the exploration of new lands, and so on. It is as if mapmakers, otherwise separated drastically by culture and language and geography itself, nevertheless share a remarkably similar group of concerns with regard to mapping the known world and its multifarious landscapes.

Another area of extensive overlap bears on the role of the human body in cartography. It is tempting to think that maps are mainly products of mentation: of the concretizing and pictorializing of attitudes and items of knowledge held within the mind. It certainly cannot be denied that "mental maps" orient much of our conduct in the lifeworld, providing condensed and schematic representations that are indispensable to habitual actions of many sorts.[43] It is indisputable that "the geography of mind can at times be the effective geography to which men adjust and thus be more important than the supposed real geography of the earth."[44] Nor can it be gainsaid that certain culturally conditioned ideas about the constitution of the natural and social world deeply inform our learned as well as our spontaneous geography. Truly so—and mental geography deserves to be studied in its own right, along with the invaluable contributions of cultural geography. But we must also consider the far less widely suspected importance of the lived body in cartography.

This body shows itself to be a potent agent of synthesis in the cartographic realm. To begin with, it helps to explain the paradoxical sense we have as map viewers that we are at once located *and* not located in relation to a given map. As located, we sense that we are directly above Yosemite Valley as we gaze at Weyss's map of this region (Figure 11.4); we feel that part of our bodily being is suspended several miles over the valley, even though the map is a representation and the suspension is virtual and not actual. It is thanks to the imaginative projection of our body outwards and upwards that this effect of suspension becomes possible in the first place. In other cases, we may have no distinct sense of being located somewhere in particular in relation to the map, but we nevertheless project our bodily self into the topography represented there: for example, in viewing relief maps on bronze Chinese incense jars, whose very tangibility invites corporeal projection onto its felt surface.

In yet other instances, it is by bodily projection of different sorts that we are able to assume various positions in relation to the dominant directionality of the map, for example, when we are invited to walk around (in fact or in imagination) a portolan chart in order to read its words and images from several points of view. This last example reminds us of the body's basic orientational capacity—a capacity first singled out in the West in Kant's observations on directionality and evident in everyone's experience of architecture and dwelling.[45] The body's equally basic motility also has implications for maps: stationary as they are as physical objects, maps nonetheless adumbrate lines of possible movement in the land (or sea) which they depict. The coastal route delineated by a portolan chart or by a Chinese navigational map is a route of movement along which the map reader may in fact be moving—or will someday move, if he or she is sailing in the region—thanks to the lived body as an actively engaged exploratory agent. It is by and with this body, after all, that we get from place to place, including those places represented on maps.[46]

Another form of synthesis effected by maps, one that is also body-based, is that of heterogeneous with homogeneous space. These two modes of the spatial are all too often regarded as incompossible, especially in the wake of the early modern attempt to convert place into site. But maps draw freely on both modes: they are often planiform in the representation of open spaces that have a paucity of landscape features, and they tend to be physiographic when it comes to representing nature in its wild state. The homogeneity at stake in the former case is not in the least incompatible with the heterogeneity that structures the latter instance, as we can see lucidly in the Peutinger Table (Figure 7.8), the medieval Roman route map that combines sheer linearity (in the form of an unremittingly flat highway drawn across unrelieved empty spaces) with detailed pictographs of prominent buildings situated along the route. Once again, maps allow us to have it *both ways*: a given map may represent to us both "spatializing space" and "spatialized space."[47] The one is essentially heterogeneous—as heterogeneous as duration in Bergson's sense of the term—while the other is just as essentially homogeneous. The map reader must be prepared to encounter both kinds of space and to assimilate them to each other in appropriate ways.[48]

And this same reader must also be prepared to experience and understand maps in a last and quite basic form of creative synthesis: as geographic *and* chorographic. The Ptolemaean distinction between these two kinds of mapping was intended to keep them so far apart from each other that only two entirely different disciplines could do justice to them. Indeed, until the middle of the nineteenth century, *chorography* was recognized in dictionaries and encyclopedias as quite distinct from *geography*. The regional focus of the former was contrasted with the global scope of the latter—as if the two enterprises were utterly divergent. Yet Ptolemy himself points in a different direction. If it is true that "the end of Chorography is to deal separately with a part of the whole" in contrast with Geography, whose task is "to survey the whole in its just proportions,"[49] neverthe-less the part in question remains a part *of* that whole, a whole that is, at the limit, the earth itself. It is a part that, though detachable for the sake of a close-up view, is in the end not separable from the earth of which it is an integral part. Chorography consists in "selecting certain places *from* the whole."[50] In Ptolemy's own favorite analogy (as cited earlier), an analogy that is revealingly corporeal in character, chorography treats "only the eye or the ear by itself," while geography depicts "the entire head";[51] but this is only to say that the chorographic is essentially related to the geographic and that it cannot stand entirely on its own. The *oikumēnē* or "known habitable earth" is "a unit in itself," and any description of "the phenomena which are contained therein,"[52] that is to say, con-tained in its various localities and regions, is ultimately answerable to that larger unit. It could not be otherwise: such is the inexorable logic of parts and wholes.[53]

It ensues that the intimate tie between art and chorography—once again: "chorog-raphy needs an artist"—cannot be unique.[54] Even if we may grant that chorography is more concerned with "the kind of places" it describes and especially with their "quality" (i.e., in contrast with their "exact position and size"), it remains the case that geography, as Ptolemy himself admits, is also in the position of "emulating the art of painting" in at least "some of its major descriptions."[55]

If we are to believe Ptolemy's own words, the very distinction between regional maps—a category including everything from austere land survey maps to colorful picto-graphic and topographic maps—and "world maps" (whether cosmographic or geograph-ic) is called into question. We are reminded of the fact that most cosmographic maps put a known point (e.g., Jerusalem, Peking) at their very center, as if to say: here I stand, in this very spot known to me, whatever else may be the case in the world unknown to me. It follows a fortiori that a map of any particular region of the earth, and thus a map that purports "to describe the smallest details of places," will be continuous in content with maps of nearby regions and ultimately with a map of all regions, even though this latter map can only convey "general features" of the earth as a whole.[56] If geography "must contemplate the extent of the entire earth, as well as its shape and its position under the heavens,"[57] chorography has to be able to do likewise, albeit part by part. The regional map will represent *part* of the earth's extent and *aspects* of its shape and position, and in this capacity is never entirely discontinuous with the world map to which it at least im-plicitly belongs.

Nor can the two kinds of map be kept separate in one final regard. The "true like-ness" that is the aim of chorography, according to Ptolemy, is surely also an aim of geog-raphy.[58] Indeed, the very idea of cartographic verisimilitude, an idea first broached at the very beginning of this part, itself exemplifies the synthetic power of maps. Those who hold, with Korzybski, that a map must be "similar to the structure of the empirical world," and those who hold, with Black, that "there is no such thing as a perfectly faith-ful model," are both right. Maps are indeed to some significant degree *like* what they represent—paradigmatically in the case of regional maps, but even in certain major re-spects in world maps as well—or else they could not serve as instruments of knowledge and orientation. But maps of both sorts are also less than strictly isomorphic with "the known habitable earth" and with its various "phenomena." They depart from this earth and these phenomena in manifold ways, some of which we have explored in preceding chapters. These departures are often themselves informative as well as aesthetically pleasing, and it behooves us to pay attention to them as much as to those representations that purport to be altogether exact.

True likenesses—veri-similitudes—come in many forms, provide many pictures, and speak many languages. This is nowhere more fully and tellingly the case than in the manifold maps of the earth and its diverse parts and regions. In such maps we find likenesses of the landscape by which this earth and these parts and regions have come into our ken in the first place. For it is ultimately landscape that bears and configures the places that hold our lives together—and that we seek to recapture in the representations of paintings and maps alike.

PART III

Re-implacement in Mapping and Painting

CHAPTER 12

Re-presenting Representation

Space is itself nothing but mere representation, and therefore nothing in it can count as real save only what is represented in it.

—Immanuel Kant, *The Critique of Pure Reason*

To represent: to set out before oneself and to set forth in relation to oneself.

—Martin Heidegger, "The Age of the World Picture"

THE WORLD AS REPRESENTED

By now we know that it was by no means accidental that mapmaking reached a new pinnacle of achievement at the very same time that landscape painting came into its own in Western art. That time, the seventeenth century, was (as we have already had occasion to note) "the age of the world picture." It was a time in which models of picturing proliferated—not only in cartography and painting, but in optics (e.g., the camera obscura and the camera lucida) and in philosophy itself (Descartes's use of the term *repraesentatio* and Locke's employment of *idea*). But more than a mere fascination with alternative modes of picturing—of the process of *pictura* as a way of describing the world—was at stake. Nor was the pictorial passion of the time, a passion for producing ever more adequate imitations of the world, what was truly new: the passion for *mimēsis* had already been evident in the Greek interest in illusionism in art, in medieval speculation about being human as *imitatio dei,* and in the Renaissance concern for the precise pictorial representation of spatial perspective. What was distinctive about the *Neuzeit,* the "era of the new" as the modern era is called in German, was its conviction that *the world itself is picture-like.* As Heidegger has put it: "Hence world picture *(Weltbild),* when understood essentially, does not mean [merely] a picture of the world but the world conceived and grasped as picture."[1]

This metaphysical premise—which is to say, the operative presumption of the modern age[2]—is dependent on four basic factors.

1. First, given the ambiguity of the word *Bild,* the world-as-picture *(Weltbild)* is at the same time the world-as-image. The world (and preeminently the perceptual world) gives itself to us not as a *factum brutum* or as an infinite Idea but in the form of an *image.* An image, unlike an eternal Idea or a brute fact, is subject to our manipulation. The

manipulation can be quite minimal, as when we attempt merely to "mirror" the presented world, for example, in the role of mirror imagery in seventeenth-century Dutch painting and mapmaking. Here the assumption is that the world is simply there to be mirrored in art or maps—or in the mind that holds iconic images of it in reserve. Or we may intervene more drastically by re-forming the *imago mundi,* as happens when a landscape painter reorganizes certain features and structures of a region and, in particular, re-implaces them: gives them a new home in a painting as a place-of-representation. Fidelity to nature in the form of genuinely naturalistic or realistic representation in painting is entirely compatible with such radical re-implacement; for in this kind of representation the world-as-image remains image as it is re-presented on the canvas. As the image of an image, a landscape painting reinstates the world, even if it does not imitate it as in a mirror. It represents the world to the second power, and thus it *represents* it in Asher B. Durand's strong sense of the word. It makes of that world not only a *Bild* but what Heidegger calls a *Gebild,* a structured image that restructures the world of which it is an image.[3]

2. The world as an imageable image is a world that comes to us as *framed* or *frameable.* Corresponding to *Gebild* is *Gestell,* the enframement of the world, its proclivity for being ordered in and through a frame.[4] As in the case of image—and as was spelled out in the Interlude—the range of framing is considerable. It can take the form of a literal picture frame (which had become established practice by the seventeenth century) or it can signify more generally that the world is grasped as arranging itself in framelike ways: for instance, in memories that we may have of it or movies we make of it. Thus, for instance, Descartes espouses the idea of memory traces that are in effect the formatting and framing of previous experiences of the natural world. Similarly, the camera obscura frames the world as seen through its lenses; or, more exactly, like a naturalistic landscape painting, it reframes the perceived world, for part of the very idea of *Weltbild* is that the world comes to us *already* framed and hence that painting, like photography, memory, or mapping, only serves to frame it again. Far from being a merely second-order act, however, such enframing re-presentations are ways of gaining new insight into what has been reframed. They are at once an ordering and a revealing, a limitation and a delimitation.[5]

3. In this dual capacity, a frame shows itself to be remarkably like a *horizon*: just as a horizon both closes in and opens out upon the receding depth of a natural scene (it is at once the culmination of this depth and its surpassing into the unseen), so the frame of the same scene as directly perceived or as represented in a painting or a map is the point where its spatiality terminates while also adumbrating something beyond the immediate presentation. Moreover, both frame and horizon present and thematize the objects that appear within their respective boundaries.

These enframed, horizoned objects constitute a third distinctive feature of the world-as-picture; in such a world, objects have a determinateness and specificity that they owe precisely to their frames and horizons. Merleau-Ponty takes up this point in relation to landscape:

> In normal vision . . . I direct my gaze upon a sector of the landscape, which comes to
> life and is disclosed, while the other objects recede into the periphery and become

dormant, while, however, not ceasing to be there. Now with [these other objects] I have at my disposal their horizons, in which there is implied . . . the [focal] object on which my eyes at present fall. *The horizon, then, is what guarantees the identity of the object throughout the exploration.* [6]

It will be noticed that in this way of putting it, the horizon is by no means limited to the distant or external horizon of an entire scene; it includes the various internal horizons that surrounding objects constitute vis-à-vis a thematized object around which they are arranged. Such discrete horizons create a nonrecessive depth in relation to this focal object. They are lateral in status and in this respect resemble the frame of a picture—only now the frame is *inside* the pictorial space as a whole and creates a special form of framed depth. Most important, however, both the external and internal horizons allow us, indeed, they actively encourage us, to pick out and to fix upon particular objects that appear within their compass. This accords well with what Heidegger says of the horizon: "what is evident [about] the horizon, then, is but the side facing us of an openness which surrounds us; an openness which is filled with views of the appearances of what to our re-presenting [activity] are objects."[7] Heidegger here proposes the paradoxical thought that it is just inasmuch as horizons transcend what they enclose that they convey our attention to the things enclosed. In other words, it is precisely because we realize that horizons outrun their contents that we come to focus on these contents as that which we can securely grasp—an observation that also applies to the frame, whose very delimitation permits it to be all the more effective as a factor of presentation.

The contents presented by horizons and frames are not merely momentary foci, ephemeral epiphenomena, of our attention. They are full-fledged *objects* that outlast any given attention span. Not only this, but such objects (as the very word "ob-ject" implies) stand on their own: indeed, they even *stand against* us and in this way live up to their designation in German as *Gegen-stände*. The evanescent and intangible character of horizons (a horizon, like the end of a rainbow, cannot be located with cartographic precision) is thus complemented by the opaque permanence of the very objects they set before us. Heidegger ties the againstness or opposition of such objects—which exhibit what Peirce would doubtless have called "Secondness," that is, resistance—to their status as represented: "to represent [*vor-stellen*] means to bring what is present at hand [*das Vorhande*] before oneself as something standing over against."[8] The literal setting-before (*vor-stellen*) of the representing subject conspires with the set-before (*vor-gestellte*) status of the represented object.[9] A metaphysical collusion between the representer and the represented occurs, with the curious result that it is precisely the *object of representation*—not the directly perceived or felt object of naive realism—that is the most impenetrable and obdurate of objects: "That which is, is no longer that which presences [i.e., is directly presented]; it is rather that which, in representing, is first set over against, that which stands fixedly over against, which has the character of object [*das Gegen-ständige*]. *Representing is making-stand-over-against*, an objectifying that goes forward and masters."[10] Here we can measure how far we are from any act of mere imitation: instead of *re*presenting an already given world by a mimetic gesture, we *bring the world to a stand*—but only within the realm of our own representation. Objects *as represented* thus come to a stand within us. Such objects are without limit as to number or type: thanks to representational activity, "everything has become an object that stands opposite us within a horizon."[11] In representing them, we bring them forward before us in their very being-there-for-us.

This scene of omnirepresentation sets a problem that, beginning with Descartes and his Dutch contemporaries (and we should not forget that Descartes wrote his most important works while residing in Holland), is still with us today: how to retain within objects of representation something that is not itself the product of representational activity. As Merleau-Ponty puts it: "We must discover the origin of the object at the very center of our experience; we must describe the emergence of being and we must understand how, paradoxically, there is *for us* an *in-itself*."[12] As we have learned from the case of Dutch painters and mapmakers, the discovering and the describing of the sort of object to which Merleau-Ponty here alludes are ultimately one and the same activity. The discovering is of something "in-itself," that is, not dependent on our representational actions. The describing, on the other hand, converts this in-itself aspect of the object into something "for us." To this we need only add that, just as the object of description thereby becomes a represented object, so the activity of description sets this object over against us within a nexus of pictorial horizons. The world-as-picture is a world that has become a totality of objects that solicit us to remake them—which is to say, to represent them in their very representedness—by our own descriptive and depictive actions. The world as imaged is ultimately made in our own image.

4. This means in turn that we must recognize as a fourth basic factor the *human being as the source of representation,* as "the representer of all representing."[13] Once all objects of experience have been drawn into the circle of representation—once the in-itself has been effectively dissolved into the for us—these objects belong to the cognizing activity of the human representer. The argument, as it was framed by Kant, goes like this: if we cannot perceive or know any objects except objects of representation ("appearances," in Kant's technical terminology), then the proper domain of such objects is in the mind of the representer—for these objects are nothing but the contents of this mind, having no standing apart from mental representations of them. In Kant's own words: "all representations, whether they have for their objects outer things or not, belong, in themselves, as determinations of the mind, to our inner state."[14]

Kant's radical conclusion—generalized in the formula that "appearances in general are nothing outside our representations"[15]—is merely the last stage in the slippery slope of the subjective idealism that is a characteristic late-eighteenth- and early-nineteenth-century development that arises from the positing of the world-as-picture in the seventeenth century. Once the world is conceived as a *Weltbild,* then the subject who represents it is ineluctably drawn into the drama of representation; for this subject is essential to the enactment of the drama: more than a mere observer of the world, the subject is responsible for representing it in ways that inhere in the subject's own mind (e.g., "forms of intuition," "categories of understanding," in Kant's nomenclature). Where observation and imitation leave the subject external to what is observed or imitated, representation cannot help but bring the representing subject back into the very picture that is the world itself: "appearances in general" inhere not in a world separate from the subject but *in this subject* as a source of representations. The world regarded as a picture (including its constituent objects) depends on—indeed, consists in—the representations we make of it, and because these representations are "our representations," they make ourselves as representers of the world part of the world picture itself. We are indispensable to the cosmic picture that we ourselves compose.[16]

From this position as supreme representational source of all that is—that is, of all objects of possible experience, as Kant would put it[17]—it is but a short step to the position of being master of the phenomenal world, at once the lawgiver of nature and the master manipulator of nature conceived as a source of energy and power to be employed by human beings. To be in the picture as the unique source of the representations that make up any picture (including the world-as-picture) is also to be capable of the calculative thinking by which one "gives the measure and draws up the guidelines for everything that is."[18] Such thinking underlies not only modern science but modern mapping of the sort which we have witnessed in the Rectangular Survey of 1785, established only four years after Kant published the first edition of *The Critique of Pure Reason.*

The four traits I have singled out—the world-as-picture, the world-as-enframed, the world as populated by represented objects that stand within horizons and over against us, and our own position as the sole source of all representations—together specify the conceptual foundations of the modern era as an age in which everything exists in order to be represented: indeed, exists only *as represented.* "Whenever we have the world picture," says Heidegger, "an essential decision takes place regarding what is, in its entirety. The being of whatever is, is sought and found in the representedness of the latter."[19]

REPRESENTATIONS OF REPRESENTATIONS

If the explicit positing of representedness as indispensable to modernity in the West and ubiquitous within it is the accomplishment of such thinkers as Descartes and Kant, the emblematic instance of this way of thinking is to be seen in the mapmaking and painting of seventeenth-century Holland. Is not mapmaking for the Dutch of this era the effort to regulate space by offering an ever more reliable pictorial representation of it? Is not a master mapmaker such as Abraham Ortelius or Jan Jansson the embodiment of the one who "gives the measure and draws up the guidelines" for the known world? Do not the Dutch cartographers render the world, the entire known world, as a picture? What is a map if not a detailed *Weltbild,* literally, a "picture of the world"? All the more so a comprehensive atlas of concatenated maps: "In 1648," we read, "Joan Blaeu compiled a large map of the world—*Nova Totius Terrarum Orbis Tabula* [New map of the entire earth]."[20] This atlas, in preparation for which Blaeu "searched the records of all nations,"[21] has been called "the highest expression of Dutch cartographical art."[22] It was from this same collection that the Dutch selected maps to offer to the Japanese government in the middle of the seventeenth century. Blaeu's atlas can be considered an epitome of the age of the world picture indeed, as its incisive cartographic expression. It exemplifies the basic premise of this age, namely, that the world is there to be represented; or, more exactly, that the world is known only in our representations of it. Just as Atlas bears the entire earth on his shoulders, so the world atlas bears the burden of representing the totality of the known world. Where better, where more graphically, than in such an ambitious atlas can we find evidence that "the picture character of the world is made clear as the representedness of that which is"?[23]

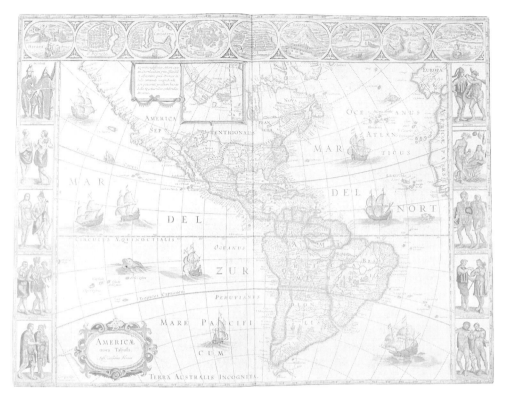

Figure 12.1. Willem Janszoon Blaeu, *Americae,* from *Theatrum Orbis Terrarum sive Atlas Novus,* 1635. The National Maritime Museum, London. Reprinted with permission.

It was Joan Blaeu's father, Willem Janszoon Blaeu (whom we first met in chapter 8), who compiled one of the first systematic atlases, titled *Theatrum Orbis Terrarum sive Atlas Novus.* True to its name, this was a veritable "theater of the earth," in which the self-proliferating character of representations was made vividly evident. Take, for instance, his plate of the Americas (Figure 12.1). Striking here is not only the central map itself, with its characteristic exaggeration of the Baja Peninsula and of Long Island. Also prominent are the vignettes in the margins, which surround the map like a theater curtain opening onto the New World. The side vignettes depict Native Americans from various countries, and at the top are tiny topographic maps of particular places, for example, "Havana," "St. Domingo," and "La Mocha in Chili." These exotic places bedeck the depiction of the map's major subject matter, North and South America. Two orders of representation are here present, two kinds of map: one large-scale and one small-scale. In an inset showing Greenland and Iceland, there is even a representation of that which cannot be represented in the map proper. Next to this inset (to the left) is an inscription in words—another form of representation—proudly anouncing the content of the map as a whole. Thus the elder Blaeu's *Americae* gives us, in one complex image, an entire theater of representations, graphically embodying the original sense of the Greek word *theatron,* "a place for viewing."[24] Here we view representations that represent in turn the

New World, its inhabitants, and its special places. We are treated to a representational spectacle, a place-of-exhibition for cartographic and pictographic representations, a place-of-representation that is representational at several levels.

The cartographic enterprise of the Blaeus and other Dutch mapmakers, seemingly so far removed from the rigorous epistemological and metaphysical analyses of Descartes, reveals not only the first three traits of the modern age—whereby the world becomes an enframed image of represented objects—but even the fourth feature: the irreplaceable presence of the representer within the represented world. In the case of maps, this presence often occurs by means of the cartouche, which (in its full form) announces the subject matter of the map, various keys to its interpretation, and the name of the mapmaker—a name that is in effect the cartographer's *signature.*[25] In the case of Willem Janszoon Blaeu's map, the cartouche is elaborately enframed and floats conspicuously in the "Mare Pacificum." It announces starkly that this is a "New Map of the Americas" that has been made by its "auctor" "Guilielmo Blaeu" (as Blaeu called himself in his later work) (see Figure 12.2). In other words, the presence of the cartographer is made known through a representation that is in effect *a representation of the representer himself.* In a world in which everything exists as (and only as) representation—even external objects, says Kant, are "nothing but a species of my representations"[26]—it is to be expected that the person who represents also exists as, and in, a representation. A cartouche is a representation of the cartographer in his or her official role as mapmaker, and in three ways. (1) As a pictorial image, it is that kind of representation which Kant would classify as an intuitively given "perception" (i.e., a conscious representation that functions entirely by temporal or, in this case, spatial figuration). (2) The cartouche qua signature functions as a *representative* of the cartographer; it stands in for him—stands surety for him, takes his place by announcing to the map viewer that it is *he,* "Guilielmo Blaeu," who made this map; it is a representative of the cartographer who, in his Atlas ambitions, represents the totality of the known world in his cosmographic chart. (3) The cartouche is also an external emblem of the "I think" that accompanies all representational activity, including that of cartographers: its singularity answers to, and literally represents, the synthetic unity of apperception that allows me, Willem Janszoon Blaeu, his son Joan, or any human being

Figure 12.2. Close-up of Willem Janszoon Blaeu's map *Americae.* Cartouche with the title of the map and the cartographer's name.

to regard a set of representations as belonging to *my* consciousness, to my own identical self, whoever may have created these representations in the first place.[27]

The same is true of the inscription "I Ver Meer" that Blaeu's contemporary added to his masterpiece *The Art of Painting* (see Plates 11A, B). The representer here represents himself not by a printed cartouche that contains his name but by an actual painted signature. Unlike Jan van Eyck in *The Arnolfini Marriage,* Vermeer feels no compulsion to represent himself explicitly by a self-portrait framed in a mirror that is itself represented within the painting. In Vermeer's case, the level of self-representation is more abstract because its medium is strictly linguistic; nor is it embedded in a cartouche containing other information. What Vermeer's simple self-assignation accomplishes, however, is essentially the same thing as van Eyck's self-portrait or Blaeu's cartouche: a representation of his own presence as the source of the representations that the painting itself provides. As if to clinch the point, he adds the word *descriptio,* that is, a verbal token of his pictorial activity of describing the world he knows. If this world is much smaller in scope than that of the total world aimed at by the geographer, it is no less a world to be described as a picture, a place-world with all that this implies of attention to the detailed structure of a single locality. In Vermeer's case, *descriptio* means not only "description" and (literally) "drawing" but also map! But where Blaeu's map is frankly cosmographic and purports to represent half of the whole known earth, Vermeer's painting gives us the chorography of known place. What is known in Blaeu's case is an expanse of open (and mostly uninhabited) space; what is known by Vermeer—and conveyed to us by his painting—is the cozy interior of an intensely inhabited domestic locale.

Divergent as the map and the painting may be, each exhibits an intertwining of representations, verbal and imagistic, which we have encountered on several occasions: from the map of Nuzi (Figure 7.7) to the notations made by Constable on the reverse side of his cloud studies (Figure 5.8)—not to mention the place-names on T-O maps and portolan charts as well as on Chinese grid maps and Japanese *ukiyoe* maps. This recurrent alliance of image and word illustrates the general principle that representations, by their very nature, give rise to recursive compoundings—to representations of representations of representations; Kant, ever prescient, had foreseen this intrinsic possibility as well. Not only did he discern numerous levels *(Stufen)* of representation—those of "perception," "knowledge," "intuition," "concept," "idea," and so on[28]—but he formulated the basic rule that "all representations have, as representations, their object, and can themselves in turn become objects of other representations."[29] The recursiveness of representations extends from the conceptual realm of judgments to the intuitive domain of maps and paintings. It makes possible not only the self-representation of the mapmaker or painter but the series of successive re-implacements that are at play in maps and paintings alike. In Kant's example, an ordinary judgment is the representation of an object (at the conceptual level) of a representation of the same object (at the intuitive level).[30] What is true of judgment is also true of other modes of representation, including the purely pictorial, whose intuitive content allies itself with verbal expression in composite structures such as dreams and rebuses, pictograms and cartograms. In particular, discursive and presentational symbols continually collude in the creation of maps, and the same is true of such scenographic paintings as *The Oxbow* or *The Art of Painting.*

REPRESENTING PLACE, REPRESENTING SPACE

The fact of the matter is that the representation of landscape—whether this occurs in the burnished vistas of Jacob van Ruisdael or in the crisp linearity of Nicolaas Visscher (whose map of the Netherlands we know to be hanging on the wall in Vermeer's *The Art of Painting*)—brings with it the reinstatement of place within the represented scene. That the place is more often local in painting than in mapmaking is only to be expected, given the frequent and literally cosmic ambition of the latter activity to represent "the total world." The locality of landscape paintings ranges from an intimate view in a Rembrandt ink drawing (e.g., Figure 8.2)—which, like Vermeer's *View of Delft*, presents itself as representing one particular place only—to the regional representations of Constable, who took large tracts of the Stour Valley as the primary topic of his paintings (e.g., Plate 10, Figures 5.1-5.8). If a region such as the Stour Valley is the most embracing natural unit of landscape painting, the earth (that is to say, the totality of known and unknown regions) is the maximal subject of maps: hence the frequently used tag *orbis terrarum* (literally, "globe of the earth") in seventeenth-century cosmographic maps.

The power of maps to incorporate contiguous and ever more inclusive parts of the earth in carefully delineated representations (e.g., in atlases of the sort that both the Visscher and the Blaeu families created) is at the same time a weakness with regard to the representation of subregional localities, which are conspicuously neglected by those maps that purport to furnish a world picture. For the most nuanced representations of the particular places that make up given "spots" and "vistas," we must turn to landscape painting. Such painting resituates these spots and vistas, these discrete *endroits* and *lieux,* in the pictorial space of its imagery, thereby giving us not a world picture but the picture of a place in all of its subtle specificity. If many maps pass over places in their zeal to represent the totality of the world—its literal "geography"—almost all landscape paintings represent places within the detotalized totality of a given locality or region, its "chorography."

It is ironic indeed that Descartes and Kant, the twin philosophical colossi of "the age of the world picture"—the one inaugurating this age, the other bringing it to its most penetrating philosophical expression—were at the same time thinkers who actively ushered in the demise of place as a significant philosophical notion. Descartes took the decisive step of dissolving place into space, reducing place to sheer volume and at the same time considering it to be a merely relational (and thus relative) entity; either way, place is only a function of a homogeneous and universal space modeled on three-dimensional Euclidean geometry.[31] To geometrize the world as did Descartes is to pursue a form of representation in which the peculiarities of local shape, the idiosyncrasies of chorography, are no longer respected.[32]

At the end of the era of representation—that is to say, by the end of the eighteenth century—Kant carried forward the Cartesian dismantling of place in his critical philosophy. In *The Critique of Pure Reason,* place figures only in the shrunken and site-specific forms of "position" and "location."[33] Place has become a subservient feature of space. As Kant says explicitly in this text, "in sensible outer intuition (in space) . . . physical locations are there quite indifferent to the inner determinations of the things [located in that space]."[34] This is just what we should expect in a philosophy for which everything, including space itself, exists in the form of representations: "there is nothing in space save what is represented in it, and whatever is in it must therefore be contained in the representation."[35] To be contained in a representation is to be contained *in the mind*; indeed, avers Kant, space "cannot exist outside our mind."[36] And if this is so, there can be no genuinely nonmental place—which is to say, the kind of place that chorographers and landscape painters alike are committed to representing. Not even Kant's incursion into the sublime in *The Critique of Judgment* saves him from this ruinous consequence so far as place is concerned: in spite of passing references to a "rugged" nature in its desolate and terrifying aspects, the source of the sublime is ultimately in the mind of the awe-struck subject.[37]

Such skepticism as to the priority of place in nature (concretely considered, *on the earth* and *in the land*) stands in the starkest of contrasts with the efforts of mapmakers and landscape painters of the same period. The eighteenth century in Germany was a time of heightened interest in mapmaking, including the creation of general atlases. Paradoxically, Kant's only full recognition of the importance of place occurs in an early discussion of orientation by means of a map of the heavens—a discussion that, despite its ingenious descriptions of the role of the body in such orientation toward "regions in space," was precritical and did not survive the demotion of place effected by the publication of *The Critique of Pure Reason* thirteen years later.[38] What was for Kant only of passing concern was a preoccupying passion for such contemporary geographers as Anich and Hueber, Koops, and Homann, who built on the tradition of cartographic excellence established by the Dutch two centuries ago.[39] Similarly, landscape painting in Germany was beginning to thrive again after the much earlier breakthrough of Dürer, ushering in the Romanticism of a painter such as Caspar David Friedrich. Friedrich, painting in the period just after the publication of Kant's *Critique of Judgment,* created his own version of the Romantic-Gothick sublime that bears resemblance to the early paintings of Cole in America and looks forward to Bierstadt's and Church's apocalyptic sublimity of the mid-nineteenth century. In all of these painters, the role of place was central and assured. Not only did place possess for them an obstinate identity that refused dissolution into homogeneous space—Blake, after all, premised his Romantic revolt on an uncompromising opposition to the Newtonian neutralization of space and time—but it also resisted dissolution into something merely mental.[40]

The same is true for a pre-Romantic contemporary of Kant's such as Gainesborough, whose early landscape paintings (dating from the same decades in which Kant was composing his critical philosophy) had such a decisive influence on Constable. Neither Gainesborough nor Constable would have admitted to the proposition that "space is itself nothing but mere representation, and therefore nothing in it [including place] can count as real save only what is represented in it."[41] For these painters, as for their fellow mapmakers, place counted as real independent of its representation in paintings or in maps—despite the fact that both groups were preoccupied with bringing representations to ever new levels of enhanced realization. Cartographers and artists refused to be confined to the circle of representation that Descartes and Kant had drawn so tightly around human experience and knowledge.

BREAKING OUT OF THE REPRESENTATIONAL CIRCLE

The question thus becomes: *are* we so confined? Can we break out of "the brightly lit circle of perfect [re]presentation"?[42] Who is right—the philosophers or the mapmakers and painters? Is there any space for place in an era of representationalism? Is there any place for place in the age of the world picture: an age in which we are still living, according to Heidegger's alarming assessment?

What is most alarming is the dire prospect of being trapped within representations themselves—which is to say, ensconced within the mind conceived as a closely confining container. When Kant says blithely (in a statement cited earlier in this chapter) that "all representations, *whether they have for their objects outer things or not,* belong, in themselves, as determinations of the mind, to our inner state,"[43] he downplays the problematicity of having "outer things" as "objects" of these representations. If representations *and their contents* are indeed "determinations of the mind," it is by no means evident how we can ever transcend the boundaries of mind and catch hold of what is nonmental: namely, things and places. As Richard Rorty asks, "How do we know that anything which is mental represents anything which is not mental? How do we know whether what the Eye of the Mind sees is a mirror (even a distorted mirror—an enchanted glass) or a veil?"[44]

Once Descartes proposed the model of the mind as a privileged inner space—*not* a place (it was left to Milton to suggest that, on an alternative poetic model, "the mind is its own place")[45]—the problem of self-enclosure and thus of epistemological skepticism loomed large, casting shadows over the entire course of modern philosophy; for the inner space of the mind in its strict self-containment consists in just two elements: representations and the scanning activity of what Rorty calls the "Inner Eye." It does not matter that these representations are of many kinds ("representation," like "idea" as employed by Locke, includes sensations, feelings, thoughts, concepts, etc.) and that the scanning activity is incorrigible in its own sphere and occurs in two major ways that Kant specified under the headings of "intuition" and "understanding." Despite the sophistication of the model, the mind remains trapped within the domain of its own representations. All it can do is to *survey* this domain, not entirely unlike a land surveyor on the open plains of western Ohio in the early nineteenth century. "The Inner Eye," writes Rorty, "surveys these representations hoping to find some mark which will testify to their fidelity."[46] The hope is in vain, however, for the fidelity sought after is a fidelity to a reality that exists only in our representations of it and thus in the domain of mind—not out there as actual in the fields being surveyed. To posit the reliability, much less the veracity, of our representations is futile in a situation in which we are trapped inside our own epistemological universe. As a consequence, a very basic premise of modern philosophy—"to know is to represent accurately what is outside the mind"[47]—is foredoomed in advance. The very representations by which we presume to know what is outside our mind foreclose access to an external world that would be able to verify them.

In such a circumstance, we can only consult our own interiority—that is, represent our own representations. That this is so is evident in the following statement of Kant's, a statement that is remarkable for its sheer circularity: "Nothing is really given us save perception and the empirical advance from this to other possible perceptions. For the appearances, as mere representations, are in themselves real only in perception, [while] perception is in fact nothing but the reality of an empirical representation, that is, appearance."[48] This claim deconstructs itself. Appearances (i.e., indeterminate objects of perception) are said to be "real only in perception"—which might seem to mean that they are real as independent objects *of* perception. Yet, because perception itself is "nothing but the reality of an empirical *representation*" and thus itself an appearance, we cannot escape the enclosure of representations—an enclosure in which *everything* has the status of an appearance. In contrast with a sheer phenomenalism, Kant allows that appearances count as *real*; but their reality is exclusively representational in status. As he adds immediately afterward, "to call an appearance a real thing prior to our perceiving it, either means that in advance of experience we must meet with such a perception, or it means nothing at all."[49] Because we cannot encounter any such perception prior to experience (perception takes place entirely within the realm of possible experience), to consider an appearance to be a real but nonperceived *thing* is meaningless; such a thing is either a figment of the metaphysical mind or it is a "thing in itself," which cannot be known at all. Once we grant that appearances are in space and time and that space and time are themselves determinations of the knowing subject, it follows that any and every appearance "consists merely of representations."[50]

The situation is desperate—much more desperate than Kant is willing to admit. Not only can we not represent anything located outside our minds, but on the representationalist model we are sealed off in an ideational labyrinth without exit. As Rorty indicates, the resulting distinction between mind and nonmind (i.e., "matter" in Cartesian language, the "real" in Kantian discourse) is uncompromising and absolute; this distinction is "more like a distinction between two worlds than like a distinction between two sides, or even parts, of a human being."[51] Between two such alien worlds precious little, if any, communication is possible; even the Cartesian imagery of a mental mirror of nature, which had been posited as a rearguard effort to hold on to a vestige of such communication in seventeenth-century thought (a form of thought that, as we have seen, privileged the mirror as a model in virtually every domain of discourse), ceases to have any application, even a metaphorical application, by the time we reach Kant's two-world hypothesis in *The Critique of Pure Reason*.

Kant's own efforts to heal the split between the representational ("phenomenal") and the nonrepresentational ("noumenal") worlds were brilliant but singularly unconvincing. When his distinction between representation and object of representation is not merely otiose, it becomes a purely functional one that is dependent on the invocation of a rule that regulates a set of representations in such a way that they can be said to be representations of one and the same object by one and the same knowing subject. The object is thus a function of representations rather than the reverse: transcendental idealism and not perceptual realism reigns. What Kant calls "objective reality"[52] is derived not from the object but from a rule supplied by the understanding—which is to say, by another source of representations—with the result that, once again, we have a circumstance in which one group of representations (those that constitute the "appearance") is related only to another group of representations (those that constitute the "rule" that makes the appearance coherent as an appearance of a self-same putative "object").[53] The circle of representations, far from being escaped, is only reinforced by this move. As Robert Paul Wolff remarks, "if the object cannot be a distinct entity from the representations of it, and if at the same time it must serve as the ground for their objective connection, then the object must be simply *a special way of organizing the representations.*"[54] The same self-enclosed circularity obtains in Kant's alternative move to posit the object of a group of representations as a "something in general = *x*"; for the object thus described is purely formal, and Kant ends by admitting that such an object "is nothing to us" and derives any unity it may possess from the unity of our own consciousness "in the synthesis of the manifold of representations."[55]

Once more, then, the object *of* representations dissolves into the mass of representations by which this object is perceived and known; and these representations in turn are nothing but denizens of our own mind, its "inner determinations."[56] If we are as mired in the morass of our own manifold of representations as Kant maintains, it is hardly surprising that he cannot effectively answer the question he raises himself: "How, then, does it come about that we posit an object *for* these representations?"[57]

How, indeed, do we? Once more we can take our cue from Descartes's contemporaries and fellow countrymen, the painters and mapmakers of his time. These latter were certainly engaged in representational activity, aiming to represent surrounding landscape space and even the space of the entire earth. In order to do so, they had to evade the fenced-in self-enclosure that we have just seen to be the fate of representation in philosophical hands. For the Inner Eye of mental inspection they substituted the Observing Eye, the outwardly glancing look, that takes in the world around them—a world composed of localities and regions in the near-space of immediate experience and of much larger geographical units in the far-space that beckons beyond the horizon. They not only *had* representations of the circumambient world: they *created new ones* in the form of maps and paintings that were genuinely representations *of* that world, a full-fledged place-world. Even if the world itself was assumed (at the deepest level of metaphysical grounding) to have the status of a picture—to *be* pictorial, to present itself in pictures, whether stemming from nature or from human beings—there was the converse conviction that pictures *of that very world* were possible and, indeed, desirable. What else is a map but a "world picture" (where *world* can signify either local landscape world or the world of the earth as such, the subject of chorography or of geography, respectively)?

And what is a painting by Vermeer but an elegant world picture, in the case of *The Art of Painting* a picture of a domestic world seen from inside its cozy enclosure? Vermeer shows us that the artist can represent an interior that is not mental only, but also architectural, decorative, human, habitational, and, above all, dramatic. The closed world he depicts opens out into the sunlight streaming in from the south window and into the larger world represented on Visscher's map on the wall. Unlike the grimly self-contained mind on Descartes's or Kant's conception, the room Vermeer paints opens onto a larger, albeit unseen, place-world. It is a *clearing* in a sense closely related to the meaning this term possesses in landscape painting proper: in the warm world of Vermeer's pictorial space things happen that exceed the narrow circle of representation. These "things" are more and other than the "mere appearances" that Kant consigned to the mind alone. If they are not things in themselves, they are at the least represented things that exceed, by their concrete externality, their "objective reality," that placeless and thingless inner space to which both Kant and Descartes confine us. Vermeer gives us a space that, although an architectural interior, is a genuine place of exterior happening—or rather, a happening that is as much outward as inward, that is in the end both at once. *In the painting,* an "object" not only *of* but also *for* my representations is presented by the painter. Standing before it, "I represent something as an event, as something that happens; that is to say, I apprehend an object."[58] These last words of Kant's are strangely appropriate as a description of our experience of Vermeer's painting and, all the more so, of paintings of landscape by his contemporaries and successors.

What maps and paintings share—and mental representations lack—is a factor that we may designate as *display.* The happening we perceive in the artwork or mapwork, that takes place in the clearing it effects, is an event of display. The place of representation is a scene of display. Such a scene arises on the other side of the windowless interiority of the mind's keep, where incorrigible in(tro)spection comes paired with lack of outlet into the extramental. Indeed, the scene of display at stake in a map or a painting defies the very distinction between mental and material, inside and outside. As the title of Vermeer's painting indicates, such a suspension of dichotomies occurs emblematically in "the art of painting;" for, in the truly realized art of painting, the world is displayed as a world-place that encompasses mind and matter, inner and outer, self and other in one coherent presentation. The world-place is shown to be part of the place-world. The art of mapmaking also exhibits world-places—ultimately as part of an entire earth-world. A map not only represents this world, it *presents* it, giving it back to us cartographically or chorographically.

In maps as in paintings, the perceived and known world comes to expression. This world is not just put *on display,* it is placed *in display*: it finds its place, a new place, in a newfound representation. Instead of representing representations in an endless spiral of self-confinement, a painting or a map re-presents knowledge and experience in the display of a translucent pictorial space. Such space breaks out of the circle of representations into the clearing of presentational re-implacement.[59]

PRESENTATION VERSUS REPRESENTATION

Thanks to the lead afforded by seventeenth-century painting and mapmaking, we are beginning to find our way out of the inner impasse of the mind into which philosophers living in the same century had led us in their zeal for irrecusable certainty. They found such certainty in the mind alone, and more particularly in the mind's inspection of its own immediately given contents, its "representations" and "ideas." But these latter, in turn, being "determinations *of the mind,*" gave no egress from the cul-de-sac of the mind itself. Only by turning to another paradigm, that offered by art and maps, can we glimpse a way out—a way out that is not simply contrary to mind but that is an ingression into the concourse of clearing, within which any opposition between mind and matter is illuminated and resolved. *This clearing is none other than place.*

To win our way back to place in this manner requires a special sensitivity to the virtues of display qua presentation. These virtues are the virtues of *Darstellung* rather than of *Vorstellung,* "presentation" rather than "representation." We have seen that the

German verb *vorstellen* ("to represent"), construed as *vor-stellen,* implies an active "setting-before" that results in the positing of objects as "standing over against us" and that encourages our exploitation of these objects as merely there for our use. To represent as *vorstellen* is thus "to set out before oneself and to set forth in relation to oneself."[60] In the early modern period, the "relation to oneself" becomes all-encompassing, because representations of everything we experience are located in the closed circuit of the mind, wherein various cognitive processes circulate without issue. In the activity of *dar-stellen* (literally, "to set there"), in contrast, we detect something distinctly different at stake: a setting out into the open of pure *presentation (Darstellung),* affording a nonexploitative relation to something other than oneself as the master representer.

This "something other" is not the represented world, the world-as-picture—that is to say, the totality of representations and the basis for manipulation and mastery. *It is place itself,* construed as the locale in which we are able to come to terms with material things and thereby to situate the concrete constituents of the place-world. Instead of taking this world back into the imprisonment of the mental, we can follow it out into the open by perceiving it and acting in it—and by painting it and making maps of it. Pursuing this exterocentric path, we escape the enforced enclosure of representations and enter the realm of presentations. In this realm, the place-world can be said to come to its self-presentation, its *Selbst-Darstellung.* It exhibits itself in its own expressions—and in the reexpressions that paintings and maps compose. By emerging and reemerging in the successive re-implacements that art and cartography provide so paradigmatically, this world regains its proper position as the cynosure of our existence. If material things are "the stars of our life,"[61] the place-world that we map and paint is the scenographic setting.

The seventeenth-century situation, which has been a recurrent focus throughout this book, is indeed an extraordinary one. In one and the same century—indeed, in the first fifty years of that century—we find two divergent directions whose interplay would determine the course of modernity in the West. The first direction dictates that we are ineluctably encased in our own mind-set of representations, which refer only to each other and to the Inner Eye that surveys them from within a purely mental space. This scene of interior spectation dominates the philosophy of Descartes, Locke, Berkeley, and Hume—and Kant, whose *Critique of Pure Reason* (1781) gives the scene an explicitly transcendental sanction, but otherwise keeps it as rigidly delimited as it is in Descartes's *Meditations on First Philosophy* (1641). For all of these thinkers, the place-world without, and all the material things that populate it, have been drawn into the mind, where they are dissolved into a collocation of closely held representations to which we have immediate access, thanks to what Wilfrid Sellars calls "the Myth of the Given."[62] To the world of things and places, however, we have no access at all, or indirect access at best—and therefore no allegiance.

In contrast with this first line of philosophical thought (whose basic metaphysical program is precisely that sketched by Heidegger in "The Age of the World Picture") is a counterdirection suggested by our considerations of landscape painting and mapmaking in the early modern era. According to this second line of thought, it is a mistake to start from the model of mental representation as an all-inclusive and all-explanatory foundation of human experience. To do this is to land us, and to keep us, stationed within ourselves construed as the "representers of all representing." As Gadamer puts it pungently, "to start from subjectivity is to miss the point."[63] It is to miss the point of being situated in the world in the first place—of being there as embodied beings whose language and thought and perception are worldly and not merely representational in scope. This means that our experience of the world is not for our singular selves alone, much less contained within our mind as a switching station of privately possessed representations. It is an experience of the world as accessible to all who dwell in it. The world is an "in-itself for us," a world that is not only filled with discrete objects that stand over against us as if to mock (while in fact colluding with) our efforts to represent them, but is also replete with places in which we are concretely located and things with which we palpably deal.

This second direction, whose full philosophical articulation does not occur until the twentieth century in the writings of Husserl and Heidegger, Merleau-Ponty and Gadamer, was already at work in the seventeenth century. Contesting the prevailing paradigm of the age, it challenged the representationalist reduction of the world to its depiction in the mirror of the mind—a mirror composed of glass so dark that no reflection of anything outside itself was possible. According to this emerging form of thought, the world, and above all the places of the world, are *presented* and not merely *represented* in paintings and maps. This presentation is for all those who can witness it—"all presentation," says Gadamer, "is potentially presentational for someone"[64]—and thus it is not limited to its representation in the minds of human subjects stranded in sequestered epistemological space. Just as it puts objects back into a nonobjective world, it puts isolated subjects back into place—the very place in which they are already located by virtue of their visual and orientational powers.

If the first direction is the direction of *disposal,* the second direction is that of *display.* The historical axis of the first direction, extending from Descartes to Kant, lies athwart that of the second direction, which reaches from Rembrandt and van Ruisdael and Vermeer to Constable and Cole and Church, and it includes many imaginative cartographers along the way. It is one thing to say that the world is "at our disposal," that it is there to be calculated and manipulated, stored and saved, and that all of this happens in the guise of cogitational "representedness," with the result that "every relation to something—willing, taking a point of view, being sensible of [an object]—is already representing."[65] It is quite another thing to hold that the world is "in display," that it exists to be presented for the sheer sake of show, to be exhibited ("dis-play" derives from *displicare*: to un-fold), to be beheld as something more (and other) than "objective reality," to be given presence and re-presence in perception and memory, paintings and maps. Unlike representation in its modernistic guise—which always brings with it, despite its mastermindfulness, a secondariness (e.g., in Kant's positing of noumenal reality as prior to phenomenal or represented reality)—presentation is in no sense secondary. As its prefix "pre-" indicates, it is always in the advance position, it is an a priori of the world,[66] it is a gift of what there is to the witness of what is there. That the world exists to *present itself,* to display itself to us, is far from being an ontological deficiency. It points to an augmentation, rather than a diminution, of being.[67]

PUTTING AND PLAYING

It is instructive to learn that the original verbal root of "disposal" is *ponere,* to put or place, which is precisely the meaning of *stellen,* the verbal root of *vorstellen* and *darstellen* alike. We have seen that whereas *vorstellen* in its Kantian acceptation (i.e., as a dis*posable* mental representation) assumes the sense of "put," as in "to put or set before," *darstellen* draws on the action of placing—as in the "placing before" of a theatrical production that *presents* itself to an audience, as in a proscenium. Hence the potential of *darstellen* as a model for artistic creation: "presentation *(Darstellung)* must be recognized as the mode of being of the work of art."[68] Unlike *Vorstellung* in its self-confining demand that the representer set out before himself only what exists "in relation to himself," *Darstellung* involves what Gadamer calls an "openness toward the spectator."[69] This openness is part of being an artistic spectacle and part of being in an aesthetic clearing, and it forms the basis for that attention to place which we have found to be intrinsic to landscape painting that is more than decorative landskip. In the seventeenth century and thereafter, it is almost as if the putting aspect of *ponere* vanquished its sense as placing, as if "disposal" came to mean "*dis*placement," as in putting-out-of-place. To the extent that this was so, it only re-inforced the primary direction of the age of the world picture, a picture in which place no longer figures—in which it has been disposed of. This early triumph of *Vorstellung* over *Darstellung,* of *repraesentatio* over *praesentatio,* is at one with the triumph of space over place as this was inaugurated by Descartes in the middle of the century and whose most complete expression is found in Newton's *Philosophiae naturalis principia mathematica* of 1687. The effect, if not the explicit agenda, was the disenfranchisement of place.

This triumph would have been complete—and the story much easier to tell—were it not for the achievement of landscape painting, which must be accorded a special originality. This originality is less evident in mapmaking, whose origins are much further back in history (e.g., in Ptolemy, whose work was rediscovered in the fifteenth century). Moreover, landscape painting possesses a feature that has little place in mapmaking. This feature is *play,* the barely concealed verbal root of *display.* The display of painting, its "presentational immediacy" (in Whitehead's term), is a matter of the play of forms and colors on a painting's presented surface, its place-of-presentation. In its complete self-presentation, that is, when there is no remainder *not* displayed on the seen surface, such display stands in stark contrast with the self-reserved character of that which is securely at our disposal: the hammer is handy, but it still has its own obdurate recalcitrance. Where we *put* the obdurate objects into the internal representations of the mind, or transform them into the external devices and tools of technology, we *place* the ingredients of a landscape painting onto the dis-play of its surface, a place-of-exhibition occupying an ambiguous location between the inner domain of representation and the outer world of perception. In creating and viewing such a work of art, we disport ourselves—and thereby, paradoxically, find our place within it and within the world it presents to us.

The idea of the work of art as a form of play is not new. It can be traced back to Kant himself. In speaking of poetry, Kant remarks that "it plays with semblance, which it produces at will, but not as an instrument of deception; for its avowed pursuit is merely one of play."[70] More generally, a genius in any art form is someone who "sets the mental powers into a [movement] that is final, i.e., into a play which is self-maintaining and which strengthens those powers."[71] The play in question is the activity of imagination in a "free harmonizing" with understanding[72]; free play also occurs in entertaining "aesthetic ideas," including the idea of the sublime. But if it is true that "the imagination here displays a creative activity,"[73] it is no less the case that this model of playful activity is limited to representations as its proper medium: "the aesthetic idea is a representation of the imagination, annexed to a given concept, with which, in the free employment of imagination, such a multiplicity of partial representations are bound up, that no expression indicating a definite concept can be found for it."[74] Kant's recourse to play in his account of art therefore remains in the shadow of his adhesion to the paradigm of representational mind; not even his invocation of the sublime—which is at once a representation (i.e., as an aesthetic idea) and yet more than a representation (i.e., as surpassing any given set of representations in its boundlessness)—saves him from confinement to the realm of subjectivity. Indeed, as Gadamer has argued, Kant inaugurates the slippery slope by which the domain of art was gradually, in the course of nineteenth-century aesthetic theory, reduced to "aesthetic experience" and thus to a pernicious subjectivism.[75]

It is Gadamer's contention that play, despite its problematic origin in Kant, is nevertheless the key to the rethinking of aesthetics as a nonsubjective enterprise; for a close examination of play reveals that the phenomenon of playing, far from bringing out the subjectivity of the players, suspends it in the common enterprise of the game: "The mode of being of play is not such that there must be a playing subject who takes up a playing attitude in order that the game may be played . . . The actual subject of play is obviously not the subjectivity of an individual who among other activities merely plays, but instead the play itself."[76] When we play in an engaged manner, we enter into the playing in such a way that we need not represent any extrinsic goal or aim, or any other explicit content, to ourselves. Our experience is that the playing "happens, as it were, by itself."[77] In other words, it happens as if our own subjectivity and its characteristic representational activity were irrelevant to its occurrence. It also happens as if the objectivity of objects—their status as resistant things standing over against us—had become just as irrelevant.

If the work of art is not merely analogous to play construed in this non-subjectivist and nonrepresentationalist sense but is itself a form of such play, then certain consequences follow immediately. First, as just intimated, it does not exist for us as an obdurate object standing over against us (*gegen-ständlich*): "the work of art is not an object that stands over against a subject for itself."[78] Indeed, the work of art is not a represented object at all. Second, the work of art does not serve to represent anything else but, like play, finally *presents itself*: it is a mode of self-presentation.[79] Third, the artwork, far from existing only in relation to its creator or to itself, exists essentially *for others,* that is, for those who admire and view it: "artistic presentation, by its nature, exists for someone."[80] This means that it opens itself up to others by including its appreciators in the scope of its own presentation. Here the work of art exceeds the realm of play in the usual sense, according to which it may take place in private. The work of art exists as a nonexclusive openness in which the spectator himself or herself becomes a player in *its* game.[81] In this last respect, the work of art can be seen as similar to a religious ritual whose enactment is shared by all who participate, including those who merely sit and listen.[82]

The contrast between Kant and Gadamer exhibits the critical difference between an aesthetic theory that, despite its grasp of the playful character of art, attempts to constrain artistic phenomena to the Procrustean bed of representational thinking and thereby to subjectify these phenomena (that is, to make them phenomena at our disposal, albeit the disposal of a productive imagination in its interplay with understanding and reason) and an approach to art that respects its character as "bringing-forth"[83] into the open clearing of its presentation the display of its own presence. In the latter case, the work of art becomes its own measure instead of being measured by the accuracy with which it represents something beyond itself. Rather than having the secondary being of a representation, "it exists absolutely as something that *rests within itself.*"[84]

For a work of art to rest within itself is for it to have found its place: to be, finally, its own place because it has cleared its own clearing. This is not a matter of the mere transposition of place: that way lies a concern with accuracy and thus with the isomorphic representation of place reduced to site. It is a matter instead of the transformation of place itself—of place beyond place, place re-created within the work itself. This is what we witness in the creative action of a successive re-implacing in landscape painting that is not obsessed with replicating the real. The effect of such re-implacement is "a wholly transformed world."[85] This world, the world of the work of art, is not a merely represented world. It is a world re-presented or, alternately put, self-presented in and through the continually transformative re-implacement effected by the work.

With this step, we move from positing the subject as the source of all representations to conceiving the artwork as the locus of all effective re-implacement. We also move from the representation of objects to the presentation of world. This is a step that, though not articulated as such until late in the history of aesthetic theory, occurred early in the history of modern painting. It was precisely the step undertaken by the Dutch contemporaries of René Descartes during the years of his reclusive residence in Holland. As Descartes was confining himself to the warmth of his stove—and to a theory of mind as the self-confinement of representations—Jacob van Ruisdael was walking the hills and mountains of Germany, taking himself bodily into the natural world, which he was to present in paint in the first full-fledged landscape paintings in oil in the West. At the same moment as well, Dutch geographers were preparing resplendent maps of the earth at large, an earth-world whose first extensive exploration had been accomplished in the preceding century: here, too, the world was being re-presented in images, its many exotic and ordinary places re-implaced in atlases that made the geographic world into a graphic spectacle for all to behold. Into this double drama—of painted landscape and mapped world—viewers were invited as active participants, as players included in the open scene of nonsubjective display.

PARTICIPATION IN PAINTING

The subsequent history of landscape painting reflects the participatory action that such Dutch painters as Rembrandt, Vermeer, and van Ruisdael had first encouraged by making this kind of painting a presentational display rather than a representational icon. Even efforts at panoramic painting (whose explicit aim was a fully comprehensive representation) could not resist the temptation to include the spectator in a quasi-participatory manner. Henry Lewis's *Panorama of the Lower Mississippi* elicited the remark of one of its viewers that "the artist has succeeded in imposing on the senses of the beholder and inducing him to believe that he is gazing, not on canvas, but on scenes of actual and sensible nature."[86] To impose on the senses and to induce belief in the actuality of depicted scenes is the work of a presentation that draws the spectator into the drama of the landscape. The same is true of Chinese panoramic scroll paintings, whose gradually unfolding scenes invite active participation thanks to their implicit diachrony: here is a theater for the eyes that overcomes the alienation that may accrue to massively panoramic "views" from a stationary point, constricted to which the spectator may feel belittled. The mere fact that the implicit position of the observer of a panoramic scroll painting may be very distant from the scene depicted (and typically high above it as well) does not, however, undermine that sense of openness which allows the same viewer to feel part of the presented pictorial space. Even a surveylike landskip such as Smibert's *Vew of Boston* (Plate 2) draws in the viewer by various devices such as the positioning of foreground figures who beckon to us to attend to the distant scene to which they are emphatically pointing. By momentarily identifying with these "players," we become participants in the scene itself, just as our projective identification with Vermeer as the painter in *The Art of Painting* (where we are literally looking over his shoulder [see Plate 11A]) encourages us to feel present in the artist's studio.

Fitz Hugh Lane retains vestiges of panorama in the strong horizontality of his Gloucester seascapes, but he supplements these vestiges with a canny inclusion of the viewer still more directly in the landscape. He places the viewer *at ground level,* as if standing on the very shore from which the painted vista extends back and out in considerable distance. The apparent demotion from a view above the scene (as is the case with many panoramic paintings)[87] is in fact a gain for presentational immediacy; and it is above all a gain with respect to participation, because the viewer now feels very much part of the landscape, especially the outstretching shore in both of the *Norman's Woe* paintings (see Figures 2.4 and 2.5). The viewer not only perceives such scenes with his or her (figurative) feet on the ground, but also feels himself or herself to be someone with a distinctly upright posture scanning the depths whose farthestmost limit is the horizon formed by earth and sky. The overall sense is not that of looking *through* the frame of such paintings—as Alberti would have it in his notion of monofocal central perspective—but of being oneself located *within* the pictorial space, continuous with it (as in the "distance point" method of perspective favored by the Dutch).[88] All of this is accomplished despite—indeed, in the very midst of—Lane's explicit ambition to record what he sees in the most faithful manner.

To this development we may add that of Thomas Cole. On the one hand, Cole places panorama itself in perspective by confining it to the background of his paintings of the 1820s and 1830s: the panoramic view is still present, but we take *a view on this view* (e.g., in the case of *The Oxbow* [Plate 9], in which we are only allowed to glimpse in the distance a panoramic presentation of a rural paradise). On the other hand, Cole dramatizes the foreground and middle ground in an engagingly theatrical manner; he makes us feel as spectators that we are standing on a precarious promontory just under our feet in the lower region of the painting; the self-represented artist (whose figure is a species of iconic signature) turns back to us at its edge, drawing us into the scene still further; and we are made to sense the danger of the abyss lurking just beyond the artist—an abyss into which he, and we too, might fall at any moment. In effect, we are treated to a melo-drama of landscape to which all of the theatrical metaphors of "staging," "acting," and (in the most encompassing sense) "playing" are applicable.[89]

By the time we come to consider Constable, the last in the select set of landscape painters whom we have taken to be prototypical in Part I, we reach a turning point in which the devices of melodrama—dependent as they are on a literary sensibility—are no longer necessary; for Constable takes us right into the region he wishes to bring forth in his paintings, albeit nonisomorphically. The spectator of his paintings, though posi-tioned by imputation a certain distance above the ground and hovering, as it were, be-tween a bird's-eye and an earth-level view, is made to feel so much at home in the picto-rial space that suggestions of theater give way to images of domesticity. The sense of spatial familiarity is such that the viewer is induced to think "I, too, belong to this re-gion, or very well could." Such sharing of a common landscape—common in senses of both "everyday" and of "held in common"—brings with it an almost anamnestic convic-tion that one has already experienced the presented scene. Constable paints out of the truth of Gadamer's maxim: "to be present means to share."[90]

Such sharing is of the essence of landscape painting that is genuinely presentation-al. The very notion of presentation entails the concomitant idea of presenting something *for* others—others who are at once perceivers and participants. Where mental representa-tions are for ourselves alone, artistic (and, equally so, cartographic) presentations are for others: for the viewers who feel included in the presentational play of images. This is so much the case that the viewers may even come to experience themselves as *placing them-selves* in the picture, entering actively into it, as often happens upon viewing the land-scapes of Ruisdael, Lane, Cole, and Constable. One reason these painters (and certain of their successors such as Cézanne and Monet) are so compelling is that they engage us as participants in the vivid display of landscape. In their case—in contrast with those paint-ings that emphasize the space of survey—"participation transforms what may appear as a distant and gentle pastoral [vista] into an active scene which requires the involvement of the viewer to render it complete."[91] Transformation is precisely what is at stake here: a transformation from the disinterested passivity of landscape as depictively represented—"landskip," in short—to an active involvement in landscape as present on its own terms. The third-person detachment of represented space gives way to the second-person inti-macy of presented place. Our own place as spectators is now included in, or is at least continuous with, the place (or places) of the landscape displayed in the painting.

The transformation at stake here can also be understood as a movement from the primacy of strict enframement in painting to a loose openness that allows the landscape itself to come forward into our clarified vision. We have seen how framing as *Ge-stell* is integral to the modern era of representation. In the instance of painting in the West, the importance of the frame is doubtless due, as I have suggested, to the Albertian injunction to think of the surface of the canvas as itself a frame that mediates between objects seen and the naked eye. In this way the painting becomes a surface for the gathering of repre-sentations of the seen world—representations that are projected onto this surface in the form of condensed iconic images. (The most famous illustration of this Albertian ideal is Dürer's 1538 woodcut of a draftsman drawing a nude through a framed and gridded window: the first Western painter of landscape as such here shows his ambivalent allegiance to a Renaissance ideal of monofocal perspective.) When Alberti says that "on the surface on which I am going to paint, I draw a rectangle of whatever size I want, which I regard as an open window through which the subject to be painted is seen,"[92] he is in effect enclosing the beholder's direct perceptual view of a scene within another view, that of a literally framed representation—with the result that a painting based on this model of redoubled viewing is a representation of a representation.

Instead of allowing themselves to become entangled in such a representational regress, those painters who contested Alberti, beginning with the Dutch masters and continuing through Cézanne, insisted that a painting is not a separate plane of projection on which representations of the visual world are collected, but is *itself a place of presentation for this world*—and for us as its direct viewers. By an active collaboration between ourselves as viewers and the painting as the place of viewing, we enter into the work of art in its full radiance. We participate in "the one great horizon"[93] that landscape painting from the seventeenth century onward provides. It provides such a horizon by unframing its formally framed view and thereby placing us as viewers within the very place of its outlook. Instead of spectating as mere onlookers of a scene, we become engaged in the clearing that the painting effects, part of its illuminative action. We look out from within—out upon landscape from within the place proffered by the painting.[94]

TRACING AND TRANSFIGURING

Landscape has served as a continual cynosure in this book, which has focused on its representation in painting and in maps. It is geographed and chorographed in maps; and it is presented and re-presented in paintings of various styles and historical periods. Landscape is traced in maps and imaged in paintings.

If the immediate perception of landscape claims to give us the thing itself in what Husserl would call a "primordial dator intuition"[95]—that is, landscape in its common, unadorned being—mapmaking and painting, both being *about* landscape, exist ineluctably at one remove from any ostensibly unmediated sensory experience of it. To this extent, each of these modes of description can be considered "secondary." But any such secondariness does not signify secondariness in import or value. For one thing, it is not at all clear that we ever have more than fleeting glimpses of landscape in the first place; and even the barest glance at it is complicated by the ingression of images, memories, or words. For another, the very idea of landscape entails a massive infusion of culture; what we consider to be "landscape" comes shaped, indeed preshaped, by human history and society, and it is indisputable that, with the possible exception of entirely untouched wilderness, "landscapes always display a fragile equilibrium between natural and human force."[96] We observe this equilibrium in John Ruskin's classic description of a delicate equipoise between the natural and the cultural in a scene in the Jura Mountains where an untamed hillside exists in graceful concord with an abandoned fortress, giving the scene a complex poignancy it would not have possessed as sheer wilderness.[97]

Still more radically, we have seen that the secondariness of many representations of landscape (i.e., their status as truth-about), far from being a defect, can be a distinct strength. To be true-about—that is, to be informative in various ways—may contribute to a deeper truth-to, as is notably the case in many maps and paintings we have considered. To the extent that these representations are genuine presentations and are not restricted to mental conjurings inspected by an Inner Eye (producing secondariness of a decidedly interior sort), they carry forward an original experience of detailed fact into a higher-order state that possesses augmented being and that is true to their subject matter. This higher-order state, whether it takes the shape of a map or a painting, is something that, in Gadamer's words, has "come to exist more fully."[98] No wonder we gaze so wistfully at a colorful and ingenious *ukiyoe* map, or look so intently at an equally colorful if somewhat subdued Cézanne watercolor. In so gazing and looking, we find access to landscapes that have been deeply transformed—that have become not only something other than the scenes from which they stem, but something *more* than these scenes.

Kant himself, to whom we have had recourse at so many points in this concluding chapter, was acutely aware of the augmenting and transforming effect of art. As he wrote in a celebrated passage in *The Critique of Judgment* (1790): "the imagination (as a productive faculty of cognition) is a powerful agent for creating, as it were, a second nature out of the material supplied to it by actual nature. It affords us entertainment where experience proves too commonplace; and we even use it to remodel experience."[99] It is a matter, in short, of "the transfiguration of the commonplace"[100] in which what is supposedly derivative—"second nature" only—proves to be uncommonly potent, more potent, in fact, than the "actual nature" that is the putative beginning point. (I say "putative" because there exists an entire genre of "imaginary" landscape paintings in which we cannot specify the precise "material supplied to [them] by actual nature." In their case, we have to do with worlds so radically transformed that we cannot recognize anything but fragmentary origins in actually perceived or remembered experiences.)[101]

Part of the power of art's secondariness resides in the very playfulness of its aesthetic surface[102]—in what Kant designates with obvious reluctance as "entertainment." But play proves to be a most serious activity. We have seen it to be an effective antidote to the subjectivism that has infected so much of modern philosophy; it draws its participants out of solipsistic self-enclosure into an impersonal activity in which the players transcend their role as the source (and the limit) of representations. In its capacity to encourage the sharing of space—or, more exactly, the exchange of place—it has a transforming effect on these players, who become one with the game and are taken up into its capacious horizons. As a result, anything that is play or playlike in this transformative scene—and this means above all a work of art—is able (in Kant's words) "to remodel experience": to change the terms of the game of life itself.

An important part of the game of life is the experience of landscape. This is especially true with regard to what J. B. Jackson calls "vernacular landscape" and John R. Stilgoe "the common landscape." This is the landscape of our ordinary lives, partly planned and partly spontaneous, possessed in common by all who share a set of contiguous places: *"common,"* says Stilgoe, "means not rude or vulgar but belonging to a people."[103] An exemplary instance of such a space held in common is the region of the Stour Valley as this was represented in Constable's paintings over a period of almost four decades.

What was experienced by Constable and his family and friends as common land—as the place of their daily inhabitation—was transformed by his paintings into what was already in the painter's own lifetime coming to be called "Constable country."[104] It was a matter of a genuine trans*figur*ation of the common places of that region—a matter of putting these places not into the representational framings of mind (where, thus enframed, they would have lent themselves to unplayful disposal), but into the fully presentational format of the artwork (where, free from formal frames, landscape becomes play and display). This is not, as I have emphasized, an act of mere trans*posit*ion, by which places are constricted to pinpointed positions with the aim of achieving a picture-perfect rendering. The "posit" of "position" goes back to *ponere* in its reductive putting-aspect, that is, to the putting of things in sites, pinning them down in quasi-geometric rigor. It is foreign to the genuine placing that the same Latin root also signifies.

Placing, in contrast with siting, is the augmentational action of transfiguration by which the common places of a perceived landscape become the extraordinary places of a landscape painting, thereby transforming actual nature into that second nature which is the work of art. This second nature can be considered the authentic sublimation of the directly experienced (but always also culturally informed) common landscape. Figurative in its presentation and playful in its presence, it is inclusive in its clearing capacity: it takes us, and the landscape we behold in common, into its captivating embrace.

The term *figurative,* the moving force in transfiguration, is of critical importance in this discussion. It derives from *figura,* shape or form (*figura* was the standard Latin translation of the Greek *schēma*) and ultimately from *fingere,* to fashion or to feign. From *fingere* come also "figment" and "fiction." The figurative factor therefore includes a factor of fantasy; it embodies the playfulness that is evident in feigning actions, in figments of the imagination, and in works of fiction. But figuration also connotes the well-shaped entity, the formation by which the work of art reaches its final form, its satisfying Gestalt, its full presentation. Landscape painting may be said to exist in the tension between these two poles of the figurative—between the fantastic and the formed, caprice and care. No wonder it is so often preceded by the sketch, as important to Dürer and Rembrandt as it is to Lane, Cole, and Constable; for the sketch is a schema that mediates between the moment of fancifully projecting a scene (i.e., the first sense of figuration) and the moment of final finish (figuration in its second sense). As the transcendental schematism takes us from sensibility to understanding in Kant's model of the mind, so the painter's schematic sketch takes him or her from "productive imagination" to the "free harmonizing" of imagination and understanding.

Trans-figuration in art is just this move from fantasy to form, from the freedom of the sketch to the constraints of the painting. What makes an artist such as Lane more topographical than transfigurative is precisely his unwillingness to treat the sketch as a schema of future (and different) possibilities; instead, he sees it as a quasi-cartographic diagram of the landscape that it is his duty to transcribe with fidelity in the finished painting of the same scene: hence his use of the grid (and for the same purposes as found in the *yujitu* map, i.e., the determination of correct area and distance). In contrast with this transpositional propensity is the way in which both Dürer and Rembrandt decide to stop at the sketch: to consider landscape fully captured in its schematic adumbration as watercolor study or ink drawing. Instead of calling for painterly transfiguration, these sketches already capture the *configuration* of the landscape.

Between these extreme solutions—the one treating the sketch as a mere prefiguration of the finished work, the other finding in the sketch the finished work itself—stand the efforts of Cole and Constable, both of whom consider the sketch to be what the word meant in its original Greek form of *skhedios,* "impromptu": done or made in a casual or offhand manner, *extempore.* Unlike the full-fledged painting, the sketch is something improvised on the spot, that is to say, *at the place* of the primary experience of landscape, when the perception of actual nature precipitates the free play of productive imagination. Rather than saying *extempore,* we might employ *exspatium* as a more pertinent term with which to describe this early moment of figuring nature in its actuality—as a prelude to what will become the second nature of the fully finished work. It is during this moment that Cole and Constable both revel in a freedom of form that is not retained in the eventual oil painting of the same scene. Standing out of doors, in *plein air* (and in *plein lieu*), they indulge their fantasy in a first set of free figurations before moving inside to finish their work in a second, more sober, moment of transfiguration. If their paintings only reach their finished state in this later moment of final shaping, the world they set forth is, paradoxically, "complete when it is yet only partial."[105] In their hands, a landscape painting has its own integrity and validity at every stage of its creation, including that of the bare sketch.

The landscape painter is thus engaged in figurative work at every moment in his transformation of the perceived world. He is preoccupied with *figura* in its full range: figure, image, picture. In this figural-pictorial preoccupation he is to be contrasted with the writer and the mapmaker, both of whom engage primarily, and not just preliminarily or exceptionally, in graphic activity—in making inscriptions on the blank surfaces of paper. *Graphos* connotes what is written and, paradigmatically, that which is written in the letters or other graphemes of a language. Naturalists such as Thoreau and Gary Snyder describe the landscape entirely in these graphic units, thereby transcribing the landscape onto the printed page. The landscape is presented, re-presented, not in pictorial images but in written signs or "logographs." Mapmakers such as Willem Janszoon Blaeu and Nicolaas Visscher also transcribe in graphic signs, but in their case such signs assume two forms, logographic and pictographic, which together inscribe the known world on the surface of the map.

But the matter is not so straightforward as the bare distinction between the graphic and the figural suggests. What we can call *chorology* includes in its purview both cartography and landscape, a combination that complicates any effort to maintain a strict division between the graphic and the figural.[106] Moreover, Alpers has reminded us that both mapmaking and landscape painting are "arts of describing" in that both capitalize on pictorial representations; and she also underlines the replacement of the Greek term *graphē* by *pictura* and *descriptio* in the late Renaissance. Both the mapmaking and the landscape painting that flourished so brilliantly in the seventeenth century were pictorial-*cum*-descriptive arts, which is to say, arts that existed on the very cusp between the graphic qua written and the pictorial qua imagistic. Even if maps favor description via inscribed lines, and landscape paintings description via images, the collusion between maps and paintings was intimate indeed in this period. As we have seen, maps of the time contained both literal writing within their cartouche and pictorial images on their margin—and certain works, such as Pieter Saenredam's *The Siege of Haarlem,* were at once maps and pictures, both together in a single construct: a map-painting. Conversely, artists of the age, anticipating Constable's cloud studies, were wont to inscribe their sketches with written words of description and identification, as in Constantijn Huygens III's *View of the Waal from the Town Gate at Zaltbommel* or in Gaspar van Wittel's *View of the Tiber at Orvieto.*[107]

We should not be altogether surprised at such hybridization of writing, mapping, and painting. The Greek *graphē* originally meant "writing or drawing or painting," and the activity of *graphein,* that is, tracing, is common to all three enterprises. *Descriptio,* as I have noted, means "world map" as well as "description"—and, again, "drawing." Muir's essay "Discovery of Glacier Bay" is no less a graphic tracing of landscape than is a map of the same region or a painting of it. In fact, Muir himself simultaneously wrote about *and sketched* Glacier Bay on his inaugural voyage of discovery there.[108] The two actions proceeded pari passu, coexisting as equably as mapping and painting in Vermeer's *The Art of Painting.*

In all such instances, the art of painting the land becomes inseparable from the science of mapping space, and both from the practice of writing about the places that make up space.[109] Thanks to their ingenious and multiple forms of tracing, each of these activities constitutes a coherent but diverse set of approaches to landscape. Each articulates in its own graphic way various aspects of landscape; and each also transfigures what is commonly experienced in landscape—what is actual in nature—into a work that presents a transformed world, a nature beyond nature, a work of culture in image and line, word and picture, chart and sketch, bequeathing to us texts and maps and paintings.

LANDSCAPE AND PLACESCAPE

But what *is* "landscape" in the end? We know what it was *in the beginning*: it was *Landschaft,* the first form of the word in any European language. A *Landschaft* is a conglomeration of the cultural and the natural, a set of dwellings constructed in the midst of cultivated fields, around which lurks an intractable and bewildering wilderness.[110] In the idea of *Landschaft,* "dwelling" and "landscape" come together. Just as the *roland,* the ancient stone or tree that served as a local *axis mundi,* stood at the center of a *Landschaft,* so wilderness was found at its outer edge.[111] Between the ritualistic center, reaching back into the ancient past, and the untamed wild land, reaching out into the future of danger, was placed the *Landschaft,* where life in the present was shared by the common people. The planting and pasture fields around the town acted as a buffer against the desolation and disorientation of wilderness; the cultivation of these fields encircled a common domestic culture, including the *cultus,* the ritualistic worship that centered on the *roland* as world-axis.[112] The wilderness itself was considered uncultivable and its inhabitants (mostly animals, though including a few outcasts and bedeviled human beings) were literally uncultured. In the Dark and Middle Ages, when the *Landschaften* were the fundamental habitational units in central Europe, there was not the least wish to explore wilderness, much less to map it: for wilderness lay beyond the known world and was not allowed to mix with this world. Like the circle of covered wagons drawn together by early American pioneers at the end of each day, the *Landschaft,* itself usually circular in shape, closed out the wildness of uncharted surrounding territory.

It is a curious fact of linguistic history that the word *Landschaft* was taken over at the end of the Middle Ages by the Dutch, who transliterated it as *landschap,* a word we have encountered before. No longer used to refer to the living arrangement of "houses surrounded by common fields and encircled by wildernesses"[113]—this division of the landscape had no point in Holland, where practically all of the land was cultivated and inhabited—*landschap* took on two specific senses: what the mapmaker charted and what the landscape painter portrayed. By the seventeenth century, "the word *landschap* was used to refer both to what the surveyor was to measure and the artist to render."[114] In other words, *landschap* reflected two crucial currents of Dutch cultural life in the early modern era—an era caught up between the science of maps and the art of painting, two modes of descriptive representation that are closely affiliated in a common praxis of graphic tracing. Where landscape painting represents the world as a set of distinctive places, mapping as practiced by the Dutch subsumed places into a projected and comprehensive world-whole.

Landschap was anglicized into *landskip* in the early years of the seventeenth century. At first, *landskip* meant "scenery," that is, painting of a decorative sort—a meaning that the word retained in the United States even into the next century. But by 1630 in England, *landskip* came to connote something much closer to what we mean today by "landscape": "large-scale rural vistas, chiefly hilltop views of woods, villages, fields, and roads, dominated by the colors of vegetation and good soil—green and brown."[115] This second meaning of *landskip* looks forward to Constable's vision of English countryside, a vision of fields and villages seen from the height of a hill or the depth of a dale and displaying the dark earth colors for which this painter was so celebrated. It is as if the painter of the Stour Valley built on and redeemed the second usage of *landskip,* not its decorative but its rural sense, by showing how the commonly beheld views of his native region could be so transfigured in landscape painting—aesthetized, as it were—as to become "a wholly transformed world."[116] In this world, we witness the very picture of "space shaped for agriculture and gently punctuated by artifice and roads."[117]

In a definition that conveys what landscape came to mean in the age of the world picture and of modern representationalism, a recent edition of the tellingly titled *Webster's New World Dictionary* states that landscape is "a picture representing a section of natural, inland scenery . . . seen by the eye in one view." This picture, unlike the (interior) mirroring picture that makes up representations in their Cartesian-Kantian acceptation, is no longer seen by the Inner Eye but by the outward-looking eye of the mapmaker and the landscape painter, and (in the nineteenth century) the naturalist writer. Each of these figures transforms what he or she sees, and often sees in more than one view.[118] Each presents, indeed re-presents, a landscape.

Or rather: a placescape. I say "placescape" partly because more than "land" in the strict sense of "inland scenery" is described by mapmakers and naturalistically minded writers and painters. Just as Muir describes the impingement of Alaskan glaciers on inlets of the Pacific Ocean in writing about his experiences there, so Constable painted seascapes while he was residing in Brighton in 1824–28; and cartographers must take account of masses of water as well as landmasses. But "placescape" also points to the fact that all three kinds of decriptor deal with the -scaping not just of land but of *place* in their respective modes of re-presentation: where *scape* implies both shape and scope. Each figure shapes and reshapes the natural world, converting the actuality of a given place into the transmuted place of the work; each is engaged in transfiguring the common place of land or sea—and of city as well—into the extraordinary place that is presented not only *in* a work but *as* a work. An enlargement of scope occurs as perceived and remembered place becomes re-presented placescape.

Such descriptive work as this extends the range of what is seen (and touched, walked over, etc.) into genuine openness. By "openness" I mean the capacity of a placescape to exceed a precisely delimited perimeter of the sort that circumscribes sites in their geometric focus or legal limits (e.g., as "building sites"). The effect is that of overflow coming from the conversion of empty site into plenary place. The openness of the work also acts to invite the participation of its viewers—to make them feel that they belong to it. Merleau-Ponty captures both aspects of openness when he writes: "it is reflection which objectifies points of view or perspectives, whereas when I perceive, I belong, through my point of view, to the world as a whole, nor am I even aware of the limits of my visual field."[119] For "world as a whole" we could substitute the word *placescape,* to which we belong thanks to its availability to us and to our own imaginative and bodily incorporation into it as a nonobjective field of experience.

It is ironic that the seventeenth century bequeathed to us the ambiguous legacy both of placescape and of sitescape. In its Cartesianism, it insisted on an objectifying rational reflection that attempted to turn place into space and, in particular, into the space of sites. In such a space, places were geometrized by a process that Husserl has described as superimposing a "garb of ideas" *(Ideenkleid)* over the qualitative features of perceived phenomena, thereby regulating and regularizing them in such a way as to reduce them to "objects," mere *Gegenstände.*[120] Such geometrizing is not to be confused with geographizing, which respects the uniqueness of landforms, their endemic geomorphology.[121] The geometrizing of natural shapes means their leveling down into discrete flat areas belonging to a homogeneous space: it means a view of the world as a sitescape. In painting, this reduction of place to site occurs in the form of obsessively descriptive "views" of natural phenomena regarded as objects located at precise points in an indifferent space. In the closed-off space of sites there is a manifest lack of openness, there is no overflow of boundary and no invitation to participate: "there is no effort to extend the landscape beyond the painting to the space of the observer; it is enclosed in its own special space."[122]

I have been arguing that a countervailing tendency was also actively at play in the seventeenth century. It resisted the geometrization of perceptual experience and the homogenization of space. It refused the idea that the pictorial space of a painting, or even that of a map, is indifferent to what is placed in it. It admitted that places may exceed their own provisional boundaries by an intrinsic generosity of scope and by the polyform shapes of particular presences. We have seen that the landscape painting of the era, especially in its Dutch origins, engaged the spectator in a way that had been unknown to the Renaissance and earlier times: "participatory landscape requires that we look into the [pictorial] space, that we enter it, so to say, and become a part of it."[123] As a result, the spectator is made to feel that he or she belongs to the place-world of the painting.

It is the world as a detotalized totality that is contained, implicitly or explicitly, in a placescape: implicitly in the case of landscape painting (which presents the world *in* and *through* the depicting of particular places and regions) and explicitly in mapmaking (which, at the limit, offers an entire *cosmographia*).[124] Into this embracing and porous placescape, whether displayed in painterly or cartographic form, we are invited to step as interested parties, joining the artist at work in *The Art of Painting* or exploring the world at large with the geographer in the *Theatrum Orbis Terrarum sive Atlas Novus*. In each of these ventures, we are asked to bracket our incredulity and our indifference—our "morbid geometrism,"[125] the pathology of shrinking place to site—and to participate as players, indeed, as full partners, in the description and discovery of placescapes whose shape and scope exceed our delimited personal experience.

Landscape is both something *to which we belong* and something that *belongs to us*: an in-itself for us. It belongs to us, and we to it, despite the efforts to disengage and displace it that were made during the age of the world picture; for when landscape, and more generally placescape, is viewed as merely another part of the world picture, it is made into sheer spectacle, a *theatrum mundi*; it is transduced into the space of sites. In the face of this alienation of place from us and us from place, perhaps only the transfiguring power of painting, in legion with creative mapmaking, is capable of restoring that primary belongingness which acknowledges our antecedent ties to landscape and those of landscape to us. Through the configurative versatility of images and words, we regain possession of that place which is no site, and of that view which has nothing to do with the Inner Eye of mental representation and everything to do with the full body of participation.

Lao Tzu said: "As for where one stays, one values the proper place."[126] In the experience of landscape and in its graphic and pictorial descriptions, one comes to stay again where one has always been—dwelling, in a second time and by a second nature, in an abode that is the transfigured re-implacement of the places one has known and valued and to which one now returns for another look.

Landscape Experienced and Re-presented

Geography [is] the eye and the light of history.

—Joan Blaeu, *The Great Atlas* (Amsterdam, 1663)

Geography is the propaedeutic for knowledge of the world.

—Immanuel Kant, lectures on "Physical Geography" (1757–97)

REPRESENTING IMAGINARY LANDSCAPES

By picking out maps and paintings as exemplary instances of representing landscape, I do not mean to imply that these should hold our attention in any exclusive manner. There is a vast plurality of ways of representing landscape—as of place generally. Landscape, as we perceive and remember it, is a decidedly polymorphic affair. Thanks to its intrinsic variety (even a desert is amazingly diverse in landforms and life-forms) and to its all-embracing and open prospect, it offers countless perspectives and vistas, all of which are subject in turn to multiple modes of representation in various media. A complete catalog of these modes and media would be an endless project.

Nevertheless, one thus far neglected form of representing landscape deserves our brief attention before we turn to more conclusive matters: literary representations of fantastic places. Although painters can certainly set forth highly imaginary scenes—as we see vividly in the work of Hieronymous Bosch or Joan Miró—the artists on whom I have chosen to focus in Part I all take their primary inspiration from the perception of actual landscapes, as is most evident in the case of members of the Hudson River School as well as in European painters such as Hobbema and Ruisdael, Constable and Cézanne. Most of the cartographers treated in Part II attempt to convey the lay of the land or the extent of the sea as exactly as possible; unexplored regions are fittingly labeled "terra incognita" and, in their very unknownness, are often filled with detailed landscape tableaux (e.g., in medieval maps of the land of Prestor John).

The fact is that landscapes that are wholly imaginary in status are comparatively rare in the practice of painting or mapping. But just such landscapes are often at stake in poetry, as in this couplet by the Chinese poet Tu Fu (A.D. 712–70):

> Stars hung down on the breadth of the plain,
> The moon gushes in the great river's current.[1]

Here is certainly a landscape representation, even though it is not the representation of any landscape that (so far as we can know) was ever actually experienced by the author of these words. It combines the generic (e.g., "stars": but we do not know *which* stars are meant by the poet) with the specific (i.e., the events of "hanging down" and of "gushing"); yet the specific itself is not sufficiently detailed in any autobiographical or historical sense for us to be able to say that the poem is the representation (in words) of an actual experience (in visual or auditory perception).[2]

An example such as this can even be seen as putting the whole idea of representation into question. Stephen Owen points out that classical Chinese poets (e.g., of the T'ang dynasty, an era that we have already visited in the case of cartography),[3] rather than presuming that their poems represented some particular reality in the landscape, regarded their works as the schematized manifestations, the "omens," of nature. Owen cites an ancient apothegm to the effect that "the true function of literature *[wen]* is to be the means by which all inherent order may come through,"[4] commenting that "literature is a gate for the latent and inarticulate to become manifest. The poem is not simply the manifest state of the world's inherent order; its movement is the process of that order *becoming* manifest."[5] If this is so—if poetry exemplifies the schematic manifestation of a burgeoning order of the world—we must be cautious in claiming that a poem is a representation of a particular place, even a particular imaginary place. Indeed, the representation of both place and time is so highly problematic in poetry that even when an exemplary Romantic poet of the West such as Wordsworth explicitly titles a poem "Composed upon Westminster Bridge, September 3, 1802," we cannot assume that the poem represents just that place at just that time.[6] Nevertheless, we are still likely to say that Wordsworth attempts to set forth an invisible significance—to represent a metaphysical meaning—whereas in the Chinese case there is a more radical rejection of the entire idea of poetic representation, because even the recourse to the representation of preexistent meaning is barred.

Western civilization since its Greek origins up to Romanticism and beyond (with the notable exception of certain avant-garde movements) is so tied to the concept of representation that it pervades not only prevailing norms of the interpretation of poetry—which is considered capable of representing emotions as well as meaning, if not particular times and places—but also literary descriptions in prose fiction of places that are overtly fantastic. These descriptions are found, for example, in myths, legends, and stories of such imaginary places as "Isaura" and "Islandia," "Mezzonaria" and "Middle Earth." I draw these latter cases from a book revealingly titled *The Dictionary of Imaginary Places.* In the same book we read the following description of "Moominpapa's Island": "An island off the coast of Moominland, covered mostly in marshes and swamps. There are no tall trees; the only vegetation is heather and thickets of dwarf spruce. On the north coast of the little island, a half-moon bay of white sand lies between two headlands."[7] This summary description of an island landscape from Tove Jansson's 1965 novel *Muuminpappa Merellä* is accompanied by a map of "Moominland & Daddy Jones' Kingdom" drawn by the editors of *The Dictionary of Imaginary Places.* The map, traced out elaborately in plan and using conventional hachure marks and established pictographic signs (though significantly without any indication of latitude and longitude, much less of exact scale), does not convince us in the least as to the independent reality of Moominland or of Daddy Jones's Kingdom. This is so even though each region is shown as bordering on "Finland" and Moominpapa's Island itself is seen as set in the "Gulf of Finland." We do not hesitate to construe both the verbal description of the island's landscape and its cartographic portrayal as entirely fictitious, so much so that we are not even tempted to speculate on the actual geographic origin of this blatantly mythical place, nor do we concern ourselves with any putative cartographic location (including any claim that Moominpapa's Island is found in the Gulf of Finland).

In other words, we assume the fictionality of the place called "Moominpapa's Island" to be even more fully invented—more a matter of literary fancy—than is the landscape setting of the Chinese couplet and, still more so, than the scene of Wordsworth's poem set on Westminster Bridge. Where we are at least tempted to posit a possible place of origin in the latter two instances—misguided as such positing may be, even in the case of Wordsworth's ostensibly autobiographical effort—we are convinced that "Moominpapa's Island," although it is given a much more complete and explicit description in words than these poems provide and is even represented in a map, is from the first and forever fictitious.

And yet the description of Moominpapa's Island remains *a representation of some sort of landscape*; we accept without question the editors' effort to represent it. The fictitious island is certainly something other than a sheer manifestation of nature, its "entelechy"[8] construed as the actualized verbal state of nature in its coming through or showing forth. But what exactly does it represent? And how does it represent it? At the very least, the verbal description given in this fictitious tale is a representation of a representation, for the figment it represents is itself the representation of a thought (here, the author's thought of what that fictive landscape, Moominpapa's Island, is like). And the thought itself? It is a *representative* if not itself a representation—a representative of an entire way of imagining nature as not delimited to its empirical, much less its scientific, description. The representation of a fantastic landscape arises by means of a special form of thought; and this thought is the *delegate* or *emissary,* the representative, of an imaginative mode of thinking about nature in certain comparatively unfettered ways. Nature so conceived, which inspires a particular mode of reflection on the part of the author of *Muuminpappa Merellä,* may be said to *face us* in the form of a literary landscape, however fabricated this landscape may be in fact. It is a matter of redoubled representation—of a possible nature and of thought about this nature—as conveyed in highly imagistic words. Purely imagined places are indeed represented, but only on the basis of this rather convoluted process of literary production. What is at first seen as a mere matter of "fiction" or make-believe—that is, the simple representation of what is unreal—ends by being a great deal more complex, thanks to the fact that the author's thought articulates the structure of an imaginary world.

GEOGRAPHY AND HISTORY

It seems a far cry from fiction to geography. Is not geography the ultimate, the most exact and complete, representation of the world *as real*? Where poets such as Tu Fu and Wordsworth and authors of fictitious prose such as Tove Jansson enjoy the freedom of imaginative figuration—being bound by very few constraints of the real, not even by the actual landscapes that may well have initially inspired their work—are not geographers compelled to transmit to us knowledge of "the known habitable world"? Do not their works reflect as precisely and fully as possible the earth's reality as a natural entity?

But these last questions, plausible as they may seem, are misleading; for it is also true that geography studies arbitrary configurations of the places and regions of the earth—above all, its division into such formal and schematic structures as parallels and meridians and into such historically contingent factors as national boundaries. In fact, the notion of "areal differentiation" is one of the most influential conceptions of geography, and can even be taken as definitive of the field.[9] Yet an "area" is often shaped by culturally and socially determinative factors, including the shifting fate of whole peoples and languages, not to mention various economic and political forces. It is evident that we need to look again if we are to grasp what is specific to geography as a form of representing places.[10]

With this aim in mind, let us contrast geography and *history.* Geography presumes the pre-givenness of the earth in its inherent spatial characteristics, above all its regionalization into the naturally configured places that are its point of departure. History, in contrast, concerns itself expressly with the way in which these same places are lived through by human beings: discovered, explored, acted on, built upon, inhabited, and memorialized. The depositions and traces of historical action modify local geography in turn—sometimes quite massively, as in the imperialist expansion of the Age of Discovery or in the effects of English Enclosure laws—with the result that geography is ineluctably cultural as well as natural. There is no adequate geography that does not somehow represent the mark of man upon the natural placescapes of the earth.

It remains the case, however, that geography, though aiming ultimately at the earth as a whole, begins with *places* as its minimal units, while history bears primarily on *events*: whereas the former is intrinsically spatial, the latter is just as basically temporal in orientation. This is precisely the perspective of Immanuel Kant in his lectures on the topic "Physical Geography," delivered yearly between 1757 and 1797. In one of these lectures, Kant said emphatically that "geography and history fill up the entire circumference of our perceptions: geography that of space, history that of time."[11] As a contemporary geographer puts it, if history may be defined as "the knowledge about events in the human past," geography is "the knowledge of the world as it exists in places."[12]

Nevertheless, even if this preliminary distinction between geography and history appeals to a concern for conceptual tidiness, it cannot withstand further scrutiny. The same geographer who proposed the formulas just cited also acknowledges candidly that "all of geography must be approached historically."[13] Place is finally inseparable from period, because the ramified character of places includes their own past history.[14] The apparently pure synchrony of geography is inevitably complicated by the vagaries of diachronic development. We witness this commixture of the placial and the historical when we study a region such as the Mediterranean as it evolves over the long term. As Braudel says of his monumental study of the Mediterranean world in the years 1550 to 1600:

> The resulting picture is one in which all the evidence combines across time and space, to give us a history in slow motion from which permanent values can be detected. Geography in this context is no longer an end in itself but a means to an end. It helps us to rediscover the slow unfolding of structural realities, to see things in the perspective of the very long term. Geography, like history, can answer many questions. Here it helps us to discover the almost imperceptible movement of history.[15]

In "the very long term" *(la longue durée),* place and event, geography and history, coalesce (Kant himself proposed a "history of nature"),[16] leaving us ill-advised to distinguish in any exclusive way between the members of the two dyads cited earlier in this sentence. On the one hand, every place that figures into geography is affected by history and is a fortiori cultural in status.[17] On the other hand, every event is place-bound; it is bound to take place in place: "every event described [in history], every series of events reconstructed, has to have a place."[18]

Yet place remains primary—not just in geography but everywhere else as well. Archytas, the Pythagorean contemporary of Plato, proclaimed that "all existing things are either in place or not without place."[19] *To be is to be in place*: perhaps this Archytian axiom is the lesson that geography, more than any other single endeavor, teaches us most poignantly. Quite apart from providing the ground of every historical and cultural nexus, in spite of its own historicity and cultural tenor, geography exhibits the irrecusably place-situated and place-situating character of life on earth, including the subhistorical and precultural life of entities other than the human.

Even within its own discipline, geography continues to insist on the primacy of place. Despite its ancient concern with the world-whole and even with its characteristically modern preoccupation with exact "locations"—that is, with the precise measurement and quantifiable differentiation of a given position from other positions in projected geographic space—geographers remain recognizant of places as the distinctive bearers of geographical realities. It would be geographical nonsense to locate, say, Chicago in an exact co-ordination of latitude and longitude without situating it in a region (e.g., Illinois, the Midwest) that in turn encompasses it and allows it to be considered *a place in that region.* There is no location without a place of location, whether this place be that of an entire region or a mere "locality." As Lukermann observes, "For possibly three thousand years the place of something has been described in terms of the internal arrangement of features (site) and of external connectivity and environs (situation). Separately or together as definitions of place, site and situation are locating *in relation to* some other place or thing. *To locate is to relate.*"[20] In geography the ultimate relationship is of places (and of aspects of places, e.g., their experiential features) to each other. If the totality of such relationships is the proper subject matter of *geographia* in Ptolemy's original sense of the term, it remains the case that any such totality is dependent on the parts of which it is composed—which is to say: geography is dependent on chorography and not the other way around; for chorography in the Ptolemaean sense is the discipline that studies the constituent units of geography. These units are places and regions, which are more encompassing than locations or positions and are integral to that totality which we call "the earth" or "the world." Middle terms in this respect, places and regions are indispensable to the pursuit of geography—to begin with, places, on which regions themselves are finally dependent. At once a mediatrix and a matrix, a place is the natural carrier of geographic knowledge; for such knowledge is "the knowledge of the world as it exists in places."[21]

TOWARD A NATURALISTIC GEOGRAPHY

If the alliance between geography and history is often implicit, thus calling for demonstration and interpretation, there is no such concealed link between geography and cartography. The link is altogether aboveboard. It is revealing that Joseph Needham titles chapter 22 of his *Science and Civilization in China* "Geography and Cartography." Geography *and* cartography: one discipline calls for another, the two form a pair. The reason for this is not difficult to determine. For one thing, maps are first of all maps of the very landforms and sea masses that it is the task of geography to describe. For another, maps may be considered the "language of geography."[22] They bring to explicit imagistic expression what is mostly treated in words and scientific formulas in geography. The intention of geography is to deliver a (mainly) verbal description of the known world; the intention of cartography is to deliver (mainly) drawn images of that world. The primarily discursive symbolism of the former is complemented and carried forward by the mostly presentational symbolism of the latter, thereby achieving the equilibrium for which I argued in chapter 7. The graphic work of geography, its basic action of delineation, is most fully achieved in the cartography of maps.

Geography as it has been pursued in the West is characteristically logocentric, as we can see in its perennial tendency to ally itself with the still more rigorous disciplines of geology, geodesy, geometry, and geomorphology. In the United States, for example, the advanced study of geography was, until recently, predicated on a previous knowledge of geology.[23] In China as well, the early advent of square grid patterns from the twelfth century onward gave to the representation of the known world a decidedly geometric cast. No wonder that, as a compensatory gesture, in East and West alike cartography was called on to give *eikon* to *logos,* image to word—in short, to furnish the iconicity and pictoriality that geography could not find from within its own resources.

In a similar vein, landscape painting also provides the imagery of land and sea that geography all too often suppresses in its haste to grasp the literally geo-metrical and geological truth of the earth and its surface. Yet the affinity between such painting and geography is as close (even if it not always as evident) as that between cartography and geography. We see this affinity in ancient Chinese landscape painting, where the energy of *ch'i* may be said to link the then-current form of geography (i.e., geomancy or *feng shui*) with the painted images of surrounding hills and mountains.[24] The same affinity is just as powerfully present in American landscape painting of the mid-nineteenth century. At the very moment when the Corps of Topographical Engineers was creating graphic depictions of the Far West—geographic descriptions in image and word that were intended to contribute to a more accurate cartography (and vice versa)—painters such as Bierstadt and Moran and Whittredge were composing painterly renditions of the same region. Their works purported to be of geographical realities, and their hyperrealism answered to the felt need to know the earth in its precise configurations. It is not surprising to learn that Frederic Edwin Church, the most assiduously naturalistic of the painters of his generation, was an avid admirer of Alexander von Humboldt, especially of the latter's five-volume *Cosmos* (first published in 1845). In this major work, the German naturalist issued a challenge to painters that acted as a clarion call to Church:

> Are we not justified in hoping that landscape painting will flourish with a new and hitherto unknown brilliancy when artists of merit shall more frequently pass [through] the narrow limits of the Mediterranean, and when they shall be enabled, far in the interior of continents, in the humid mountain valleys of the tropical world, to seize, with the genuine freshness of a pure and youthful spirit, on the true image of the varied forms of nature?[25]

Church followed the advice of the great geographer literally: he traveled to Central and South America, where he made studies for several of his most renowned paintings (e.g., *Cotopaxi* [Plate 7], *Heart of the Andes* [Plate 6]). These studies, to which I have alluded earlier, are unmistakably naturalistic in intent and execution; they compose the pictorial equivalent of a geographer's detailed field notes.[26] What Emerson wrote of Humboldt in a journal entry of 1845 could apply as well to Church: "The wonderful Humboldt, with his extended centre [and] expanded wings, marches like an army, gathering all things as he goes."[27] Church himself gathered everything he saw into images—virtually everything that Humboldt had earlier described in words. If nature for Humboldt could "only be vividly delineated by thought clothed in exalted forms of speech,"[28] the same nature was for Church clothed in equally exalted forms of image. The relationship between geography and painting was made even more intimate by the fact that Humboldt included a section in *Cosmos* titled "Landscape Painting in Its Influence on the Study of Nature." Here he emphasized that "colored sketches, taken directly from nature, are the only means by which the artist, on his return, may reproduce the character of distant regions in more elaborately finished pictures."[29] In other words, the landscape painter operates like a geographer while in the midst of nature and only fully assumes the role of painter in the sanctity of the studio. The painter's detailed sketches, created in *plein air,* rejoin the geographer's verbal accounts, each being made *sur place.*[30] Either way (and *both* ways in the case of Humboldt's closest American counterpart, John Muir), the wildness of nature comes to articulate expression in images and words that convey its incarnate essence.

RE-PRESENTING EARTH IN WORLD

Just as geography and history call for each other—and even merge in "the very long term"—so does geography affiliate itself with cartography and landscape painting. But, in the latter two instances, there is an additional elective affinity to be found in the fact that both terms paired with geography, its sister specialties, as it were, occur in an expressly imagistic format. Each supplies images that are lacking in geography taken as a logocentric enterprise; both, as I have been suggesting, give something pictographic to the geographer immersed in the logographic. Each provides images, cartographic or painterly, by which Gaea, Goddess of the Earth, is given pictorial description. For earth (*gē* in Greek) is very much at stake here; the common prefix *geō-* ties together such diverse but logocentric terms as geometry, geodesy, geology, geomorphology, and geography itself. *Only the earth* conjoins these disparate disciplines, just as *the earth alone* subtends the edifice of human history and culture. If it is true (as Sauer states) that "geography is [the] spatial differentiation of nature and culture,"[31] the basis of this differentiation is the earth that stands beneath and beween the natural and the cultural as their common ground.

But the earth—what is this? It is not merely the visible landscape, or even the land itself, much less its sheer surface. It is all this but also more than this: it is the *deepening of land,* the land's end as it disappears outward into the horizon and downward beneath the palpable soil. Rilke expresses this distinctive deepening in a celebrated passage from the ninth *Duino Elegy*:

> Earth, isn't this what you want: an invisible re-arising in us?
> Is not your dream to be one day invisible?
> Earth! Invisible![32]

The earth that furnishes depth to things on its surface itself possesses invisible depths. It is "self-secluding" in the term of Heidegger's to which we have already had recourse: ingathering, as we might also say.[33] In its self-secluded depths, earth gathers the depths of things and keeps them there in a deeply enracinated state. No wonder, then, that *geo-* as a shared root holds together the various discursive and scientific disciplines (e.g., geodesy and geology) by which the earth has been studied under the general heading of "geography." (Only geomancy, the ancient and still current Chinese practice, is resolutely nondiscursive.)[34]

Not only is it the case that "Geography [is] the eye and the light of history."[35] It is arguable that geography underlies *all* scientific disciplines. John Dewey, writing in 1899, put it this way: "The unity of all the sciences is found in geography. The significance of geography is that it presents the earth as the enduring home of the occupations of man."[36] Earth, in its very withdrawal from bright visibility, alone bears the brunt of the various sciences pursued on its surface; it alone bears up under the pursuit of *logos* and *morphē* (i.e., structure and form) and of *metron* and *graphē* (i.e., measure and trace). This is so even though the various sciences of the earth—from geometry to geology and geography itself in its modern guises—have come to detach themselves from their origins in the earth, losing touch with what Husserl called "the origin of geometry" in basic natural forms and shapes. But the "geognost"—the one who truly *knows* the earth—recognizes that any knowledge of the earth, no matter how refined it has become, is supported by the earth itself and is rooted in its secret structures.

What is ultimately supported by earth is the *world*—the abstract world of research as well as the concrete world of looking and touching. World, as Heidegger also reminds us, connotes illumination, clearing, culture, history, language, the "self-disclosing."[37] The world is a cosmos, a world of appearances, where "cosmos" connotes both order and decoration: Humboldt speaks of "the dignity of the word *Cosmos* in its signification of *universe, order of the world,* and *adornment* of this universal order."[38] The cosmos is a world order in which appearances are allowed free play. If the aspect of *order* leads to science, the aspect of *adornment* leads to imagery and play (especially the play of light): both are Apollonian in contrast with the earth's Dionysian depths. But Apollo and Dionysus, self-revealing surface and self-concealing depth, invoke each other, indeed, require each other. Given this coimplication, we should not be surprised by the alliances we have discovered between the science of geography and the art of the image that underlies maps and landscape paintings alike. All proceed by "morphic resonance,"[39] by the inculcation of forms, whether these be the strictly geomorphic forms of geography or the pictorial shapes of cartography and painting.

This cosmic/telluric alliance casts a new light on landscape representation. Far from such representation being a merely secondary matter, something of second-order status in comparison with an aboriginal nature, it exists in parity with its own putative "origin." Even more radically, to represent landscape in maps or paintings is to achieve an *advance in being* with respect to the earth from which they take their departure. Instead of being a mere attenuation of earth—its sheer repetition or replacement—these two modes of representation carry forward the being of nature-as-encountered into forms of articulation that describe a new order of being, a renewed cosmos. This is why I have designated them as modes of *re-presentation*: they present again, at another and more sublime level, what is first of all presented in perception.

In place of translation, that is, of mere conveyance of the self-identical over time or space, such redoubled presentation effects a thorough transformation. It is this transformation that we first met in Lane's subtle but masterful movement from the allure of iconographic depiction (as exemplified in his careful pencil drawings, taken as they were "directly from nature") to the alembic of a painted scene. This was a movement from the mere transposition of site to the transplacement of place itself; it constitutes both the conversion of site into place and the enhancement of place itself. The site of the actual coastal configuration of the cove called Norman's Woe, faithfully inscribed on gridded paper that uncannily resembles the format of an early Chinese map, became the transfigured place of the painting *The Western Shore with Norman's Woe* (Figure 2.5). The same process of transplacement was even more evidently at work in the landscape paintings of Lane's compatriots Cole and Church. All three painters brought about an advance in being; by re-presenting the earth, they created worlds in their works—worlds that were the effective transformations of the very ground on which they stood. If each began by the meticulous observation of the geomorphology inherent in this ground, each ended by reconfiguring it in significantly sublime ways.

THE TRUTH OF RE-PRESENTATION

Place will out. It will come out in the world of the work from the earth of its own ground. In re-presentation, topography taken literally transcends itself.[40] The tracing of topos becomes place-in-the-work; but it does so only by remaining in lived synchronism with its own soil—with *chōra*, the preorganized and preformal earth, Gaea. Thanks to the synchronism, the false narrative of secondary being, of "representation" as a derivative imitation of the originary, gives way to the true tale of placial primacy. In this account, the re-presentations afforded by maps and paintings stand more than proxy for their origins. The very idea of proxy or substitute, although richer than the notion of sheer simulacrum, here cedes place to the notion of augmented being; for the conception of a proxy, and even more so that of simulacrum, is dependent on the Cartesian assumption that there must be at least as much formal reality in the cause as there is objective reality in the effect.[41] By "formal reality" Descartes means a preexisting substance, and by "objective reality" he means precisely such a reality *as represented*. But contrary to Descartes's assumption, certain effects are capable of *exceeding* their own causes in force, power, and meaning. Preeminent among these effects are cartographic and painterly creations that display the earth in the world of the work.

Thus exhibited, thus transformed, place is indeed "the first of all things" in Archytas's prophetic phrase.[42] It is first not only in *being*, as Archytas maintains, but first also in the order of *representation*: it is *what* is represented—directly, in the case of landscape paintings; directly and sometimes indirectly, in maps. Moreover, in being represented, and even more so in being re-presented, place comes into its own. It acquires the superordinate being that it lacks in the realm of bare perception, in "firsthand" experience. Place as re-presented in the world of the work takes precedence; it acquires, at last and for the first time, the truth of its own being.

The question of representation moves us back into the domain of truth. The truth at stake is not the truth-about of exact discernment in a facsimile. In the case of mapping, not even the geomorphic accuracy of photometric mapping[43] gives us the kind of truth that matters most when it comes to representing landscape: that is, *truth-to*. Transformative re-presentations furnish us with images that are true to the experience of landscape from which they proceed—and ultimately true to the earth on which they are based. Despite their salient differences, portolan charts and medieval *mappae mundi* are both true to the experiences of their creators or appreciators, whether these experiences be of particular Mediterranean coasts or of a theocentric universe. However deficient or removed these maps may be from the standpoint of purportedly "objective" standards of measurement—standards that we have seen to be highly problematic because of the distortions introduced by the necessity of projection—they remain true to two aspects of human experience: its sheer *that* (i.e., the fact that the earth has been experienced in these ways) and its concrete *how* (i.e., in what guise the earth has thereby presented itself).[44]

Landscape paintings also attain this double truthfulness, as we see vividly in the works of Constable and Cézanne, or Marsden Hartley and Milton Avery, Chaim Soutine and Willem de Kooning. In other works, notably those of the Hudson River School and Luminism on which I have placed such emphasis in Part I, a decided confluence of truth-about and truth-to (in both of its forms) manifests itself. As in Humboldt's writings about nature, the detailed and the exact (the criteria of truth-about) are combined with the aesthetic and the imaginative (the main dimensions of truth-to). In these complex combinations of two alethic modalities, we witness position and site (the normal objects of truth-about) undergoing transmutation into place and region (the proper subjects of truth-to).

THE DOUBLE ASPECT OF LANDSCAPE

Place is the module of landscape—indeed, its very element. Landscapes are, in the final analysis, placescapes; they are congeries of places in the fullest experiential and re-presented sense. *No landscape without place*: this much we may take to be certainly true. Not only is it difficult to imagine or remember an actual landscape devoid of places; it is not possible to come upon a landscape that does not contain them in some significant fashion. When John Ruskin described the landscape in the Jura Mountains to which I referred in chapter 12, he found neither a congeries of teeming detail nor a single whole mass, but a coherent group of interarticulated places, built and natural alike. Even in the midst of the most remote glacial wilderness, John Muir encountered the tropics of particular places, their peculiar turnings.[45] Everywhere we turn in the presence of landscape, whether the landscape itself be cultivated or wild, we find the inscape of place, its ingression into the underlying land. To be in a landscape is to be in the midst of places.

But what is landscape? The question returns insistently one last time. What we now know is that, at the very least, landscape is something situated at the intertwining of earth and world: at (and as) their "common outline."[46] *Between earth and world is landscape.* Neither as deepgoing or reclusive as earth nor as ascendant or illuminated as world, neither self-secluded nor self-shown, landscape is the pivot of the two together. It is where earth and world meet, their shared surface. No wonder it is so variegated in form and so diversified in content! It brings the dark earth to bear at the very places where the world begins to manifest clarity and order. Stretched out like the Tibetan demoness Srin-mo over an entire countryside, landscape articulates and specifies the earth/world interface. Like such a demoness as well, the landscape is pinned down at particular *points de capiton.*[47] These points are the places that punctuate and populate the landscape as a whole.

If geography is the tracing of earth's landforms, landscape is the worlding of the earth—a duplex surface on both sides of which (as on a Möbius strip) earth takes on shape without having to pass over the edge of the world. What Wallace Stevens said of imagination we can now say of landscape: "I am the necessary angel of earth, / Since, in my sight, you see the earth again."[48] In landscape the earth becomes visible; it is the most graphic moment in the earth's history: it is geo/graphy writ large. Such writing is strictly cosmic: it constitutes the world of the work. And it occurs as re-presentation in image and word, both of which delineate place in its plenipotentiary powers.[49]

Yet landscape is not only a matter of visibility. The etymology of "scape" suggests that landscapes *cut into* the earth's surface, marking it as age lines inscribe a human face and give it a distinctive character. Landscapes possess ley-lines that exhibit the precise directionality of the earth on which they are laid down as coruscating presences.[50] Such lines, often merely implicit in the landscape (and thus lying *between* the visible and the invisible), can be said to trace the figures of the land, to delineate its character in the guise of landforms: "landscape," as I suggested earlier, is very much a "landshape."[51] *Skep-,* a likely root of "scape," means to cut, to scrape, and even to hack. Landscape possesses the sculpting power of a *hapjo,* Germanic for "cutting tool" and a close relative of *skep-*; for landscape cuts out from the earth the figure of a world.

Landscape is also *what is cut out,* the *skopo*—another Germanic word, one that means "container," as does Middle Dutch *schope,* a cousin of English "scoop." As cut out from the earth, scraped within its surface, a landscape becomes a container: the scrape is also a scoop. Thus telluric violence gives way to conservation, allowing landscapes to hold and retain things (and memories of things). Figuration becomes configuration as cosmographic script betokens a lasting landscape.

The -*scape* of *landscape* thereby points us in two complementary directions. On the one hand, the *scope* implicit in *scape* indicates that aspect of landscape which has to do with its ex-tended and laid-out character: its tendency to arrange itself *around* us as a circumambient spectacle. "To have scope" means to possess breadth or span, *envergure*, as the French say. As scopic, a landscape opens up and opens out upon a vista; it gives us a "prospect" on the land. It offers a prospective view even as it encourages us in viewing it to look and move and explore within it.[52] A landscape in its scopic aspect is an arena of open spectacle, a mise-en-scène of the world as it spreads out before and around us, giving us its layout.

On the other hand, what we can call the *scoopic* aspect of a landscape presents to us the same spectacle under the mode of reclusion. As scooped-out or cut into, a landscape possesses an implicit concavity that contrasts with the convexity of its scopic outreach. Exhibiting this concavity, landscape exemplifies Husserl's idea of the "near-sphere" in its capacity to shelter and surround the viewer, who experiences herself as *within* its ambience rather than looking out onto it as a prospect.[53] In this capacity, landscape serves as a "refuge" for us, a place to which to retreat and in which we may be protected; such a refuge is the second basic directionality that a landscape affords. Rather than expanding outward over the earth and across its very surface, landscape here sinks down into the earth's ingathering depths. As one geographer has put it, "visible landscapes are like icebergs: only a small proportion of their real substances lies above the surface."[54] If landscape as prospect constitutes a world on the earth—on its own double-sided surface—landscape as refuge draws us into the earth itself.

The scopic and scoopic aspects of landscape, its prospective and refugial dimensions, are formative and pervasive. In being drawn into the drama of a glaciated landscape, John Muir sought its refugial being, its capacity to charm and shelter; but in drawing back from the same landscape, viewing it from afar and sketching and writing about it, he also took his distance from it and regarded it as a powerful prospect to be remembered and recounted. Similarly, Thomas Cole created landscape paintings that are scopic in their sweeping vision—most patently in the distant panorama of *The Oxbow* (Plate 9), in which a domain of prosperity is offered in prospect—while being at the same time scooped out in the near distance, where an arena of momentary reclusion is offered (e.g., shelter before the forthcoming storm as well as protection from the abyssal falling off of the earth in the middle distance). Cole's early paintings, especially those set in the Catskill Mountains, are also constituted by a dialectic of refuge and prospect.

Maps participate in this same dense dialectic, at once drawing us into the intimacy of local space (e.g., in the topographic insets on the margins of early Dutch maps [as depicted in Plate 11A, Figure 12.1], or in the detailed coastlines of a portolan chart [e.g., Plate 14]), while also proferring a more comprehensive worldview: if the coastline or inset constitutes a literal refuge, the representation of the open land and sea provides, just as literally, a prospect of the *oikumēnē*. The strange spatiality of Mercator maps—which oscillate between concavity and convexity in their efforts at projection—illustrates the same double aspect of landscape. So too does an early-nineteenth-century *ukiyoe* map from Japan (e.g., Plate 16), whose oblique vista is both extensive (i.e., toward the external horizon) and contractive (i.e., as representing the in-depth contours of particular landscape features).

In each of these instances we observe an intimate interplay between prospect and refuge, scope and scoop. When the two aspects are taken together, like the two infinite attributes of Substance in Spinoza's metaphysics, they furnish us with the full picture of landscape, its most fully augmented being. Landscape as set forth in this bivalent way gives us not less than landscape-as-experienced but *more*. It gives us the transformation of landscape as "the assemblage of real-world features"[55] into a detotalized totality of re-presented attributes in an intensified state.

RE-PRESENTING THE PLACE OF LANDSCAPE

To borrow a phrase from William James (a phrase James himself applied to the lived body): landscape is "the palmary instance of the ambiguous."[56] The literally ambi-guous means *being (or having) something both ways*. This is what we have seen landscape to ex-hibit at every remove. Beyond its power to combine earth and world as well as refuge and prospect, landscape is also a creature of surface as well as depth, of visibility as well as in-visibility, of image as well as word, of nature as well as culture. It can be just as well paint-ed as mapped. In addition to being perceived, it can be actively imagined. In every case, it offers something scenographic to us; but it also calls for interpretation and under-standing: Constable said that "we see nothing until we truly understand it."[57] Con-stable's words apply above all to landscape as experienced and sensed. Precisely because of its deep-seated ambiguity, we need to scrutinize landscape continually—again and again, as Husserl would put it, until the "thing itself" is finally grasped. And when we do understand landscape itself, we realize that it is capable of keeping together all of the otherwise divisive terms in which its ambiguity consists.[58]

One such set of terms is that of prose fiction and history. The former is held to be about unrealities, the latter is said to concern real events. Yet both are bound to narrative, thus to sequential time—in contrast with lyric or episodic poetry, in which the poet's in-tense thought or feeling at a given moment is at play. Beyond their shared dependence on narrative temporality (history being more fully dependent on such temporality, given its aim of presenting a seamless web of happenings), history and prose fiction also rely on at least a tacit reference to landscape, which provides the larger scene in which narrated events, historical or imagined, take place. We witnessed such a reference to an imaginary landscape in the striking instance of Moominpapa's Island—a fictitious entity that was nevertheless given quasi-cartographic representation by the editors of Jansson's novel. A given landscape, whether actually experienced or entirely imagined, *holds narrated events together,* furnishing for them a common matrix of interconnected places.

Another ambiguous commixture concerns the combination of history and space set forth by a given landscape. We have seen that Kant's effort to hold these categories apart in his discussion of geography foundered, leading to a recognition on his and (still more so) Alexander von Humboldt's part that events are inseparable from places—that time and space are intertwined in a truly comprehensive geography. *It is in landscape that this intertwining of place and historical event is most intimately and completely realized.*

Not only is each landscape that we experience and represent indelibly temporal in the ultimate but abstract sense of belonging to geological and cosmological orders of time (which entail their own modes of narration). A given landscape is also concretely historical. As W. G. Hoskins has pointed out, each generation can be said both to inherit and to change the landscape in which its fellow members live.[59] Further, a particular landscape bears within itself the marks of historical life: "an eighteenth-century hedge-row, itself now incorporating nineteenth-century trees, twentieth-century barbed wire, and, sometimes, a Victorian bedstead, may [in turn] reflect a boundary line two thousand or more years old."[60] Looking into the refuge of the countryside from the prospect of his study, Hoskins remarks that "one is reaching back, in a view embracing a few hundred acres at the most, through ten centuries of English life, and discerning shadowy depths beyond that again."[61] These historical depths rejoin, even as they extend, the earth's depths that we have found to be so essential to the constitution of the perceived and remembered, as well as the variously represented, landscape.[62]

But if there is indeed an inextricable commingling of history and landscape, such that we must acknowledge their indissoluble ambiguity, it is no less true that even here landscape is a primus inter pares. As Hoskins himself says, "all studies of the past, in fact, draw their evidence from three primary terms—documents, archeology, and the landscape."[63] Of these three sources, landscape is the most fundamental: documents and archaeological research finally refer back to its vast holdings, its unending resources, its archaic archive. For landscape is the world-text on which the much more discrete texts of history—texts preserved in written or printed documents—are inscribed; and it is within landscape as well that the remains of past periods of dwelling on the earth, the traces sought by archaeologists, are to be found.

These texts and traces are *found in place*: the place of landscape itself. A given landscape retains and presents the evidences of history that come to enter its generous embrace; more exactly, it both withholds these evidences and renders them visible. The writing of history, like the drawing of maps and the creation of paintings, would not be possible without the implacement provided by landscape: all three re-implace what is already found in the first figuration, the primary perception, of landscape. Indeed, the happening of history itself would not be possible without landscapes in which to occur. Every event happens in a landscape that straddles the uneasy boundary between earth and world, thereby engendering history as their middle term.

The intertwining of the earth and its multifarious worlds, including the worlds of painting and mapmaking as well as those of fiction and history, is the accomplishment of the places afforded by landscape, that is to say, by placescapes. To enter into history or fiction and, all the more so, such forms of re-presentational display as paintings and maps provide is certainly to be in time. But it is also, and much more tellingly, to be in place.

Notes

PROLOGUE

1. "I do not know much about gods; but I think that the river / Is a strong brown god—sullen, untamed, and intractable" (T. S. Eliot, opening lines of "The Dry Salvages," *Four Quartets*).

2. Maurice Merleau-Ponty, *The Visible and the Invisible,* trans. Alfonso Lingis (Evanston, Ill.: Northwestern University Press, 1968), p. 253.

3. On "the imaginative experience of certain complicated muscular movements," see R. G. Collingwood, *Principles of Art* (Oxford: Clarendon Press, 1938), pp. 147ff.

4. On the Dymaxion Sky-Ocean World Map, see R. Buckminster Fuller, *Critical Path* (New York: St. Martin's Press, 1981), pp. 163–71, esp. Figure 31 on p. 169. Fuller writes that viewers of such a map—which can be projected onto a three-dimensional model of the earth in the form of a geodesic dome with a diameter of 100 feet—witnessed "the whole of the Earth's surface simultaneously without any visible distortion of the relative size and shape of the land and sea masses" (p. 168).

5. Svetlana Alpers, *The Art of Describing: Dutch Art in the Seventeenth Century* (Chicago: University of Chicago Press, 1983), p. 124.

6. See Michael Aris, *Views of Medieval Bhutan: The Diary and Drawings of Samuel Davis (1783)* (Washington, D.C.: Smithsonian Institution Press, 1982). Davis, like his compatriot Turner, conceived it as his bound duty to transmit to his fellow countrymen precise topographical representations of a foreign land. Still another Turner, Samuel Turner, published a work in 1798 in the same tradition: *An Account of an Embassy to the Court of the Teshoo Lama in Tibet.* (This latter is discussed in Aris's book, pp. 26ff.)

7. For a very different view of the necessity of representation—one that emphasizes its biological origins—see Ellen Dissanayake, *What Is Art For?* (Seattle: University of Washington Press, 1998), esp. chapters 1 and 5.

8. See Denis Cosgrove, *Social Formation and Symbolic Landscape* (London: Croom Helm, 1984); Ann Bermingham, *Ideology and Landscape: The English Rustic Tradition, 1740–1860* (Berkeley: University of California Press, 1986); Linda Nochlin, *The Politics of Vision: Essays on Nineteenth-Century Art and Society* (New York: Harper & Row, 1989); and David C. Miller, ed., *American Iconology: New Approaches to Nineteenth-Century Art and Literature* (New Haven: Yale University Press, 1993).

9. See Edward S. Casey, *Getting Back into Place: Toward a Renewed Understanding of the Place-World* (Bloomington: Indiana University Press, 1993), pp. 150–55.

10. See Edward S. Casey, "The Ghost of Embodiment: Is the Body a Natural or a Cultural Entity?" in Donn Welton, ed., *Body and Flesh* (Oxford: Blackwell, 1998).

11. See Edward S. Casey, *Getting Back into Place* and *The Fate of Place: A Philosophical History* (Berkeley: University of California Press, 1997). A fourth volume, dealing with the confluence of mapping and painting in the work of certain twentieth-century artists, is in preparation. A fifth and final volume, bearing on senses of place in non-Western cultures, is projected.

1. FROM LANDSKIP TO LANDSCAPE

1. Kenneth Clark argues for the priority of the *Hours of Turin*, "the first modern landscape," in his *Landscape into Art* (New York: Harper & Row, 1976), pp. 33ff. Constable's claim concerning Titian is contained in a lecture on the history of landscape painting that he delivered in June 1833, and which is reprinted in *Memoirs of the Life of John Constable*, ed. C. R. Leslie (Oxford: Phaidon, 1980), pp. 293–94. The date of composition of Titian's painting is currently regarded as closer to 1530.

2. Clark, *Landscape into Art*, p. 229. What Clark asserts here regarding the seventeenth century (he has Dutch landscape mainly in mind) is not inconsistent with his claim concerning the van Eyck masterpiece of the early fifteenth century: the latter is quite exceptional, being without precedent or successor, and in any case does not allow us to "systematize the rules."

3. Constable, *Memoirs of the Life of John Constable*, p. 303. The full statement is: "Landscape is the child of history, and though at first inseparable from the parent, yet in time it went alone, and at a later period (to continue the figure), when history showed signs of decrepitude, the child may be seen supporting the parent."

4. On this topic, see Osvald Sirén, *The Chinese on the Art of Painting* (New York: Schocken, 1963), as well as L. Sickman and A. Soper, *The Art and Architecture of China* (Harmondsworth, England: Pelican, 1956). For a more complete statement, see chapter 5 in this volume.

5. Petrach was "the first man to climb a mountain for its own sake, and to enjoy the view from the top" (Clark, *Landscape into Art*, p. 10). Leonardo followed suit by climbing Mount Monboso.

6. The (false) premise that actual experience of landscape must precede its representation in art is stated lucidly by Barbara Novak: "only the man practised in reading nature's text could appreciate paintings dealing with that experience" (*Nature and Culture: American Landscape and Painting, 1825–1875* [Oxford: Oxford University Press, 1980], p. 20). The same premise is just as false as applied to artists themselves. Doubtless, *some* experience of nature is required for the painting of landscape. But this experience need not be of any *particular* part of nature, not even in the case of a given landscape painting of that part. Certainly no comprehensive experience is required: Willem de Kooning found all that he needed for creating an entire series of paintings was a few furtive *glances* at the Long Island landscape as viewed from his speeding car. (See his "Sketchbook I: Three Americans.")

7. Clark, *Landscape into Art*, p. 231.

8. Constable, contra Clark, holds that landscape and imitative painting arose hand in hand: "It was, however, at Venice, the heart of colour, and *where the true art of imitation was first understood*, that landscape assumed a rank and decision of character that spread future excellence through all the schools of Europe" (*Memoirs*, p. 293; my italics). From this claim, Constable's view of Titian as the first great European landscapist follows as a matter of course. Any complete assessment would have to include a treatment of Giorgione as well.

9. On the notion of "detotalized totality," see Jean-Paul Sartre, *Being and Nothingness: An Essay on Phenomenological Ontology*, trans. Hazel E. Barnes (New York: Washington Square Press, 1966), p. 563: "the first phenomenon of being in the world is the original relation between the

totality of the in-itself or world and my own totality detotalized." On "the encompassing" *(das Umgreifend),* see Karl Jaspers, *Von der Wahrheit* (Munich: Piper, 1947), pp. 53ff., 85ff., 138ff., 624ff. It is evident that both notions owe much to the earlier idea of a perceptual gestalt that cannot be reduced to the mere addition of its component parts.

10. See J. J. Gibson, *The Ecological Approach to Visual Perception* (Hillsdale, N.J.: Erlbaum, 1986).

11. I say "panperceptual" and not "paraperceptual," because I have elsewhere used the latter term to refer to the various ways in which imagining supplements perceiving (see *Imagining: A Phenomenological Study,* 2d ed. [Bloomington: Indiana University Press, 2000], p. 140). Clark hints at the idea of panperceptuality when he remarks that "landscape painting depends so much on the unconscious response of *man's whole being to the world which surrounds him*" (*Landscape into Art,* p. 232; my italics). "Man's whole being" I am construing as the totality of panperception, and "the world which surrounds him" I take as landscape qua encompassing detotalized totality.

12. Clark, *Landscape into Art,* p. 229.

13. For further discussion of the frame of painting, see Interlude, "Material Conditions of Representing Place in Landscape Painting" (in this volume).

14. Is this why Cézanne was convinced that "a picture must exist as a design of flat patterns even before it creates an illusion of depth" (Clark, *Landscape into Art,* p. 216)? If so, this suggests that the painter should attend above all to the peculiar properties of the picture plane—to a painting's "frontality" (ibid.)—instead of trying to capture *directly* the dimension of depth, which refuses any straightforward encapsulment.

15. On the genre of panoramic painting, see John Francis McDermott, *The Lost Panoramas of the Mississippi* (Chicago: University of Chicago Press, 1958); Barbara Novak's account in *Nature and Culture,* pp. 20–27; and, most recently, Stetan Oetermann, *Panorama: History of Mass Medium* (New York: Zone, 1997).

16. Cited in David Robertson, *West of Eden: A History of the Art and Literature of Yosemite* (Yosemite National Park: Yosemite Natural History Association and Wilderness Press, 1984), p. 11. Claveau's work, like that of Lewis, has been lost, making it all the more difficult for us to imagine the powerful impact of such efforts at achieving an encompassing view of landscape.

17. Cited in Novak in *Nature and Culture,* p. 24, from James Jackson Jarves's *The Art-Idea* (Boston: Hurd and Houghton, 1864), p. 205.

18. "Thus the public was already experiencing a kind of 'motion art' with Cole's cycles, albeit the spectators were the ones in motion [i.e., in contrast with the panoramas presented in theaters]. The kinetic or cinematic aspects of such art had of course a popular counterpart in the panorama, which, however, permitted the public to stand still while the canvas unrolled. The overlap between Cole's serious cycles, which represented, for him at least, his most profound philosophical thought, and the popular art of the panorama is an important juncture of the high art of history painting, appreciated by an intellectual elite, and public or popular art" (Novak, *Nature and Culture,* p. 20).

19. On the panoramic aspect of arcades, especially those of Paris, see Walter Benjamin, *The Arcades Project,* trans. H. Eiland and K. McLaughlin (Cambridge: Harvard University Press, 1999), pp. 31–61, 527–36, 871–84.

20. Cited in Novak, *Nature and Culture,* p. 71, from Alexander von Humboldt's *Cosmos,* trans. E. C. Otté (New York: Harper & Brothers, 1850), vol. 2, p. 98.

21. Novak, *Nature and Culture,* p. 71.

22. "The gradual unrolling of a panorama took several hours, generally affording the audience only a glimpse of each of the many sites and objects depicted. As an experience, it was cumulative and extended in time" (ibid., p. 23).

23. Cited in ibid., p. 27.

24. Ibid., p. 23.

25. Joseph S. Czestochowski, *The American Landscape Tradition: A Study and Gallery of Paintings* (New York: Dutton, 1982), p. 5.

26. Cited in Wolfgang Born, *American Landscape Painting: An Interpretation* (New Haven: Yale University Press, 1948), p. 78; quoted by Czestochowski, *The American Landscape Tradition,* pp. 14–15.

27. "Even those works of Bierstadt's that can be identified with a specific geographical location often have erroneous titles. 'Island Lake, Wind River Range, Wyoming' was originally identified as a scene in the Wabash Mountains owing to the artist's confusion about western geography" (Patricia Trenton, "Purple Mountains' Majesty," *Portfolio* [July–August 1983]: 32). In the nineteenth century, the term *portrait* was applied very broadly to other than human subject matter (ships, mountains, etc.).

28. Cited in Robertson, *West of Eden,* p. 23, from Twain's review in the *San Francisco Alta,* August 4, 1867.

29. Robertson, *West of Eden,* p. 27. This comment is all the more revealing in that only a few pages earlier Robertson had defended Bierstadt against Twain's charges: "But why, we should ask, is a 'glorified atmosphere' out of place in a painting of Yosemite? Why should Yosemite on Bierstadt's canvas resemble as closely as possible the way it looks to the eye? The answer is, it need not" (p. 23).

30. I take up Eve Ingalls's work in more detail in a forthcoming monograph on contemporary painting regarded as a form of mapping.

31. Clark, *Landscape into Art,* p. 65.

32. Ibid., p. 105. Clark maintains that this degeneration applies more generally to much painting of the eighteenth century, "that winter of the imagination" (ibid.).

33. *American Paintings,* a guide published by the Metropolitan Museum of Art (New York, 1962), p. 3.

34. On landskips in this early sense, see Nina Fletcher Little, *American Decorative Wall Painting, 1790–1850* (Sturbridge, Mass.: Old Sturbridge Village, 1952). The word *landskip* is of Dutch origin and derives in turn from *land* and *skepen* (to create); *landschap* was the standard Dutch word for landscape from the sixteenth century onward. On the landskip versus landscape distinction, see J. B. Jackson, "The Meanings of 'Landscape,'" *Kulturgeographi* 88 (1965): 47–50; Kenneth J. Myers, "On the Cultural Construction of Landscape Experience: Contact to 1830," in David C. Miller, ed., *American Iconology: New Approaches to Nineteenth-Century Art and Literature* (New Haven: Yale University Press, 1993), pp. 58–79. Myers points out (pp. 70–71) that *landskip* was used for maps and verbal descriptions as well as paintings and decorative panels.

35. From the placard at the Metropolitan Museum of Art, where the Marmion Room has been transported and reconstructed in the American wing.

36. An excellent example of such a chair, made in Boston in 1820–25, sports a riverscape on its crest panel and a writing tablet on its arm. It is in the collection of early American furniture at the Metropolitan Museum of Art.

37. On the close connection between these two terms, via a shared "pictoriality," see John Barrell, *The Idea of Landscape and the Sense of Place, 1730–1840* (Cambridge: Cambridge University Press, 1972), pp. 1–3.

38. Czestochowski, *The American Landscape Tradition,* p. 5.

39. I speak here of the "decorative"—not the *ornamental,* which is integral to an art work to begin with. For this distinction, and for its full ramifications in art and architecture, see Kent

Bloomer, *The Nature of Ornament: Rhythm and Metamorphosis in Architecture* (New York: Norton, 2000).

40. For a discussion of the more general significance of site, see Edward S. Casey, *Getting Back into Place: Toward a Renewed Understanding of the Place-World* (Bloomington: Indiana University Press, 1993), pp. 258–70, 302–3. Concerning its philosophical antecedents, see also Edward S. Casey, *The Fate of Place: A Philosophical History* (Berkeley: University of California Press, 1997), pp. 183–84, 201, 204, 299–302.

41. This is to leave aside the distinct likelihood that cartography in its modern forms *preceded* the constitution of nature as an extended space of sites, for example, in the exemplary instance of Cartesian *res extensa*.

42. See note 26. I here underline "and."

43. Robertson, *West of Eden*, p. 21.

44. Cited in ibid., p. 23.

45. Ludwig Wittgenstein, *Zettel*, ed. G. E. M. Anscombe and G. H. von Wright (Berkeley: University of California Press, 1967), sec. 681. Concerning surface as a nonobjective place of "singularities" (and not mere "individuals"), see Gilles Deleuze, *The Logic of Sense*, trans. M. Lester with C. Stivale, ed. C. V. Boundas (New York: Columbia University Press, 1990), pp. 103–4. Deleuze also remarks that "what is most deep is the skin . . . [an event is] all the more profound since it occurs at the surface" (p. 10).

46. Arthur Danto writes that "a representation is something that stands in the place of something else, as our representatives in Congress stand proxy for ourselves" (*The Transfiguration of the Commonplace: A Philosophy of Art* [Cambridge: Harvard University Press, 1981], p. 19).

47. I am not forgetting those legendary cases (such as that of Harnett) in which viewers have been alleged to take a painting—or a movie—for "the real thing." It is striking, however, that these cases often occur at the dawn of an art form: for example, the Athenians who mistook Zeuxis's grapes for real grapes, or the early audiences who thought that the oncoming steam engine in a movie was going to crush them. The fact that these occurrences arose during a first "naive" experience suggests precisely that they are not normative for subsequent experiences of the same representation. I suspect, moreover, that they are often posited in retrospect as mythical events that reflect one's sense of what one *wishes* or *imagines* had been the case—where "wishes" and "imagines" connote a desire to trespass limits of representation that are not in fact trespassable.

48. Both Peirce and de Saussure define a sign as something that stands for something else. The same basic definition is found in Husserl's treatment of signs in the first of his *Logical Investigations*. Despite this striking convergence of initial conception on the part of the three pioneers of modern semiology, later researchers rarely focus on the basic relation of standing for.

49. These forms of landscape are described in succession in the first five chapters of Clark's *Landscape into Art*.

50. The interplay between experience and concept of experience is the overriding theme of E. H. Gombrich's *Art and Illusion*, which explores the nuanced relationship between "making" and "matching" in art. See E. H. Gombrich, *Art and Illusion: A Study in the Psychology of Pictorial Representation* (Princeton, N. J.: Princeton University Press, 1960), pp. 29, 73, 186–89, 356–58.

51. "Taking the place of" is not to be confused with "standing in place of," much less "standing in for": the former entails the strict replacement of subject matter by its surrogate, the latter do not.

52. A notice in the *National Advocate* of April 21, 1818, says that "preparations for the rotunda [to be erected] by Mr. Vanderlyn, for panorama views, have commenced . . . [Vanderlyn has

decided] to devote his time to the more humble, though more profitable, pursuit of painting cities and landscapes . . . [P]anorama views of our battles, such as Chippewa, Erie, New Orleans, Lake Champlain, and so forth . . . would not only be highly national and popular, but exceedingly profitable" (cited in John McCoubrey, ed., *American Art, 1700–1960* [Englewood Cliffs, N.J., Prentice Hall, 1965], p. 44).

53. Clark says of Vermeer's *View of Delft* that "This unique work is certainly the nearest which painting has ever come to a coloured photograph" (*Landscape into Art*, p. 65).

54. I refer in this sentence to Plato's distinction in *The Sophist* between *eikastikē* and *phantastikē*: in the exact *eikon* a faithful replica is produced that is the same as its original in size as well as proportion—thus can take its place—whereas a "phantastic" representation allows for liberties in size and proportion. Thus, a *re*presentation as I here use the term is intrinsically "eikastic" in character, or at least aspires to being as exact an *eikon* as possible.

55. On Thomas Hill and William Keith as painters of Yosemite Valley, see Robertson, *West of Eden,* pp. 27–34.

56. I take the terms in quotes approximately in the sense of Martin Heidegger's 1935 essay, "The Origin of the Work of Art," in *Poetry Language Thought,* trans. Albert Hofstadter (New York: Harper & Row, 1971), pp. 17–87.

2. FINDING PLACE FOR THE ELEMENTAL

1. Peter Palmquist, "Views to Order," *Portfolio* (March–April 1983): 87.

2. David Robertson, *West of Eden: A History of the Art and Literature of Yosemite* (Yosemite National Park: Yosemite Natural History Association and Wilderness Press, 1984), p. 62.

3. For an analysis of the effects of frames on paintings, see the Interlude, "Material Conditions of Representing Place in Landscape Painting" (in this volume).

4. For an analysis of this effect, see Robertson, *West of Eden,* p. 65.

5. Ibid.

6. Ibid., p. 73.

7. Asher B. Durand, "Letters on Landscape Painting" (1855), as cited by Joseph S. Czestochowski, *The American Landscape Tradition: A Study and Gallery of Paintings* (New York: Dutton, 1982), p. 88; Durand's italics.

8. Husserl speaks of "phantom things" *(Phantomdinge)* in his later writings. These are positive epistemic presences that are irreducible either to mental images on the one hand, or to physical things *(Körper)* on the other.

9. The drawing measures 8½ × 25½ inches. The first painting, usually referred to as *Norman's Woe, Gloucester,* is 28 × 50 inches; the second painting, titled *The Western Shore with Norman's Woe,* has dimensions of 21½ × 35¼ inches.

10. Lisa F. Andrus, "Design and Measurement in Luminist Art," in John Wilmerding, ed., *American Light: The Luminist Movement, 1850–1875* (New York: Harper & Row; National Gallery of Art, 1980), p. 40. I would question, however, the attribution of "uninspired" to the first phase of Lane's career. A close look at his lithographic productions suggests otherwise. The word *Luminism* does not designate an official school or movement; it is a term of convenience that collects together painters who, despite manifest differences of technique, exhibit a commitment to conveying what I shall call the "contemplative sublime."

11. Lane's drawings were not only exact as to a specific location; they were also comprehensive with respect to the entire shoreline of Gloucester: "from his surviving drawings it is apparent that he knew intimately the details of the entire Cape Ann shoreline. Between 1850 and his death

in 1865 he made drawings of every prominent point and cove, offshore island or intown building. Sometimes he did several sketches of the same spot, changing his viewpoint slightly" (John Wilmerding, *Fitz Hugh Lane* [New York: Praeger, 1971], p. 46).

12. Andrus, "Design and Measurement in Luminist Art," p. 40.

13. On this question, see Barbara Novak, *Nature and Culture: American Landscape and Painting, 1825–1875* (Oxford: Oxford University Press, 1980), pp. 28–29 and 231–38.

14. Andrus, "Design and Measurement in Luminist Art," p. 40. For the close link between Luminism and the panorama, see Novak, *Nature and Culture*, p. 29: "Significantly, the luminist artists duplicated the horizontal extensions of the panorama in their pictures' proportions. I say significantly because I am suggesting that they had a profound understanding of the structural means whereby the popular panorama could be transformed back into high art."

15. Andrus, "Design and Measurement in Luminist Art," p. 40.

16. On the grid of recession, see J. J. Gibson, *The Perception of the Visual World* (Boston: Houghton Mifflin, 1950). As Novak comments, "mensurational control of the parts between the planes . . . is often aided by subtle horizontal alignments" (*Nature and Culture*, p. 23). An even more extreme instance of such horizontal alignments, leading to a still more radical recession into depth, is found in Martin John Heade's studies of Newbury Marsh. For example, his *Sunset on the Newbury Marshes*, painted the same year (1862) as that in which Lane painted his two versions of *Norman's Woe*, is paradigmatic in this respect: "The lines of recession are established by discreet placements of haystacks that recede into the distance with dramatic and compelling effect . . . In no other American landscapes does spatial recession play such an important role or is it developed with such careful geometric precision" (Earl A. Powell, "Luminism and the American Sublime," in Wilmerding, *American Light*, p. 83).

17. Andrus, "Design and Measurement in Luminist Art," p. 40.

18. Ibid., p. 42; my italics.

19. On the critical difference between identity and sameness, see Martin Heidegger's essay "Identität und Differenz." The same theme is developed by Jacques Derrida in his early essay "Différance," in *Margins of Philosophy*, trans. A. Bass (Chicago: University of Chicago Press, 1982), pp. 3–27.

20. Andrus, "Design and Measurement in Luminist Art," p. 40.

21. Wilmerding, *Fitz Hugh Lane*, p. 31. Wilmerding is in fact describing Lane's last extant lithograph, *Castine from Hospital Island* (1855), but I take his words to apply with uncanny accuracy to the first painted version of *Norman's Woe*.

22. Ibid., p. 35.

23. I owe this last observation to Andrus, "Design and Measurement in Luminist Art," p. 40. Notice that the removal of the ripples on the beach amounts to a re-implacement *within* represented space—as does the changed position of both ships.

24. Wilmerding, *Fitz Hugh Lane*, p. 31.

25. But not to the artist, who has become absent to his own creation, realizing an egoless luminescence. As Novak says with explicit reference to *The Western Shore with Norman's Woe*: "in actual practice, only in luminist quietism does the presence of the artist, his 'labor trail', disappear. Such paintings, in eliminating any reminders of the artist's intermediary presence, remove him even from his role of interpreter" (*Nature and Culture*, p. 44).

26. What Powell says of Heade's *Sunset on the Newbury Marches* holds true as well of *The Western Shore with Norman's Woe*: the foreground factors of the painting "appear to extend uninterrupted and unimpeded beneath the viewer" (Powell, "Luminism and the American Sublime," p. 83).

27. John I. H. Baur, "Early Studies in Light and Air by American Landscape Painters,"

Bulletin, Brooklyn Museum 9 (winter 1948): 7; cited by Barbara Novak, *American Painters of the Nineteenth Century* (New York: Praeger, 1969), p. 115. Baur was the first to identify Luminism as a distinct style of painting and to appreciate Lane's contribution to this style. Concerning the importance of view for Lane, Charles Olson has written: "Lane painted true color, and drew true lines, and [had] 'view' as a principle" ("An 'Enthusiasm,'" *Gloucester Daily Times,* October 9, 1965).

28. As Novak remarks, Luminist paintings such as those of Lane "reach to a mystical oneness above time and outside of space" (*Nature and Culture,* p. 44).

29. Novak, *American Painters of the Nineteenth Century,* p. 117.

30. Ibid., p. 116.

31. "The ideal emanated from the core of the artist's sensibility, and out of that ideal core the real took shape. Actuality quietly encysted a nucleus of abstract idea . . . luminism was the most genuine answer to the demands of the age for a synthesis of the real and the ideal" (ibid., p. 117).

32. See Edward S. Casey, "Wild Places," Part IV of *Getting Back into Place: Toward a Renewed Understanding of the Place-World* (Bloomington: Indiana University Press, 1993), esp. pp. 202–26. The elemental also has deep affinities with Bachelard's examination of the material imagination in poetry (not surprisingly, given the turn toward the poetic that we have observed in the place-for aspects of landscape painting). For Bachelard's treatment, see his series of books on the material imagination: *The Psychoanalysis of Fire, Water and Dreams, The Earth and Reveries of the Will,* and *The Earth and Reveries of Repose.*

33. I say "phenomenal" in order to stress that we are dealing here with felt or sensed factors of the landscape, not its ultimate physical constituents as these are posited in scientific accounts.

34. Karl Jaspers, *Philosophy,* trans. E. B. Ashton (Chicago: University of Chicago Press, 1969), Part I, 105; translation slightly modified.

35. Novak, *Nature and Culture,* p. 76. Novak is here speaking of Church, but her remark applies with equal validity to Lane, Heade, Gifford, Kensett, and other Luminists.

36. Ibid., p. 41. Bachelard, who was also much taken by alchemical transformation, would have said that light allows us to imagine matter more sensitively.

37. Andrus, "Design and Measurement in Luminist Art," p. 40.

38. Novak, *Nature and Culture,* p. 42.

39. Ibid., pp. 41–42. Novak remarks rightly that with Church "light is more closely attached to what we generally call atmosphere" (41).

40. Ibid., p. 42.

41. Ibid., p. 43.

42. Ibid., p. 41.

43. It is striking that Claude Lorraine's magisterial landscape *The Ford* was painted in 1636, exactly two hundred years before Cole's *The Oxbow.* Cole himself, like a number of other nineteenth-century American painters, studied in Lorraine's still-existing studio in Rome before he returned to America in 1832. From Lorraine he learned the value of undertaking landscape painting on a grand scale and unsullied by extraneous considerations that would subordinate such painting to other concerns.

44. Novak, *Nature and Culture,* p. 40. For a very different reading of the significance of water in nineteenth-century painting, especially that of Niagara Falls, see David Nye, *The American Technological Sublime* (Cambridge: MIT Press, 1994). I owe this reference to Edward Dimendberg.

45. Henry David Thoreau, *Walden* (New York: New American Library, 1942), pp. 128 and 129, respectively; cited in Novak in *Nature and Culture,* p. 41.

46. On the idea of interplace and, more generally, on inhabitation, see Casey, *Getting Back into Place,* Part III, "Built Places," esp. chapter 5, "Two Ways to Dwell."

47. Ralph Waldo Emerson, "Nature," in *Essays: Second Series* (New York: AMS, 1968), p. 171.

48. Speaking of bodies, places, and motions, Husserl writes that "all of that is relative to the earth-basis ark *[die Arche Erdbodens]* and 'earthly globe' . . . the earth-ark *(die Arche Erde)*" (Edmund Husserl, "Foundational Investigations of the Phenomenological Investigations of the Phenomenological Origin of the Spatiality of Nature," trans. F. Kersten, in *Shorter Writings,* ed. Peter McCormick and Frederick A. Elliston [South Bend, Ind.: Notre Dame Press, 1981], p. 228). On the felt immobility of earth, see ibid., pp. 224ff., esp. p. 225: "the earth does not move—perhaps I may even say that it is at rest."

49. Novak is only an apparent exception. She devotes chapter 4 of *Nature and Culture* to "The Geological Timetable: Rocks." But the primary focus of this chapter is on the impact on painters of the scientific conception of the earth, especially its geological conception as this emerged in early-nineteenth-century science. There is very little consideration of "earth" in the felt phenomenological sense I have been developing in my remarks here.

50. On this notion, see Edward S. Casey, *Remembering: A Phenomenological Study* (Bloomington: Indiana University Press, 1987), p. 294.

51. On the primal significance of the "wherein" *(das Worin),* see Martin Heidegger, *Being and Time,* trans. Edward Robinson and John Macquarrie (New York: Harper, 1962), pp. 119 ff.

52. Novak, *Nature and Culture,* p. 29.

53. This is not to say, however, that the significance of brushstrokes is limited to their indication of the artist's ego. They can very well represent certain structures in the natural world, or set up visual pathways for the viewer of the painting, or embody a certain life force (e.g., *ch'i* in Chinese landscape painting). (I owe this cautionary note to Eve Ingalls; conversation, June 28, 1999.)

54. As Novak remarks, "absorbed in contemplation of a world without movement, the spectator is brought into a wordless dialogue with nature, which quickly becomes the monologue of transcendental unity" (*Nature and Culture,* p. 29). It will be noticed that this statement of Novak's tacitly acknowledges both silence ("a wordless dialogue with nature") and the void ("a world without movement").

55. Powell, "Luminism and the American Sublime," p. 71. Powell also says that "Fitz Hugh Lane and Martin Johnson Head, working independently, created luminist pictures that exemplified the transcendentalist fascination with the contemplative sublime" (ibid.).

3. APOCALYPTIC AND CONTEMPLATIVE SUBLIMITY

1. Thomas Cole, "Essay on American Scenery," cited in K. McShine, ed., *The Natural Paradise: Painting in America 1800–1950* (Boston: New York Graphic Society, 1976), p. 77.

2. Edmund Burke, *A Philosophical Enquiry into the Origin of Our Ideas of the Sublime and the Beautiful,* 4th ed. (Dublin: Cotter, 1767), p. 104.

3. Ibid., p. 105.

4. Ibid., p. 222; Burke's italics.

5. Ibid., p. 119. Darkness is treated again on pp. 225–29.

6. Ibid., p. 82. Burke adds that "when we know the full extent of any danger, when we accustom our eyes to it, a great deal of the apprehension vanishes" (ibid.).

7. Ibid., p. 81. On p. 80, Burke says similarly, "whatever therefore is terrible, with regard to sight, is sublime too."

8. I mention "self-preservation" because in Part I, section 6 of his *Enquiry* Burke classifies terror as a passion of self-preservation, adding that "the passions . . . which are conversant about the preservation of the individual, turn chiefly on *pain* and *danger,* and they are the most powerful

of all the passions" (p. 49) The weaker passions, most of which deal with pleasure, are termed "passions of society" by Burke.

9. Ibid.; Burke's italics. Burke adds that "I say the strongest emotion, because I am satisfied [that] the ideas of pain are much more powerful than those which enter on the part of pleasure" (ibid.).

10. Ibid., pp. 79–80; my italics. We are reminded of Emerson's remark that "in every landscape the point of astonishment is the meeting of the sky and the earth" (Ralph Waldo Emerson, "Nature," in *Essays: Second Series* [New York: AMS, 1968], p. 176).

11. Ibid., p. 80.

12. Ibid., p. 84. William Blake implicitly acknowledges Burke's point as it applies to the graphic arts but concludes more radically that "Obscurity is neither the Source of the Sublime nor of anything else" (cited by Morton D. Paley, *The Apocalyptic Sublime* [New Haven: Yale University Press, 1986], p. 100).

13. Burke, *Enquiry*, p. 123.

14. Ibid., p. 118.

15. Ibid.

16. Ibid.; my italics. The unacknowledged premise in this line of thought is that ideas can operate as impressions—that is, "impressions of reflection" in Hume's terminology. Burke's invocation of light in its impressional state contrasts revealingly with Lane's use of light in its lingering and quiescent state.

17. Barbara Novak, *Nature and Culture: American Landscape and Painting, 1825–1875* (Oxford: Oxford University Press, 1980), p. 35: "This older romantic-Gothick sublime endured well into the American nineteenth century." By "romantic-Gothick sublime," Novak means the impetuous and tumultuous sublimity found paradigmatically in the paintings of Salvator Rosa.

18. Maximal circumambience, a criterion I invoked earlier in first describing the sense in which landscape wraps around the viewing subject, cannot, taken by itself alone, account for sublimity. For an ordinary (e.g., pastoral) landscape to become sublime in prospect there must also be present certain extremities of spatial dimension, obscurities of vista, special effects of lighting, and so on.

19. Novak, *Nature and Culture*, p. 36. The version of *Cotopaxi* here under description is that painted in 1862, the last in a series of sketches and paintings of the same mountain done from 1853 onward. It hangs at the Detroit Institute of Arts.

20. *Cotopaxi* has been called a painting of a "natural Armageddon" (David C. Huntington, *The Landscapes of Frederic Edwin Church: Visions of an American Era* [New York: Braziller, 1966], p. 54).

21. Burke, *Enquiry*, p. 93.

22. On the apocalyptic sublime in British art of the late eighteenth and early nineteenth centuries, see Morton D. Paley's thorough study *The Apocalyptic Sublime* (New Haven: Yale University Press, 1986). But I disagree with Paley's judgment that "No such development [i.e., of the apocalyptic sublime] occurred outside of England" (p. 1). On the contrary, nineteenth-century American art was a fertile field for such sublimity.

23. "It is important to remember that the American landscape had long been identified as the site of the promised millennium . . . [T]he American landscape was interpreted as a metaphor for biblical landscape" (Earl A. Powell, "Luminism and the American Sublime," in John Wilmerding, ed., *American Light: The Luminist Movement, 1850–1875* [New York: Harper & Row; the National Gallery of Art, 1980], pp. 90, 93). As Novak puts it, in this vision "nature is both sublime and sanctified" (*Nature and Culture*, p. 38). For further treatment of the apocalyptic

sublime, see ibid., pp. 34–38, where Novak emphasizes as well the democratic and bourgeois—the specifically nationalized—aspects of the apocalyptic sublime as it pervaded American landscape painting of the nineteenth century.

24. Burke, *Enquiry,* p. 91.

25. Ibid., p. 93.

26. Cole, "Essay on American Scenery," p. 77.

27. Ibid.

28. Burke, *Enquiry,* p. 120.

29. Novak, *Nature and Culture,* p. 43; Novak's italics.

30. The latter alternative is suggested in Powell's statement that Church's painting "is a compelling work which combines two aspects of the sublime, the traditional interest in nature as object [i.e., the Burkean sublime] and the transcendental [i.e., Kantian] concern for nature as experience, through color, space, and silence" ("Luminism and the American Sublime," p. 90).

31. Burke, *Enquiry,* p. 103; Burke's italics.

32. See the chapter titled "The Arc of Desolation and the Array of Description," in Edward S. Casey, *Getting Back into Place: Toward a Renewed Understanding of the Place-World* (Bloomington: Indiana University Press, 1993).

33. Burke, *Enquiry,* p. 120.

34. Ibid., p. 119.

35. Powell, "Luminism and the American Sublime," p. 85.

36. For a detailed discussion of such meditation and its consequences, see Paul J. Griffiths, *On Being Mindless: Buddhist Meditation and the Mind-Body Problem* (LaSalle, Ill.: Open Court, 1986).

37. This dichotomous classification is adopted both by Novak and by Powell, despite their own nuanced accounts of significant exceptions in the form of certain paintings by Cole and Church.

38. I counted no less than eight twilight scenes in the exhibition at the Metropolitan Museum titled "American Paradise: The World of the Hudson River School" (fall 1987). Two of these scenes, including the original of *Twilight in the Wilderness,* were by Church.

39. Burke, *Enquiry,* p. 133.

40. Immanuel Kant, *The Critique of Judgment,* trans. J. C. Meredith (Oxford: Clarendon Press, 1952), p. 90.

41. Ibid; Kant's italics.

42. Ibid., p. 94; Kant's italics.

43. Ibid., p. 97.

44. Burke, *Enquiry,* p. 104.

45. Ibid., p. 106. The phrase "delightful horror" is not Burke's alone. It also occurs in John Dennis's account of his trip to the Alps in 1688. (See Louis Hawes, *Presences of Nature: British Landscape 1780–1830* [New Haven: Yale Center for British Art, 1982], pp. 2–3.) And it reappears, in slightly altered form, in Freud's account of the Rat Man's perverse pleasure in recounting the horror of a certain Chinese technique of torture.

46. Kant, *Critique of Judgment,* p. 109.

47. "The sublime [as absolutely large] is that, the mere capacity of thinking which, evidences a faculty of mind transcending every standard of sense" (ibid., p. 98).

48. Ibid., p. 97.

49. Ibid., p. 104.

50. The sublime "cannot be contained in any sensuous form, but rather concerns ideas of

reason" (ibid., p. 92). These ideas, although never adequately presentable in sense experience, "may be excited and called into the mind by that very inadequacy itself" (ibid.).

51. On this accord, see ibid., p. 104.

52. Ibid., p. 105.

53. Ibid.

54. Ibid., p. 94.

55. Ibid.

56. Ibid., p. 109. Kant italicizes "dynamically sublime."

57. Ibid., p. 115.

58. Ibid., pp. 110–11; my italics.

59. Ibid., p. 160. On the distinction between being threatened in a *fearful* situation and the actual *fear* we might feel, see ibid., pp. 120–21.

60. The phrase "pre-eminence over nature" occurs at ibid., p. 111. The moral character of supersensible freedom, in keeping with Kant's doctrine of practical reason, has the interesting (and controversial) implication that "without the development of moral ideas, that which, thanks to preparatory [moral] culture, we call sublime, merely strikes the untutored man as terrifying" (ibid., p. 115). Thus it is our capacity for moral action that is the ultimate source of resistance to the ostensibly terrifying aspect of nature on which Burke, in other writings such an ardent moralist himself, placed such emphasis.

61. Ibid., p. 114. On this superiority, see also p. 121.

62. Maurice Merleau-Ponty, *Phenomenology of Perception,* trans. C. Smith (New York: Humanities Press, 1962), p. ix: "I am the absolute source."

63. Kant, *Critique of Judgment,* p. 114.

64. Ibid., p. 100.

65. Ibid., p. 92; my italics.

66. Ibid., p. 106.

67. Ibid., p. 92.

68. Ibid., pp. 91–92. Kant adds: "although we may with perfect propriety call many such objects beautiful."

69. Ibid., p. 121.

70. Ibid., p. 91.

71. Concerning interest in mensuration, see Barbara Novak, "On Defining Luminism," in Wilmerding, *American Light,* pp. 23–28.

72. On passions of self-preservation (versus passions of society), see Burke, *Enquiry,* Part I, section 6 (pp. 48–49).

73. See ibid., pp. 79–80. Kant treats respect in *Critique of Judgment,* pp. 96 and 105, and he expressly contrasts astonishment with admiration at p. 125.

74. Kant, *Critique of Judgment,* p. 100; my italics.

75. Henri Bergson, *Matter and Memory,* trans. N. M. Paul and W. S. Palmer (New York: Zone, 1988), pp. 186 and 220. Bergson systematically contrasts concrete "extensity" *(étendue)* with abstract, homogeneous "extension" *(extension)* in the last part of chapter 4 (pp. 211ff.).

76. See ibid., pp. 202–3.

77. Both phrases are from Freud, *The Ego and the Id,* trans. J. Strachey, in *The Standard Edition of the Complete Psychological Works of Sigmund Freud* (London: Hogarth Press, 1967), vol. 19, chapter 2.

78. I have treated the question of the complex implacement of the sublime in my essay,

"The Place of the Sublime," in A.-T. Tymieniecka, ed., *Analecta Husserliana* (Amersterdam: Kluwer, 1997), pp. 71–85.

79. From a letter of 1821; cited by Claude Marks in a lecture on landscape painting at the Metropolitan Museum of Art (November 4, 1987).

80. Kant, *Critique of Judgment*, p. 97.

81. Ibid., p. 93: "for the beautiful in nature we must seek a ground external to ourselves . . . [while] for the sublime [we must seek a ground] merely in ourselves."

82. Ibid., p. 92. I have changed "power" to "might" as a translation of *Macht*.

83. Ibid., p. 93. I have changed "gives a representation" to "yields a representation." The German is *"vorgestellt . . . macht."*

84. Ibid. Indeed, for Kant the sublime is in the end only an "appendix" *(Anhang)* in a treatise on aesthetic judgment.

85. This is Kant's somewhat contemptuous word for Burke's approach to the sublime and the beautiful, despite the fact that he is "the foremost author in this way of treating the subject": see ibid., "General Comment on the Exposition of Aesthetic Reflective Judgments."

86. On "le préjugé du monde," see Merleau-Ponty, *Phenomenology of Perception,* pp. 5ff.

87. It is a matter of overlooking a basis in a subjectivity that itself has crucial social and cultural determinants. Kenneth John Myers, remarking that nineteenth-century American attitudes toward landscape promote "the culturally powerful fiction that landscape appreciation is a natural ability available to all uncorrupted men and women," suggests that "the mental practices necessary to the objectification of particular environments as picturesque landscapes became so commonplace among leading elements of the northeastern elites that they were reconceptualized as natural rather than learned abilities" (Kenneth John Myers, "On the Cultural Construction of Landscape Experience: Contact to 1830," in David C. Miller, ed. *American Iconology: New Approaches to Nineteenth-Century Art and Literature* [New Haven: Yale University Press, 1993], p. 59).

88. Kant, *Critique of Judgment*, p. 92.

89. Ibid.

90. Ibid.

91. "The feeling of the sublime involves as its characteristic feature a mental *movement* combined with the estimate of the object" (ibid., p. 94; Kant's italics). Cf. also p. 107: "the mind feels itself *set in motion* [*Bewegung:* also translatable as "agitation"] in the representation of the sublime in nature; whereas in the aesthetic judgment upon what is beautiful therein it is in *restful* contemplation" (Kant's italics).

92. See Aristotle, *De Anima*, book 3, chapter 3, chapters 7 and 8.

93. Kant's word for "mind" is *das Gemüt,* an ancient German word that has connotations of heart and feeling and memory as well as thought: and all of these as loosely assembled in one place, as the collective *Ge-* prefix indicates. The German equivalent of what I have been calling "psyche"—the root of my neologism "psychotopics"—would be *Seele,* conventionally translated as "soul."

94. On "purposiveness without purpose" *(Zweckmässigkeit ohne Zweck),* see Kant, *Critique of Judgment,* sections 10–11.

4. PURSUING THE NATURAL SUBLIME

1. Immanuel Kant, *The Critique of Judgment,* trans. J. C. Meredith (Oxford: Clarendon Press, 1952), p. 94; Kant italicizes "restful" *(ruhig).*

2. For comprehensive treatments of the beautiful/sublime relation, see Walter J. Hipple Jr., *The Beautiful, the Sublime, and the Picturesque in Eighteenth-Century British Aesthetic Theory* (Carbondale: Southern Illinois University Press, 1957); and Samuel H. Monk, *The Sublime: A Study of Critical Theories in XVIIIth-Century England* (Ann Arbor: University of Michigan Press, 1960).

3. This is not to deny that the natural sublime also inheres in the sea of seascape, as we see in the case of Lane and even more conspicuously in the seascapes of Turner. But, for the sake of simplification, I here restrict comment to the land and landscape.

4. Oswaldo Rodriguez Roque, "The Exaltation of American Landscape Painting," in *American Paradise: The World of the Hudson River School* (New York: Metropolitan Museum of Art, 1987), p. 29: "For Cole, the experience of the Italian land contrasted sharply with that of the American wilderness, whose sublime rawness could put him in touch with God."

5. Thomas Cole, "Essay on American Scenery" (lecture of May 9, 1835), reprinted in J. McCoubrey, ed., *American Art, 1700–1960* (Englewood Cliffs, N.J.: Prentice Hall, 1965), p. 100. The identification of God and Nature goes back, of course, to the seventeenth century, when God's attributes of infinity and immensity were considered physical, that is, belonging to Nature, "Deus Sive Natura" in Spinoza's formula. See Edward S. Casey, *The Fate of Place: A Philosophical History* (Berkeley: University of California Press, 1997), chapters 6–8. See also the classic essay by Ernest Tuveson, "Space, Deity and the 'Natural Sublime,'" *Modern Language Quarterly* 12 (1951): 20–38. For its effect on American art in particular, see Barbara Novak, *Nature and Culture: American Landscape and Painting, 1825–1875* (Oxford: Oxford University Press, 1980), chapter 1, esp. p. 3: "In the early nineteenth century in America, nature couldn't do without God, and God apparently couldn't do without nature."

6. Novak, *Nature and Culture*, p. 3.

7. William Cullen Bryant, "Funeral Oration," in McCoubrey, *American Art, 1700–1960*, p. 95. Bryant also remarks that Cole's "mode of treating his subjects was not bounded by the narrow limits of any system; the moral interest he gave them took no set form or predetermined pattern; its manifestations wore the diversity of that creation from which they were drawn" (pp. 96–97).

8. Henry T. Tuckerman, *Sketches of Emiment American Painters* (New York, 1849), p. 80; cited by Roque, "The Exaltation of American Landscape Painting," p. 31.

9. Kant, *Critique of Judgment*, section 59.

10. Quoted by Louis Legrand Noble, *Life and Works of Thomas Cole* (New York, 1853), p. 148.

11. Bryan Jay Wolf, *Romantic Re-Vision: Culture and Consciousness in Nineteenth-Century American Painting and Literature* (Chicago: University of Chicago Press, 1982), p. 178. Wolf underlines the word *American*. Part III of Wolf's book, "Romanticism and the Unconscious: Thomas Cole," offers a remarkable interpretation of Cole's work on which I draw freely in what follows. I am also indebted to Alan Wallach, "Making a Picture from Mount Holyoke," in David C. Miller, ed., *American Iconology: New Approaches to Nineteenth-Century Art and Literature* (New Haven: Yale University Press, 1993), pp. 80–91.

12. Wolf, *Romantic Re-Vision*, p. 177.

13. Ibid., p. 229; my italics. However, when Wolf claims that "Cole subordinates historical narrative to the experience of the landscape" (p. 228), I think that he needlessly dichotomizes the situation. Part of Cole's genius is to offer to us narration *and* spatial structure in a single coherent mixture—from which, nevertheless, we can prescind purely spatial elements. We can also prescind from the psychological matrix of Cole's work that Wolf makes his own special focus in chapter 4 of *Romantic Re-Vision*—prescind from it in order to concentrate on the character of place in Cole's representation of landscape. This is what I shall undertake in what follows.

14. Roque, *American Paradise*, p. 125. Roque also calls *The Oxbow* (as it is normally referred

to) "one of the established icons of American art" (p. 127). See also Roque's article "*The Oxbow* by Thomas Cole: Iconography of an American Landscape Painting," *Metropolitan Museum Journal* 17 (1984): 63–73.

15. In fact, Cole himself remarked in a letter of March 1836 to his patron Luman Reed that, in undertaking *The Oxbow*, "I thought I would do something that would tell a tale" (cited in Roque, *American Paradise*, p. 127).

16. In any case, the story here intimated is quite banal: to be out from under a thunderstorm is hardly exciting enough to hold our lasting attention. Cole wrote narrative accounts of his sketching trips in his journals—accounts whose details often overlap with the paintings of this early period of his work. But the accounts themselves leave much to be desired from a literary point of view and fall far short of the paintings themselves in terms of sheer drama.

17. Wolf, *Romantic Re-Vision*, p. 228.

18. See Aristotle, *Physics*, book 4, chapters 1–9.

19. The depth in this portion of the picture is not only spatial; it is temporal insofar as the promise of increased prosperity on the land is held out. In his "Essay on American Scenery," Cole stressed the idea that the American landscape invokes associations less to the past than to the future: "Seated on a pleasant knoll, look[ing] down into the bosom of that secluded valley . . . the mind's eye may see far into futurity" (p. 108).

20. For a convincing account of *The Oxbow* as a combination of a panoptic perspective (with all its implications of a powerful Foucauldian super-vision as well as panoramic layout) with a teloscoping of detail—in short, of extensity with intensity—see Alan Wallach, "Making a Picture of the View from Mount Holyoke," in Miller, *American Iconology*, pp. 80–91. Wallach posits a specific "panoptic sublime" as the most distinctive achievement of this celebrated painting: "There is, in other words, a spatial conflict in *The Oxbow* in which contradictory perspectives and modes of seeing, along with other sets of visual and symbolic oppositions—storm and sushine, wilderness and pastoral landscape (the proverbial 'Garden')—produce a type of optical excitement, a cacophony of vision that can be taken as a pictorial equivalent to the exhilaration of the panoptic sublime" (p. 90).

21. As does Wolf too frequently in his otherwise admirable treatment of Cole in *Romantic Re-Vision*, a book that draws heavily, and sometimes clumsily, on Freudian and Lacanian readings of Cole's early Catskill paintings. What I here call the "actively imagining psyche" is equivalent to the term "imaginal psyche" as employed by James Hillman, *Re-Visioning Psychology* (New York: Harper & Row, 1975).

22. "Ye Presences of Nature in the sky / And on the earth! Ye Visions of the hills! / And Souls of lonely places!" (William Wordsworth, *The Prelude*, book 1, lines 464–66).

23. This book appeared as a companion volume of illustrations to Hall's *Travels in North America*, 2 vols. (London, 1829). Cole's tracing, now in the Detroit Institute of Arts (where the second sketch is also found), is analyzed by Roque in *American Paradise*, p. 126. It is also studied by Wallach, "Making a Picture of the View from Mount Holyoke," pp. 85–88, who avers that "Hall's etching demonstrated [to Cole] the possibility of representing what had hitherto seemed unrepresentable" (p. 86). It had seemed "unrepresentable" because no one in the United States had tried to represent such a panoramic view in a single sketch or painting, even though panoramic paintings per se had existed since 1788, when Robert Barker exhibited the first large-scale panorama in London.

24. Cole had begun the practice of using both pages of his sketchbook to expand the angle of vision in sketches he made of the Bay of Naples in 1832; he continued the practice in other sketches he made in the Catskills in 1833. Concerning the significance of this technique—designed

to make panoramic views representable in paintings made from sketches—see Wallach, "Making a Picture of the View from Mount Holyoke," pp. 88–90.

25. Cited from Cole's letter of March 2, 1836, in Roque, *American Paradise,* p. 126.

26. "Cole represents the equivalent of a section of a panorama without the panorama's consistent and apparently seamless transition from point to point within the visual field. Instead, two related vistas are compressed or jammed against each other, so that a viewer scanning the landscape experiences both a feeling of panoramic breadth and a sense of imminent split or breakdown" (Wallach, "Making a Picture of the View from Mount Holyoke," pp. 89–90).

27. Wolf, *Romantic Re-Vision,* p. 177.

28. "Although writers like John Dennis and Lord Shaftesbury had been discussing the sublime, or the sublime and the beautiful, for some years, it was Addison's 'Essay on the Pleasures of the Imagination' which formulated the problems of aesthetics in such a fashion as to initiate that long discussion of beauty and sublimity—and later of the picturesque—which attracted the interest and exercised the talents of philosophers, men of letters, artists, and amateurs until well into the nineteenth century" (Hipple, *The Beautiful, the Sublime, and the Picturesque in Eighteenth-Century British Aesthetic Theory,* p. 13). Hipple also remarks that "the work which, after Addison's essays, was most influential on the course of British aesthetic speculation in the eighteenth century, was Edmund Burke's *A Philosophical Enquiry into the Origin of Our Ideas of the Sublime and the Beautiful*" (ibid., p. 83).

29. I except from this assessment of Cole several works that he produced immediately after his return to the United States in 1832 and that are largely unsuccessful efforts to paint in the Claudian mode: for example, *Landscape Composition, Italian Scenery* (1833; now at the New York Historical Society). Cole never lost touch entirely with his own resources, as is revealed by the fact that in 1833 he also painted the remarkable *View on Catskill Creek* (also at the New York Historical Society), which further explores his own distinctive approach to landscape as it was first realized in his Catskill paintings of 1825–27.

30. On Claude Lorraine as exemplary of the beautiful in painting (and in contrast with Salvator Rosa as the first painter of the sublime), see Monk, *The Sublime,* pp. 194–95.

31. I have in mind such paintings as *Snow Squall, Winter Landscape in the Catskills* (c. 1825–26), *Sunrise in the Catskills* (1826), and especially *Sunny Morning on the Hudson* (1827)—the last of which includes a slowly winding river in the far distance not unlike the Connecticut River in *The Oxbow.* These three paintings of the mid-1820s are the subject of intensive analysis in Wolf's *Romantic Re-Vision,* pp. 185–214.

32. Edmund Burke, *A Philosophical Enquiry into the Origin of Our Ideas of the Sublime and the Beautiful,* 4th ed. (Dublin: Cotter, 1767), p. 174. Cf. also his rhetorical question: "If the qualities of the sublime and beautiful are sometimes found united, does this prove, that they are the same, does it prove that they are any way allied, does it prove even that they are not opposite and contradictory? Black and white may soften, may blend, but they are not therefore the same" (pp. 195–96).

33. Ibid., pp. 181–82.

34. Ibid., p. 172.

35. Kant, *Critique of Judgment,* p. 93.

36. Cole, "Essay on American Scenery," p. 101.

37. Ibid., p. 105.

38. The case of Niagara Falls illustrates the close connection between tourism and landscape painting in the United States after 1820. Concerning this connection—less evident in Cole than in others (though even Cole painted *View of the Catskill Mountain House* [1831], engravings

from which became popular among tourists in the region)—see John F. Sears, *Sacred Places: American Tourist Attractions in the Nineteenth Century* (New York: Oxford University Press, 1989), and especially Elizabeth McKinsey, *Niagara Falls: Icon of the American Sublime* (Cambridge: Cambridge University Press, 1985). The larger issue is that of the rising middle class, whose prosperity allowed it to travel freely to see the natural wonders of the United States and to appreciate their representation in landscape paintings. For the collusion between the new bourgeoisie and landscape painters—both of whom had a vested interest in presuming that the beauty and sublimity of the natural world are intrinsic, if not actually God-given, properties, thereby overlooking "the labor of admiring" and the inculcation of certain social and cultural values—not to mention epistemic models according to which painters could disinterestedly render Nature as it exists in itself, see Kenneth John Myers, "On the Cultural Construction of Landscape Experience: Contact to 1830," in Miller, *American Iconology,* pp. 72–79.

39. Burke, *Enquiry,* p. 84. Milton is cited on p. 104.

40. Kant, *Critique of Judgment,* p. 107; my italics. For further discussion of the abyss, see p. 115.

41. Ibid., p. 121.

42. About the infinity of the sublime, Kant and Burke are in express concord. (Cf. Burke, *Enquiry,* pp. 106–8; Kant, *Critique,* pp. 101–12). For Kant's explicit (and mainly affirmative) treatment of Burke, see *Critique,* pp. 130–33.

43. Cole, "Essay on American Scenery," p. 104.

44. Ibid., p. 103.

45. Ibid.

46. Kant, *Critique of Judgment,* p. 121; my italics.

47. Cole, "Essay on American Scenery," p. 109. Cole makes it clear in his immediately succeeding remarks that he is thinking of just such scenes as he paints in *The Oxbow:* "May we at times turn from the ordinary pursuits of life to the pure enjoyment of rural nature?" The "ignorance and folly" to which human beings fall prey are exemplified in the increasing industrialization lamented so keenly by Cole. On the other hand, Cole is not invoking an entirely bucolic and precultural scene; he engraves Hebrew letters on the central hill in the distance—as if to suggest that even the idylically natural is never wholly independent of the cultural, and that there is a place for civilization of a higher visionary order within the sphere of the natural itself. For bringing these letters to my attention, I am grateful to Eve Ingalls.

48. Kant, *Critique of Judgment,* p. 89. Kant underlines "grasps."

49. Margaret Fuller Ossoli, "A Record of the Impressions Produced by the Exhibition of Mr. Allston's Pictures in the Summer of 1839," reprinted in McCoubrey, *American Art, 1700–1960,* p. 60.

50. Wolf, *Romantic Re-Vision,* p. 214.

51. Ibid., p. 212.

52. On the significance of vaporization, see ibid., pp. 205ff. The early Catskill paintings often feature a vapor trail that plays a comparable physically sublime role.

53. I refer to Carl Gustav Jung's interpretation of alchemy as having both material and psychological significance in his *Psychology and Alchemy,* trans. R. F. C. Hull (London: Routledge & Kegan Paul, 1968).

54. Cole twice uses *sublime* in its active verbal form (i.e., "to sublime") in his Lyceum lecture. See his "Essay on American Scenery," pp. 98 and 110.

55. This is the phrase of Richard Ray in his address of November 17, 1825, to the American Academy of the Fine Arts: "To the Artist . . . our country affords peculiar advantages: I mean the Landscape-painter. For him extends an un-appropriated world, where the glance of genius may descry new combinations of colors, and new varieties of prospect" (cited in Joseph S.

Czestochowski, *The American Landscape Tradition: A Study and Gallery of Paintings* [New York: Dutton, 1982], p. 12).

5. REPRESENTING A REGION

1. See Edmund Husserl, *Ideas: General Introduction to Pure Phenomenology,* trans. W. R. Boyce Gibson (New York: Macmillan, 1931), Part I, chapter 1.

2. For a treatment of the protoregionalization of *chōra* in the *Timaeus,* see chapter 2 of Edward S. Casey, *The Fate of Place: A Philosophical History* (Berkeley: University of California Press, 1997). *Chōrai,* the plural of *chōra,* means "regions" in ancient Greek.

3. For a sociological confirmation of the special satisfaction afforded by landscape painting, showing the spontaneous preference of most Americans for landscape paintings in their homes, see David Halle, *Inside Culture: Art and Class in the American Home* (Chicago: University of Chicago Press, 1986).

4. A show in Paris of painters of the "Midi," that is, the southernmost region of France, in the summer of 1995 was notable for its proliferation of approaches and styles. (The exhibition, titled "Peintres du Midi," was held at the Palais du Luxembourg.)

5. Michael Rosenthal, *Constable: The Painter and His Landscape* (New Haven: Yale University Press, 1983), p. 2. This is not to say, however, that Constable could not be ambivalent toward his own region. Before he met his future bride-to-be, Maria Bicknell, in 1809, he passed through a phase of rebellion against his father (who discouraged him from becoming a painter)—a phase in which he took both psychological and physical distance from the East Bergholt world in which he grew up. As Ann Bermingham remarks, from 1800 to 1809, Constable, "virtually ceased to paint not only his father's property but even (with a few exceptions) the whole East Bergholt landscape . . . [During this time] he engaged in a vigorously critical apprenticeship in his art" (*Landscape and Ideology: The English Rustic Tradition 1740–1860* [Berkeley: University of California Press, 1986], p. 100). Bermingham interprets Constable's resolute recovery of the East Bergholt landscape after 1809 as an act of reappropriation of the paternal world, thanks to the much more benevolent associations he now had with it through his experiences with Maria in the same landscape. By 1811, this landscape "no longer represented a culture that Constable did not want to possess; it embodied a nature that was his own" (p. 105).

6. Rosenthal, *Constable,* p. 2. The fact that Constable began this topographical practice as early as 1808 shows that he was already getting ready to appropriate his native ground just before he met Maria, following the period of ambivalence to which I refer in note 5.

7. Ibid., p. 6; Rosenthal's italics.

8. Both Constable and Cézanne—and, for that matter, Lane—are to be contrasted in this respect with such itinerant seekers of the exotic as Turner, Church, and Bierstadt. For these latter, a brief but intense immersion in one place such as Venice, Cotopaxi, or Yosemite was sufficient to inspire a series of landscape paintings based on an isolated, albeit exalted, moment of epiphany.

9. Rosenthal, *Constable,* p. 6.

10. Letter of May 27, 1812; cited in ibid., p. 49; my italics.

11. According to Freud, "the pertinacity of early impressions" is based in "the preponderance attaching in mental life to memory-traces in comparison with recent [i.e., later] impressions" (*Three Essays on the Theory of Sexuality,* in Sigmund Freud, *The Standard Edition of the Complete Psychological Works of Sigmund Freud,* trans. J. Strachey [London: Hogarth Press, 1953], vol. 7, p. 242).

12. "The subjects of all the plates are from real scenes" (from a draft of an address written

in May 1832, on the occasion of the publication of twenty-two prints on the subject "English Landscape," as reprinted in *John Constable's Discourses,* ed. R. B. Beckett [Suffolk, England: Suffolk Records Society, 1970], vol. 14, p. 84). Similarly, in his Introduction to *English Landscape* Constable says that most of his subjects "are taken from real places, and are meant particularly to characterize the scenery of England" (in ibid., p. 10).

13. In the Introduction to his volume *English Landscape* Constable wrote that "the immediate aim of the Author in this publication is to increase the interest for, and promote the study of, the Rural Scenery of England, with all its endearing associations, its amenities, and even in its simple localities" (in ibid., p. 9). One is tempted to replace "even" with "especially." Note also that Constable is here alluding to a much earlier theme of landscape painting—that of the *locus amoenus,* the "pleasant place." For a detailed treatment of this theme, and "the supersession of the pleasant" in eighteenth- and nineteenth-century painting and maps, see Kenneth John Myers, "On the Cultural Construction of Landscape Experience: Contact to 1830," in David C. Miller, ed., *American Iconology: New Approaches to Nineteenth-Century Art and Literature* (New Haven: Yale University Press, 1993), pp. 60–68.

14. This letter is cited in C. R. Leslie, *Memoirs of the Life of John Constable* (Oxford: Phaidon, 1951), p. 15; Constable's italics. The word *painture* signifies either the action of painting or a particular way of painting. The testimony of this letter is all the more remarkable in view of Constable's ambivalence toward his father during this early period of his career.

15. From the same letter of May 29, 1802.

16. As Rosenthal brings out in chapter 2 ("Early Works") of *Constable.* Claude Lorraine was, in Constable's eyes, "the most perfect of all masters of real chiaroscuro" (from his lectures on landscape painting of 1836; cited in Beckett, *John Constable's Discourses,* p. 63). Constable was also impressed by the work of Rubens, one of whose landscapes was in the collection of Constable's friend and patron George Beaumont. For the relationship between Constable and Gainesborough, see Bermingham, *Landscape and Ideology,* pp. 110–14.

17. For a comprehensive treatment of the picturesque, see Walter J. Hipple Jr., *The Beautiful, the Sublime, and the Picturesque in Eighteenth-Century British Aesthetic Theory* (Carbondale: Southern Illinois University Press, 1957), pp. 192–201; Bermingham, *Landscape and Ideology,* pp. 57–87, and especially the treatment of "the picturesque legacy" in the early work of Constable at pp. 105–114.

18. Rosenthal, *Constable,* p. 21. In fact, Constable could often be "interpreted as having painted a scene [during this period] which reflected the concerns of the farming interest in deliberate opposition to those of the aficionados of the picturesque" (ibid.).

19. Cobbett and Gainesborough, among others, had praised the scenery of the Orwell. The same point applies to the choice of the Stour over the Thames, which was even more sanctified at the time as a paintworthy river.

20. A remark made to Lucas as reported in I. Fleming-Williams, ed., *John Constable: Further Documents and Correspondence* (Ipswich and London: Tate Gallery and Suffolk Records Society, 1975), p. 55. The concern over the hanging of the painting is reported in a letter of Farington's of April 23, 1811: "Constable called, in much uneasiness of mind, having heard that his picture . . . was hung very low in the Anti-room of the Royal Academy. He apprehended that it was proof that he had fallen in the opinion of the members of the Academy" (cited in Rosenthal, *Constable,* p. 57).

21. Constable, cited from a conversation with Farington on November 5, 1814.

22. Rosenthal, *Constable,* p. 56.

23. Letter of 1812 cited in ibid., p. 57.

24. See Figures 53 and 54 in ibid., pp. 52–53.

25. From the section titled "Summer," lines 1408–9 (cited by Rosenthal, *Constable,* p. 54). Constable knew Thomson's poem quite well and quoted it on a number of occasions.

26. Thus I cannot agree with Rosenthal's summary judgment that "With 'Dedham Vale, Morning' [Constable] defined the scene's potential for himself, recognizing the capacity of reality to fit a Claudean pattern, and by *imposing* this [pattern] on his topography revealed that his attitude to landscape had changed and become more sophisticated" (*Constable,* p. 52; Rosenthal's italics). Quite to the contrary: the Claudian pattern is here in radical retreat, being only echoed at best. The "scene's potential" is thereby allowed to come forward on its own terms, not those dictated by the dominant landscape tradition of the previous century and a half.

27. Ibid., p. 5.

28. From the letter of May 29, 1802, to John Dunthorne, cited earlier.

29. Concerning the place-of-the-surface, see the Interlude in this volume.

30. Letter of July 22, 1812 (cited in Rosenthal, *Constable,* p. 60); Rosenthal's italics.

31. Ann Bermingham avers that "the oil sketch came to determine the formal language of Constable's late style, in which the difference between a sketch and a finished exhibition piece is almost nonexistent" (*Landscape and Ideology,* p. 128). But she goes on to point out that nonetheless, in the finished works, Constable "transformed the emotional appearance of the sketch into a more detached, naturalistic description of the scene" (ibid.) The sketch, she adds, "functioned for Constable primarily as a way for him to collect his feelings and associations with regard to a particular subject" (p. 129).

32. Ibid., p. 129. She adds: "Many of [the oil sketches] serve no preparatory purpose, and those that do connect to finished works have no consistent rationale, as they would if they really were *études* . . . This paradox of Constable preparatory oil sketch, the preparatory sketch that appears to do almost no preparatory work, suggests that for Constable the relationship between the sketch and the finished work was never a simple case, as it was for the academic artist, of process and product, of analysis and synthesis, or of finished and unfinished" (ibid.).

33. For an account of these influences, see E. H. Gombrich, *Art and Illusion: A Study in the Psychology of Pictorial Representation* (Princeton, N. J.: Princeton University Press, 1961), pp. 176ff. Elsewhere in the same book, Gombrich discusses the role of preexisting schemata in Constable's painting of landscape, most notably that of *Wivenhoe Park* (1816), ibid. (pp. 376ff.). Gombrich also briefly discusses the pencil drawing *Dedham from Langham* in relation to another pencil drawing of 1813 and one of the oil sketches of 1812: about these three works, he remarks that "the artist . . . cannot transcribe what he sees; he can only translate it into the terms of his medium" (p. 36). In other words, the translation of pencil into oil—and, indeed, of one kind of pencil sketch into another—presents us with another case of free variation.

34. Bermingham hints at this last combination of the sameness of land under a highly mutable sky: from 1820 onward, what Constable himself called "change of weather & effect" became "no less important to the work than its subject; and storms, clouds, rainbows, and other transitory aerial effects came to carry as much meaning as the familiar sites over which they played" (*Landscape and Ideology,* p. 148). More generally, "Constable's art in this period is organized by the contrast between a principle of repetition and a principle of difference, between the sameness of the subjects and the infinite diversity of the conditions under which they are seen" (ibid.).

35. Cf. Constable's admiring remark concerning Jakob van Ruisdael: "he *understood* what he was painting" (cited from Constable's 1836 lectures on landscape in Beckett, *John Constable's Discourses,* p. 64; Constable's italics). Part of this understanding is of the *people* who inhabit a given region, and Constable is no exception. Many of his Stour River paintings, especially after

1819 (when he exhibited *The White Horse* at the Royal Academy), show farmers and millers and other laborers intently at work on the land or river. Prominent as they sometimes are, these figures rarely stand out but instead merge with the larger landscape: "Absorbed in their work, they are also absorbed (as Barrell has noted) into the landscape" (Bermingham, *Landscape and Ideology*, pp. 138–39, with reference to John Barrell, *The Dark Side of the Landscape: The Rural Poor in English Painting, 1730–1840* [Cambridge: Cambridge University Press, 1980]). Whether these absorptive figures merely reflect Constable's conservative political stance—as Barrell claims—or are caught up in "a system of production that they neither control nor see to its end" (Bermingham, *Landscape and Ideology*, p. 142), they are shown in foregrounded actions of silent labor whose heaviness (opening water locks, pulling ferries, gathering hay) rejoins the earth on which they take place. In their explicit and large-scale depiction, such actions contrast with those suggested by Cole in *The Oxbow*, where the few ethereal signs of labor (e.g., a solitary ferry crossing the Connecticut River, thin trails of smoke rising from workers' cottages) are subsumed into the paradisiacal prospect of a productive future hovering above and beyond the land on which labor is effected. The macroscopic, muscular actions of laborers conspicuously located on the land stand in stark contrast with the microscopic traces of workers who have vanished into its midst.

6. REPRESENTING PLACE ELSEWHERE

I wish to thank Professor Samuel C. Morse, Professor of Fine Arts at Amherst College, for his close reading of an earlier draft of this chapter.

    1. Speaking of Han art (202 B.C.–A.D. 220), Alexander Soper writes: "In highly elaborate form in the inlaid bronzes, and more simply in textiles, the silhouette of repeating peaks may be transformed into an extraordinary abstraction . . . This readiness to substitute more congenial—because more abstract—versions of the mountain theme shows the general indifference of Han art to landscape as anything but an element of decoration, with only the slightest connection with the natural world" (Alexander Soper, "Early Chinese Landscape Painting," *Art Bulletin* 23 (1941): 148).

    2. This stands in contrast with the poets of the same period, who were immersed in the description of landscape: for example, Ch'ü Yüan and Sung Yü. The work of these latter, as compiled in the anthology *Elegies of Ch'u*, "reveal a sensitive, even passionate response to nature" (Michael Sullivan, *The Birth of Landscape Painting in China* [Berkeley: University of California Press, 1962], p. 8). Sullivan also remarks on the curious discrepancy between poetry and painting in the Chou era, when "painting lagged behind poetry in its range and power of expression" (p. 24). Notable in this connection is the remarkable essay "Introduction to Painting Landscape" by Tsung Ping (375–443). This text, arguably of Buddhist inspiration, sets forth a doctrine of *kan-lei* or "response to kind" whereby landscape painters are enjoined to provide viewers of paintings with an experience comparable to that of actually being in the midst of sacred mountains. Painters provide icons of the essence *(lei)* of such mountains, inducing a karmic connection between the spirit of the viewer and that of the depicted mountain. Munakata has proposed that certain Han scenes inlaid in gold on bronze tubes embody this doctrine. But it remains that in these scenes landscape, however vividly realized, continues to surround thematic human and animals figures. (See Kiyohiko Munakata, "Concepts of *Lei* and *Kan-lei* in Early Chinese Art Theory," in S. Bush and C. Murck, eds., *Theories of the Arts in China* [Princeton, N. J.: Princeton University Press, 1983], pp. 105–31; with special reference to Figure 2 on p. 126. See also Susan Bush, "Tsung Ping's Essay on Painting Landscape and the 'Landscape Buddhism' of Mount Lu," in ibid., pp. 132–64, esp. her remark that "in [such] texts, as in contemporary scenes of nature, the human image seems to loom over the landscape setting" [p. 152].) Overall, it remains the case that

theory of landscape—theory especially evident in Tsung Ping's essay and others of its kind—was far more advanced than the *practice* of landscape painting and poetry.

3. Susan Bush and Hsio-yen Shih, eds., *Early Chinese Texts on Painting* (Cambridge: Harvard University Press, 1985), p. 18. The authors add: "Hence painting merely signified decoration in the *Lun–yü* (Confucian Analects)" (ibid.).

4. See Sullivan, *The Birth of Landscape Painting in China,* chapter 2, "The Han Dynasty," and Bush and Shih, *Early Chinese Texts on Painting,* p. 19.

5. Sullivan, *The Birth of Landscape Painting in China,* p. 42.

6. Kuo Hsi (after A.D. 1000–1090), "The Significance of Landscape," in Bush and Shih, *Early Chinese Texts on Painting,* p. 151. This is an instance of the advanced state of theory of landscape to which reference was made in note 2.

7. Ibid. Sullivan comments on this text: the person who is "tied to his desk, immersed in the world and its troubles, finds in the contemplation of a landscape painting a refreshment of mind and heart as compelling as though he were to wander among the mountains themselves. The Chinese painter who in his pictures satisfies this longing depicts not merely the outward and visible forms of nature, but the inner life and harmony that pervade them" (*The Birth of Landscape Painting in China,* p. vii).

8. Sullivan, *The Birth of Landscape Painting in China,* p. 2.

9. Ching Hao, *Hua Shan Shui Lu* (essay on landscape painting, tenth century A.D.), cited in Osvald Sirén, *The Chinese on the Art of Painting* (New York: Schocken, 1963), p. 40.

10. See Sullivan, *The Birth of Landscape Painting in China,* p. 7 and notes 15, 16, for further specification, as well as John Hay, "Surface and the Chinese Painter: The Discovery of Surface," *Archives of Asian Art* 38 (1988): 108: "This universe of t'ai-chi and wu-chi polarity represented itself materially as yin and yang, structurally as the achievement and generation of form, geologically as valleys and mountains, consistently throughout all levels of existence."

11. I take the term "senses of place" from the volume *Senses of Place,* ed. Steven Feld and Keith H. Basso (Santa Fe: SAR Press, 1996), esp. the Introduction by the editors.

12. *I Ching,* trans. R. Wilhelm (Princeton, N. J.: Princeton University Press, 1976), p. 324. *Hsiang,* here translated as "images" and elsewhere as "emblematic images," is a quite ambiguous word, having the same kind of range as the Greek word *phantasmata.*

13. "We know little of the religion of China before the Chou Dynasty, except that the Shang people believed in a being whom they called Shang-ti (Supreme Emperor). This deity presided over a hierarchy of spirits; these included the ancestors of the nature spirits and *genii loci* which have been prominent in Chinese popular belief to this day" (Sullivan, *The Birth of Landscape Painting in China,* p. 3).

14. Kuo Hsi, as translated in Bush and Shih, *Early Chinese Texts on Painting,* p. 179. Note that this "advice" is both theoretical (i.e., metaphysical) *and* practical.

15. *I Ching,* p. 280.

16. Sullivan, *The Birth of Landscape Painting in China,* p. 7.

17. Bush and Shih, Introduction, in *Early Chinese Texts on Painting,* p. 4.

18. On this axiom and its ramifications, see Edward S. Casey, *Getting Back into Place: Toward a Renewed Understanding of the Place-World* (Bloomington: Indiana University Press, 1993), pp. 14–15, 98, 313; and *The Fate of Place: A Philosophical History* (Berkeley: University of California Press, 1997), pp. 101–3, 122–23, 338–39. For the surviving fragments of Archytas, see S. Sambursky, ed., *The Concept of Place in Late Neoplatonism* (Jerusalem: Israel Academy of Sciences and Humanities, 1982), pp. 170ff.

19. Thus Han Cho (c. 1095–c. 1125) says that "Generally, in using the brush, the first thing

painters must seek is spirit resonance. Next, they must decide upon stylistic essentials and afterwards upon subtleties. If, before the completion of the formal structure, the painter applies cleverness and subtleties, then he will perforce lose the spirit resonance. If he but strives for spirit resonance, then, as a matter of course, formal likeness will be present in his work" (cited in Bush and Shih, *Early Chinese Texts on Painting,* p. 183). Prior to this, the situation was quite different: "Early Chinese pictorial representation was at first concerned with *hsing-ssu,* or 'formlikeness', that is, formal resemblance of the image to what the eye sees in reality" (Wen C. Fong, *Beyond Representation: Chinese Painting and Calligraphy, 8th to 14th Century* [New York: Metropolitan Museum of Art; New Haven: Yale University Press, 1994], p. 5).

20. On these connotations, see John Hay, "Values and History in Chinese Painting, I," *RES* 6 (1983): 98.

21. Alexander Soper, "The First Two Laws of Hsieh Ho," *Far Eastern Quarterly* 8 (1949): 411.

22. Hsieh Ho as quoted by Kuo Jo-hsü (c. 1080), in Bush and Shih, *Early Chinese Texts on Painting,* p. 95. The translation follows closely that of Soper, "The First Two Laws of Hsieh Ho," pp. 412–13. I have substituted "resonance" for "consonance." Other versions place *ch'i* as such first, followed by resonance *(yün),* for example, that of Ching Hao, who cites an "old man": "The first [Essential] is called 'Spirit *(ch'i)*'. The second is called 'Resonance *(yün)*'" (translated in Bush and Shih, *Early Chinese Texts on Painting,* p. 146). But this seems to be a case of a distinction that does not make an important difference, because *ch'i-yün* combines both factors. Concerning the systematic ambiguity of *fa* ("law," "canon," "principle," etc.), see Hay, "Surface and the Chinese Painter," pp. 76ff.

23. Kao Jo-hsü, in Bush and Shih, *Early Chinese Texts on Painting,* pp. 95–96. I again employ "resonance" rather than "consonance." This statement makes it clear that *ch'i-yün* belongs primarily to the represented content of a painting—that is, "the sentient objects depicted within a picture" (Soper, "The First Two Laws of Hsieh Ho," p. 418)—rather than to the artist.

24. Ching Hao, "Technical Secrets," translated in Bush and Shih, *Early Chinese Texts on Painting,* p. 170.

25. Kuo Hsi, cited in ibid., p. 180.

26. Soper comments that the first Law of Hsieh Ho signifies that "the artist must first of all seek out and stress the ultimate, quintessential character of his subject, the horseness of horses, the humanity of man; on a more general level, the quickness of intelligence, the pulse of life, in contrast to brute matter" ("The First Two Laws of Hsieh Ho," p. 422). To this we need to add: the placiality of place.

27. Li Ch'eng-sou, cited in ibid., p. 163; my italics.

28. Kuo Hsi, ibid., p. 156.

29. We cannot say the same of painting in the West until at least 1480, when Dürer painted sketches of the Italian hillside on his return to Germany. Shortly before his death in 1528, Dürer was making etchings in which landscape is the main topic, for example, his *Landscape with the Cannon* (1518). Albert and Erhard Altdorfer, Dürer's disciples, made pure landscape etchings in the early 1520s.

30. For an acute analysis of the emotionality of this pictorial drama, see Fong, *Beyond Representation,* p. 93: "to re-create through landscape mental, or 'idea', images *(i-ching)* and emotional states *(ch'ing-ching)* rather than merely to describe landscape realistically, Kuo moves beyond 'principles of nature'."

31. Maurice Merleau-Ponty, "Eye and Mind," trans. C. Dallery, in *The Primacy of Perception,* ed. J. Edie (Evanston, Ill.: Northwestern University Press, 1964), p. 180.

32. Jao Tzu-jan (active A.D. 1340), cited in Bush and Shih, *Early Chinese Texts on Painting,* p. 266.

33. Kuo Hsi, cited in ibid., pp. 168–69. Hsi adds: "Without deep distance [a mountain] seems shallow; without level distance it does not recede and without high distance it stays low" (p. 169). For a somewhat different articulation of these basic three distances, see Huang Kung-wang (A.D. 1269–1354), in ibid., p. 263. Concerning three additional kinds of distance, see Han Cho's statement in ibid., p. 170: "When there is a wide stretch of water by the foreground shore and a spacious sweep to distant mountains, this is called 'broad distance'; when there are mists and fogs so thick and vast that streams in plains are interrupted and seem to disappear, this is called 'hidden distance'; when scenery becomes obliterated in vagueness and mistiness, this is called 'obscure distance'."

34. Here I differ from Wen Fong, who maintains that *Early Spring,* "though it suggests great depth, remains a composite of compartmentalized pockets of space" (Fong, *Beyond Representation,* p. 93). In my view, the pockets to which Fong refers are subordinated to the dialectic of the three kinds of depth just discussed.

35. I owe this observation to Samuel C. Morse.

36. Ching Hao (c. A.D. 870–c. 930), cited in Bush and Shih, *Early Chinese Texts on Painting,* p. 172. I have added the exclamation mark.

37. Han Cho (active c. A.D. 1095–c. 1125), cited in ibid., p. 186.

38. Shen Kua (A.D. 1031–1095), cited in ibid., p. 119.

39. Li Ch'eng (d. 967), cited in ibid., p. 177.

40. Merleau-Ponty, "Eye and Mind," p. 180.

41. The term "leaping scale" is that of Fong in *Beyond Representation,* p. 86: "The use of a leaping scale exponentially heightens the impression of size and distance."

42. Cited in ibid.

43. Kuo Hsi again: "If one wishes to make a mountain appear high, one must not paint every part of it or it will seem diminished. It will look tall when encircled at mid-height by mist and clouds" (cited in ibid.).

44. "A mountain has the significance of a major object" (Kuo Hsi, cited in Bush and Shih, *Early Chinese Texts on Painting,* p. 166).

45. Kuo Hsi, *An Essay on Landscape Painting,* trans. S. Sakanishi (London: Murray, 1949), p. 38.

46. Ching Hao, cited in Bush and Shih, *Early Chinese Texts on Painting,* p. 164.

47. The ideograph for mountain, *shan,* presents three triangles grouped closely together— that is, in effect a host peak surrounded by two guest peaks. (On this point, see Fong, *Beyond Representation,* p. 72.) The term "master peak" is Kuo Hsi's term: "In landscape, first pay attention to the major mountain, called the master peak. When this one is established, you can turn to others, near and far, large and small. We call it the master peak because, in this manner, it is sovereign over the entire scene" (cited in ibid., p. 178). Not only are certain mountains hosts vis-à-vis other mountains as guests, but more generally "the mountains and water are the hosts; clouds, mists, trees, rocks, figures, birds, animals, and buildings are all guests" (T'ang Hou, cited in Bush and Shih, *Early Chinese Texts on Painting,* p. 247).

48. Cited by Michael Sullivan from the *Po-hu-t'ung,* in *The Birth of Landscape Painting in China,* p. 183 n. 1.

49. Concerning *li,* Su Shih has this to say: "men and animals, buildings and utensils, all [had] constant forms, and as for mountains, trees, water, and clouds, although they lacked constant forms, they had constant rightness (*li,* 'right ways of being'). If constant forms are lost everyone knows it, but when constant rightness is not there then even the connoisseurs may not realize it" (cited in Susan Bush, *The Chinese Literati on Painting* [Cambridge: Harvard University Press, 1971], p. 42). More generally, *li* is a pervasive principle of organization—"a hierarchy of wholes

forming the cosmic pattern" (Bush and Shih, *Early Chinese Texts on Painting,* p. 5). But it is a pattern precisely of *ch'i*: "In the neo-Confucian system all pyenomena were *ch'i* as patterned by *li*" (Hay, "Surface and the Chinese Painter," p. 103).

50. Sullivan, *The Birth of Landscape Painting in China,* p. 1. Fong remarks that "the sparsely inhabited mountains are both a reminder of a simpler past, of truth unquestioned and nature unspoiled, and a spiritual refuge, where moral values can be cultivated and harmony with nature restored" (*Beyond Representation,* p. 71). For a general study of the significance of mountains in Chinese culture, see Kiyohiko Munakata, *Sacred Mountains in Chinese Art* (Urbana: University of Illinois Press, 1991).

51. "To the Chinese poet, painter, or philosopher . . . to wander in the mountains is an act of meditation, even of adoration . . . By climbing the hills and looking out over range upon range of peaks he discovers man's true place in the scheme of things. When the sun first strikes the high, bare eastern slopes at dawn, while the cloud-filled hollow lies dark and hidden, he observes the workings of the cosmic dualism of *yang* and *yin,* which, forever interacting yet forever held in balance, set in motion the due process of nature" (Munakata, *Sacred Mountains in Chinese Art,* pp. 1–2).

52. Kuo Hsi, cited in Bush and Shih, *Early Chinese Texts on Painting,* p. 152. Cf. also the advice of Li Ch'eng-sou: "Those who paint landscapes must travel everywhere and observe widely, only then will they know where to place and move the brush" (cited in ibid., p. 163).

53. Kuo Hsi puts it this way: "It is generally accepted opinion that there are those through which you may travel, those in which you may sightsee, those through which you may wander, and those in which you may live. Any paintings attaining these effects are to be considered excellent, but those suitable for travelling and sightseeing are not as successful in achievement as those suitable for wandering and living" (cited in ibid., pp. 151–52). By this strict criterion, Fan K'uan has not yet provided us with the ultimate experience of journeying! He presents us with the fruits of his own wandering, but (according to his title) he restricts these fruits to the representation of possible travel, not to wandering as such.

54. Sullivan, *The Birth of Landscape Painting in China,* p. 2.

55. See Sigmund Freud, "Mourning and Melancholia," in *The Standard Edition of the Complete Psychological Works of Sigmund Freud,* ed. and trans. J. Strachey (London: Hogarth Press, 1953–74), p. XIV.

56. Kuo Hsi, *An Essay on Landscape Painting,* p. 35. By "signification" Kuo Hsi means *essence* in the sense discussed earlier: "In painting a scene, irrespective of its size or scope, an artist should concentrate his spirit upon the essential nature of his work. If he fails to get at the essential, he will fail to present the soul of his theme" (p. 33).

57. Shen Kua, citing the *Chiu T'ang shu,* book 192 (the biography of T'ien Yu-yen, seventh-century painter), as translated in Bush and Shih, *Early Chinese Texts on Painting,* p. 120.

58. T'ang Hou, cited in ibid., pp. 247–48; my italics.

59. On encrypting as the baneful consequence of incorporation that keeps the other *other* rather than assimilating that other to oneself, see Jacques Derrida, "Fors," Preface to Nicolas Abraham and Maria Torok, *The Wolf Man's Magic Word: A Cryptonymy,* trans. Nicholas Rand (Minneapolis: University of Minnesota Press, 1987). In their book, Abraham and Torok distinguish systematically between "incorporation" and "interiorization." Only the latter succeeds in assimilating the other to oneself in nonpathological identification.

60. On the notion of the redifferentiated other, see Hans Loewald, "On Internalization," in *Papers on Psychoanalysis* (New Haven: Yale University Press, 1980), p. 83.

61. Tun Yu (active in the first quarter of the twelfth century A.D.), cited in Bush, *The Chinese Literati on Painting,* p. 58.

62. Kuo Hsi, cited in Bush and Shih, *Early Chinese Texts on Painting,* p. 166. The reference to the body as "dry wood" is found in Chang Yen-yüan's statement about the process of painting: "Object and self both forgotten, [the artist] departs from forms and leaves knowledge behind. When the body can truly be made to be like dry wood, and the mind can truly be made to be like dead ashes, is this not to have attained mysterious principles? It is what can be called the true way *(Tao)* of painting" (cited in Bush, *The Chinese Literati on Painting,* pp. 41–42).

63. Cited in ibid., p. 59, from *Chuang-tzu* 19 in Legge's translation.

64. The phrase "spirit of the landscape" comes from this observation of Tung Ch'i-ch'ang: "If one has read ten thousand books and travelled for ten thousand miles, in one's breast one can cast off all impurities and spontaneously form hills and valleys within. When one has established protective barriers [for the inner self], and sketches as one's hand moves, then all will convey the spirit *(ch'uan-shen)* of the landscape" (cited in Bush, *The Chinese Literati on Painting,* p. 45).

65. Cited from the *Hsüan-ho hua-p'u* in ibid., p. 47.

66. Paul Valéry, cited by Merleau-Ponty, "Eye and Mind," p. 162.

67. Sullivan, *The Birth of Landscape Painting in China,* p. 1.

68. Here we may detect a dissymmetry in the situation, not a simple balance between mind and nature—as is implied in the following lines from Chang Ts'ao (active c. 766–78): "A reaching outward to imitate Creation, / And a turning inward to master the mind" (cited in Fong, *Beyond Representation,* p. 76). The inward occurs only once, while the outward occurs twice: in the original perception and in the painterly representation.

69. In contrast with traditional Western paintings, which in the interest of illusion try to hide the tracks of the artist's actual gestures, "a Chinese painting projects a painter's physical movements" (Fong, *Beyond Representation,* p. 5). Fong cites Norman Bryson's claim that "painting in China is predicated on the acknowledgement and indeed the cultivation of deictic markers [which contain references to the painter's bodily action]" (Bryson, *Vision and Painting,* p. 92; cited by Fong, p. 10n). As to the proximity of painting to calligraphy, Fong says outright that "the key to Chinese painting is its calligraphic brushwork" (p. 5). For a discussion of brushwork in its bodily bearing, with special attention to the complex relation between calligraphy and painting, see Hay, "Values and History in Chinese Painting, I," pp. 86–93, as well as "Surface and the Chinese Painter," pp. 99ff. Hay discusses the divining of landscape in the latter essay at p. 92. Such divining is tantamount to getting at the essence of landscape.

70. Fong, *Beyond Representation,* p. 87.

71. "Instead of the three distinct technical steps of continuous and complete contour lines, incisively descriptive texture strokes, and graded ink washes, as seen in the earlier works [of the Northern Sung era], the freer and broken contours and modeling strokes now merge and blend with ink washes to suggest, rather than describe tactilely, the sense of volume and texture of the forms in space" (Wen C. Fong, *Summer Mountains* (New York: Metropolitan Museum of Art, 1975), n.p. In representing depth of recession—"deep distance," in Kuo Hsi's nomenclature—the artist makes use of overlapping volumes explicitly: his "concern for volume and space appears to be a conscious choice; even the mountains trailing off toward the right are carefully modeled to suggest movement into space. Receding mountain ranges are built with overlapping triangular motifs" (ibid.).

72. Contributing to this hospitality is the fact that "the painting is in every way the product of an opulent culture, its very air breathing a well-endowed contentment" (ibid., n.p.)

73. Ibid., n.p.

74. "The depiction [of Fan's mountains], with its nervously charged contours, deep crevices,

and pointillistic texture dots describing the gritty, soil-covered surfaces, conveys a powerful and immediate sense of tactile realism" (ibid., n.p.). The realism is soundly based, because the scene here depicted "vividly captures the landscape peculiarities of Shensi in northwest China, where trees and brush grow in wind-deposited soil on rocky mountaintops" (Fong, *Beyond Representation,* p. 83).

75. Fong, *Summer Mountains,* n.p.

76. In both cases, however, "the depths are on the surface," as Wittgenstein puts it in *Zettel.* The temporal depth of the horizontal axis is wholly laid out before us in one sweep that we can take in with a glance, and the depth of recession in the vertical axis is likewise fully available to us, despite the occlusion of one object by another. For further on the surface/depth relation, see Hay, "Surface and the Chinese Painter," pp. 96ff. and esp. p. 109, where the *Chung-yung* is quoted: "There is nothing more visible than what is hidden and nothing more manifest than what is subtle" (chapter 1). Hay comments: "The particular significance of *t'ien* [i.e., of truth] is that it is to be found within the depths. The degree of 'authenticity' is therefore *manifested* by the placidity of the surface" (Hay's italics). Hay puts this point differently elsewhere: "the structural pattern of the figure is actualized from within, according to its inherent system of order. The exterior is still an aspect of the *ku* [i.e., 'bone']" ("Values and History in Chinese Painting, I," p. 95). The inner, the "bone," is actualized in the outer, the "surface."

77. I owe this observation to Dan Rice, who also points out that while the primary diagonality is created by the rightward-tending slopes of mountains subordinate to the major central mountain, there is a mimetic but inverse directionality on the part of the smaller mountain group whose leftward slopes are accentuated. Examples of truly harsh vertical/horizontal intersections are found in Cartesian analytic geometry and in the linear perspective systems of the Renaissance as these latter were applied to the representation of architecture receding in depth. But there are instances of Northern Sung paintings that make use of unremitting verticals and horizontals, for example, *Fishermen* by Hsü Tao-ning, painted c. A.D. 1040.

78. Fong, *Beyond Representation,* p. 83.

79. "The rectangular format of the Northern Sung landscape [painting] does, of course, confirm the vertical and horizontal axes that are fundamental to the classic cosmology" (Hay, "Surface and the Chinese Painter," p. 113). In contrast, the oval frame of fan painting refers "not to the universe, only to the precise location and moment. The conversation between oval fan and the landscape it contains is pure aestheticism" (ibid.). For an excellent contrast between Northern and Southern Sung painting in terms of the "paradigmatic" tendency of the former and the "syntagmatic" tendency of the latter, see ibid., pp. 105ff., 112ff.

80. In the case of mountains, the distinction between "thing" and "place" is moot: a thing of this magnitude counts as a place. For further discussion, see Casey, *Getting Back into Place,* p. 216, where I speak of a mountain as a "thing-place."

81. This episode of Ni Tsan (A.D. 1301–75) is reported both by Fong, *Beyond Representation,* p. 5, and by Hay, "Surface and the Chinese Painter," p. 98.

82. "[Kuo Hsi] told the wall plasterer not to use the trowel, but instead to throw the plaster on the wall . . . After the plaster had dried, applying ink to modify the forms, he created mountain peaks, forests, and valleys . . . making them look as if they had been realized in heaven" (cited in Fong, *Beyond Representation,* p. 95).

83. Ching Hao, cited in Bush and Shih, *Early Chinese Texts on Painting,* p. 146.

84. Skillful work "may pretend to follow the basic principles of nature, but in actuality it heedlessly copies the outer appearances of things, and more and more diverges from the true images filled with vital force. Works of this kind may be said to lack 'reality' but possess an excessive 'outward beauty'" (Ching Hao, cited in ibid., p. 171).

85. Tung Yu, cited in Bush, *The Chinese Literati on Painting*, p. 58.

86. For a different sense of "truth to self" *(t'ien chen)* that comes closer to what I prefer to call "truth to place," see Lawrence Sickman, *Eight Dynasties of Chinese Painting: The Collections of the Nelson Gallery-Atkins Museum, Kansas City, and the Cleveland Museum of Art* (Cleveland: Cleveland Museum of Art in Co-operation with Indiana University Press, 1980), p. 15: "The rhythmic animation and naturalism of these trees leave the viewer with a conviction of their rightness, their 'truth-to-self' *(t'ien-chen)*." Mi Fu (A.D. 1051–1107) says expressly that in Tung Yüan's painting, "Peaks and ranges emerge and disappear; clouds and mists now reveal, now obscure. He does not adorn himself with ingenious tricks; in everything he is completely true to nature *[t'ien-chen]*" (Mi Fu, *Marriage of the Lord of the River*, cited in Hay, "Surface and the Chinese Painter," p. 109).

87. Whether being true to nature means going "beyond representation" is an issue addressed by Fong in his book of the same name: see especially the Introduction, where Fong argues that Chinese painters "had no need to create a nonobjective art" (*Beyond Representation*, p. 5) insofar as they were already beyond any imperative to create a perfectly illusionist, realist art. If so, this can only be the case for painters from the T'ang dynasty onwards—when the question of conveying the spirit of something became more important than representing its detailed infrastructure.

88. Cited in Fong, *Summer Mountains*, n.p.

89. Yang Wei-chen (A.D. 1296–1370), cited in Bush and Shih, *Early Chinese Texts on Painting*, p. 246.

90. Ibid.

91. This is Soper's considered translation in "The First Two Laws of Hsieh Ho," p. 423. Note that *sheng* is "a common and extremely powerful term that signifies not simply 'living' but the unceasingly generative transformations constituting the living universe" (Hay, "Surface and the Chinese Painter," p. 99). *Tung* means "move." Thus *sheng-tung* is an intensified, redoubled expression for "animation." Hay translates the same phrase as "the generating of a process" (ibid.).

92. "To paint the whole extent in one [work] would simply produce a map. All works which do so suffer from the fault of not discovering the quintessential" (Kuo Hsi, cited in Bush and Shih, *Early Chinese Texts on Painting*, p. 182). Han Cho insists on the degraded status of a map versus a genuine landscape painting: "Those who understand painting methods will comprehend the vitality of spiritual perfection, but those who study copying methods will possess the defects found in geography book illustrations" (ibid., p. 184). In Part II, we shall see that the relationship between maps and paintings is considerably more intimate and more productive than this exclusivist model allows.

93. Han Cho cited in ibid., p. 182. Compare Ching Hao's remark that "[a picture that attains] likeness achieves the physical form but leaves out the life breath of the subject, while in [a picture that attains] truth the life breath and inner qualities of the subject are fully present" (cited in Fong, *Beyond Representation*, p. 76). Tung Yu said similarly: "How can it be only likeness that is valued? For we know that one who does not consciously try to paint penetrates to what is there before created things. In general, the process of painting issues from a sense of life *(sheng-i)* and comes through naturalness *(tzu-jan)*, and one has to wait for the forms to appear in one's mind: it is like the unfolding and blooming of leaves and flowers. Only after this is it externalized by making use of the hand and colors. One never first seeks for likeness and then lodges one's conceptions in it" (Tung Yu, cited in Bush, *The Chinese Literati on Painting*, p. 56).

INTERLUDE

1. Cited in Kevin J. Avery, "*The Heart of the Andes* Exhibited: Frederic E. Church's Window on the Equatorial World," *American Art Journal* 18 (1986): 58 and 70 n. 41. A replica of Church's

original flamboyant frame and curtains—the painting itself ensconced within it—was created at the Metropolitan Museum of Art in the fall of 1994.

2. Jacques Derrida, "Parergon," in *The Truth in Painting,* trans. Geoffrey Bennington and Ian McLeod (Chicago: University of Chicago Press, 1987), p. 63.

3. It is this very possibility that Kant regretted: the decorative frame amounts to mere "adornment" or "finery" *(Schmuck)* and "takes away from the genuine beauty" of a work. See Immanuel Kant, *The Critique of Judgment,* trans. J. C. Meredith (Oxford: Clarendon Press, 1952), p. 68. For Kant, "ornamentation" of any kind is "only an adjunct, and not an intrinsic, constituent in the complete representation of the object" (p. 65). It is significant that Kant takes up the phenomenon of *parerga* near the conclusion of the "Analytic of the Beautiful."

4. I owe the substance of these last two sentences to comments by Ann Cahill on an earlier version of this text.

5. On the annexation of the frame—and other forms of surrounding—see Mikel Dufrenne, *The Phenomenology of Aesthetic Experience,* trans. Edward S. Casey, Albert Anderson, Leon Jacobson, and Willis Domingo (Evanston, Ill.: Northwestern University Press, 1973), pp. 152, 153.

6. "The background is that through which the world manifests itself as the background of all backgrounds . . . The background is the guarantee of the form because the world is the guarantee of the object" (ibid., p. 150).

7. "[Frames as] *parerga* have a thickness, a surface which separates them not only (as Kant would have it) from the integral inside, from the body proper of the *ergon,* but also from the outside, from the wall on which the painting is hung" (Derrida, "Parergon," pp. 60–61). In this very capacity, the frame should be considered as merging with its two grounds *(fonds)*: "the parergonal frame stands out against two grounds, but with respect to each of these two grounds it merges *(se fond)* into the other. With respect to the work . . . it merges into the wall . . . With respect to the background . . . it merges into the work . . . [Thus] it disappears, buries itself, effaces itself, melts away at the moment it deploys its greatest energy" (p. 61). With this claim, however, the separative power of the frame is lost: a power underlined in Dufrenne's remark that "the necessary separations by which [the aesthetic object] attracts my attention are as much a way of embodying as of detaching—witness the frame around the painting which is not seen as an intermediary between it and the wall" (Dufrenne, *Phenomenology of Aesthetic Experience,* p. 155).

8. For further discussion of "boundary" versus "border," see Edward S. Casey, *Getting Back into Place: Toward a Renewed Understanding of the Place-World* (Bloomington: Indiana University Press, 1993) pp. 14–16, 150–51, 165–68, 240–41. Concerning the interplace, see ibid., chapters 4 and 6.

9. "The object's outer limit—its 'frame'—is never completely isolated. The wall is always present around the picture's frame, and the city surrounds the theater . . . In any case, it is difficult for perception to isolate the aesthetic object [entirely] from the perceptual field" (Dufrenne, *Phenomenology of Aesthetic Experience,* p. 152).

10. Derrida, "Parergon," p. 56; Derrida's italics. Derrida designates *parergon* as a "philosophical quasi-concept" (p. 55).

11. I borrow this term from the landscape painter Eve Ingalls, who also suggested in conversation with me the example of the shed transformed by its inclusion in a painting.

12. See Martin Heidegger, *The Origin of the Work of Art,* trans. A. Hofstadter (New York: Harper & Row, 1979), pp. 39–50.

13. Charles Olson, *The Maximus Poems,* ed. F. G. Butterick (Berkeley: University of California Press, 1983), p. 3.

7. FIRST CONSIDERATIONS

1. See Clarence M. Bicknell, *A Guide to the Prehistoric Rock Engravings in the Italian Maritime Alps* (Bordighera: G. Bessone, 1913).

2. Catherine Delano Smith, "Cartography in the Prehistoric Period in the Old World: Europe, the Middle East, and North Africa," chapter 4 in J. B. Harley and David Woodward, eds., *The History of Cartography: Cartography in Prehistoric, Ancient, and Medieval Europe and the Mediterranean* (Chicago: University of Chicago Press, 1987), p. 67. Cf. also Smith's earlier essay, "The Emergence of 'Maps' in European Rock Art: A Prehistoric Preoccupation with Place," *Imago Mundi: The Journal of the International Society for the History of Cartography* 34 (1982): 9–25.

3. For this analysis, see Smith, "Cartography in the Prehistoric Period," p. 74. In addition to the minimum of six signs, Smith also invokes three other general criteria for determining a simple map: "that the artist's intent was indeed to portray the relationship of objects in space; that all the constituent images are contemporaneous in execution; and that they are cartographically appropriate" (p. 61).

4. Bicknell, *A Guide to the Prehistoric Rock Engravings in the Italian Maritime Alps,* p. 39. Smith doubts this particular claim in view of the very high position of the Mont Bégo petroglyphs, situated far from cultivated fields. She thinks they more likely served a "primarily symbolic purpose" that entailed " a now unknown abstract context or message" (Smith, "Cartography in the Prehistoric Period," p. 80). Elsewhere, however, Smith is more partial to a representationalist view of the same maps, allowing that "the bird's eye views *could* be called images" even if "it is probably more accurate to consider the rectangles as schematic representations of the hut roofs as seen from above and the large sub-circular features as indicating rather than portraying the neighboring pastures" ("The Emergence of 'Maps' in European Rock Art," p. 15; my italics). At stake here is the question of accurate representation, to which we shall return at the end of this section.

5. Smith, "Cartography in the Prehistoric Period," p. 62.

6. These terms are those of P. D. A. Harvey, *The History of Topographical Maps: Symbols, Pictures, and Surveys* (London: Thames and Hudson, 1980), chapter 1.

7. Walter Blumer, "The Oldest Known Plan of an Inhabited Site Dating from the Bronze Age," *Imago Mundi* 18 (1964): 10. Blumer cites Emmanuel Anati's claim that this map is "la più antica carta geografica ritrovata in Europa, che risale all'età del Bronzo" (p. 10n).

8. Thus (as I can attest from personal observation of this map in situ) certain of the pictographs, for example, those of the houses in the lower part of the map, are awkwardly situated vis-à-vis the views in plan of the "fields" positioned just above: awkward insofar as the two sets of objects are set in two quite different perspectives, as occurs also with respect to the animals and human figures throughout. Still, the shift in perspective is not so abrupt as to lead to perceptual disarray: we take it in as somehow making spatial sense.

9. For the "progressivist" view, see Norman J. W. Thrower, *Maps and Civilization: Cartography in Culture and Society* (Chicago: University of Chicago Press, 1996), pp. 3–4. Thrower regards the Bedolina map, the "oldest known plan of an inhabited site," as representing a "progression" from "abstract symbols" (e.g., rectangles and circles) to pictographic elements (e.g., figures, houses, animals). Similarly, Walter Blumer observes that in the Bronze Age carvings of Bedolina "the execution is more schematic . . . abstract and symbolic in character" versus the Iron Age additions that are "rendered realistically in perspective" ("The Oldest Known Plan," p. 9). It is now generally accepted that the huts and a number of the human figures were added in the later period of the map's composition. But this fact alone does not constitute a case for "progression." In many respects, the earlier nonpictographic elements are more informative as well as more economic; they do not require "realistic" representation to become more cartographically complete;

and they are already "in perspective," albeit plan rather than elevation or profile. Moreover, the schematic representations themselves are not wholly lacking in pictographic significance: the squares filled with points, the circles with dots in them do depict parts of the landscape. They are topographically effective even if not pictographically explicit. For a detailed study of the phases of actual composition—of which there may be as many as four—see Miguel B. Lloris, "Los Grabados rupestres de Bedolina (Valcominica)," *Bulletino del Centro Camuni do Studi Preistorici* 8 (1972): 122. The map proper, according to Lloris, belongs to the second phase; but this is not to deny that the expressly pictorial elements, whether added earlier or later, contribute importantly to the overall topographic force of the work.

10. Smith, "Cartography in the Prehistoric Period," p. 79, with reference to the earlier work of Lloris.

11. Ibid., p. 80.

12. This analysis is undertaken in Smith's essay "The Emergence of 'Maps' in European Rock Art," pp. 15–16.

13. Smith, "Cartography in the Prehistoric Period," p. 92.

14. Harvey, *The History of Topographical Maps,* p. 45. Here Harvey rejoins and reinforces Thrower's view as set forth in note 9 above.

15. Ibid.

16. Ibid., p. 46.

17. Ibid. It needs to be noted that the identification of the earlier map (i.e., the one in pure plan) is given a different description by the two authors: Smith calls it "a simple topographic map from Seradina, Italy," whereas Harvey calls the same map a "carving from the Valcamonica in north Italy" (Smith, "Cartography in the Prehistoric Period," p. 75; Harvey, *The History of Topographical Maps,* p. 45). For Smith, the Seradina carving, which has no explicitly pictographic elements, is truly a *map*—a claim that Harvey adroitly avoids making.

18. Smith, "Cartography in the Prehistoric Period," p. 80. In her earlier essay, she notes that "all who have published [on] the Bedolina complex as a map must have tacitly accepted that [its figures and elements] are representational signs" ("The Emergence of 'Maps' in European Rock Art," p. 15). But she goes on to say that the depiction of huts and fields in rock art may "have been associated less with the 'objective' recording of a spatial distribution for purposes of direction or reference than with invoking the various controlling forces of those aspects of life these topographical features were used to represent or with attempts to appease them" (p. 18). In her later article, Smith attributes the greater pictoriality of the Bedolina map to its having "a significantly different context" from the Mont Bégo maps she examines; for the Bedolina map "overlooks what would have been, even then, a cultivated valley and a route across the Alps" (ibid.). But does a closer overlook onto a particular landscape necessarily entail an augmented pictoriality in the mapping of this landscape? The Seradina map overlooks the *same* valley and route below; indeed, it does so from a still closer viewing place, as any visitor can discover; and yet it is resolutely nonpictorial in character. In a third example, that found at Giadighe in the same immediate region, in a place that overlooks the valley below from still higher up, there is a highly schematic view in plan, yet one that includes meandering double lines that almost certainly represent the river Oglio, along with two anthropomorphic figures in the lower part of the map. See Figure 4.29 in ibid., p. 79. Raffaello Battaglia, the discoverer of the Giadighe map, said that the "broad winding band" in the petroglyph "represents the course of a river" (Raffaello Battaglia, "Incisioni rupestri di Valcamonica," *Proceedings of the First International Congress of Prehistoric and Protohistoric Sciences, London, August 1–6, 1932* [London: Oxford University Press, 1934], p. 236). Even so, the depiction of this river, as of the fields and huts and streams in the Bedolina map, may

not be sufficiently interpreted in purely pictographic terms: all of these items may also exhibit what Smith calls "the sacred aspects of daily life" ("The Emergence of 'Maps' in European Rock Art," p. 18).

19. Smith, "The Emergence of 'Maps' in European Rock Art," p. 80. She adds that "when the arguments and examples [from Valcamonica] are weighed, it remains doubtful that even these two examples [of Bedolina and Giadighe] can in fact be seen as marking the introduction of the use of maps as factual records in the prehistoric era" (ibid.).

20. Harvey, *The History of Topographical Maps,* p. 46. He adds: "The two forms of mapping correspond to distinct stages of cultural or artistic development. To this extent, and to this extent only, we can speak of topographic mapping as progressing from symbols to pictures" (ibid.). The phrase "to this extent only" is, in my view, evasive and self-deconstructing.

21. This is not to exclude the real possibility that a complex combination of both interpretations obtains for certain prehistoric maps; but it does suggest that where we put the emphasis will matter greatly. In a similar spirit, each such map can be said to exhibit all three sorts of sign in accordance with Peirce's semiology: they are at once iconic (in their pictographic elements), indexical (as indicating where one must go in order to find certain destinations), and symbolic (insofar as certain items are purely conventional in origin). But in any given case, one form of sign will be more prominent than the others.

22. Max Black, "Models and Archetypes," in *Models and Metaphors: Studies in Language and Philosophy* (Ithaca, N.Y.: Cornell University Press, 1962), pp. 232–33; my italics.

23. See Jean Piaget and Bärbel Inhelder, *Mental Imagery in the Child: Study of the Development of Imaginal Representation,* trans. P. A. Chilton (New York: Basic Books, 1971), Introduction.

24. Alfred Korzybski, *Science and Sanity* (Lancaster, Pa.: International Non-Aristotelian Library, 1941), p. 58; Korzybski's italics. Korzybski also says that "any map or language, to be of maximum usefulness, should, in structure, be similar to the structure of the empirical world" (p. 11).

25. Amando Cortesào, *History of Portuguese Cartography,* vol. 1 (Lisbon: Coimbra, 1969), p. 3. Compare Max Black: "there is no such thing as a perfectly faithful model; only by being unfaithful in *some* respect can a model represent its original" ("Models and Archetypes," p. 220; Black's italics). To be truthful in *every* respect is to *be* the original, or else its mere replication. For representation to occur, difference must exist. On this last point, see Descartes, *Optics,* Discourse 4.

26. Smith, "Cartography in the Prehistoric Period," p. 92. By "contemporary" Smith means "original," that is, the landscape of origin. That Smith leaves this "primary task" to the archaeologists does not diminish, in her view, its urgency for cartographers and geographers.

27. Bicknell, as we have seen, considered the figures on his maps to be likenesses of huts and pathways as viewed from high up on Mont Bégo; Smith, as we have just witnessed, holds to the in-principle possibility of reconstructing the "real-world localities" on which maps at Mont Bégo and Valcamonica are based; Anati maintains that the Bedolina map "represents the landscape of the valley as seen from the very spot where it was found . . . [it conveys] a sure sense of the real" (Emmanuel Anati, *Camonica Valley: A Depiction of Village Life in the Alps from Neolithic Times to the Birth of Christ as Revealed in Thousands of Newly Found Rock Carvings,* trans. Linda Asher [New York: Knopf, 1961], p. 105); Blumer speaks of the same map as "the actual plan of a definite district drawn by the artist from nature" ("The Oldest Known Plan," p. 10); Ausilio Priuli proposes that the very undulations of the rocks on which maps in Valcamonica were carved represent the landscape shapes of the region depicting, for example, that of the Castelliere del Dos dell'Archa in the case of the Bedolina map (*Incisioni rupestri della Val Camonica* [Ivrea: Priuli and Verlucca, 1985], p. 24 n. 98). Priuli has also constructed, in his Museo Didattico D'Arte e Vita

Preistorica in Capo di Ponte, a three-dimensional model of the landscape as it is represented in the Bedolina map, linking specific figures in the map with particular parts of the original landscape as here reconstructed. This is isomorphism with a vengeance!

28. Juan Luis Borges, "On Exactitude in Science," in *Collected Fictions,* trans. A. Hurley (New York: Viking, 1998), p. 325. Subsequent generations, adds Borges, "saw that that vast Map was Useless" (ibid.) and altogether neglected it.

29. On the inherent vagueness of "morphological essences, " see Edmund Husserl, *Ideas: General Introduction to Pure Phenomenology,* trans. W. R. Boyce Gibson (New York: Collier, 1962), sections 67–69.

30. Harvey, *The History of Topographical Maps,* p. 45. The examples I cite in the preceding sentence are analyzed by Harvey on pp. 39–44.

31. Ibid., p. 43.

32. The Nuzi map is to be compared with the celebrated Babylonian World Map of 600 B.C. In the latter, the words are placed not only on the map—mainly place-names identifying Babylon, Assyria, and Urartu, along with numbers denoting distances between the seven outer regions signified by triangles affixed to the circle of the surrounding ocean—but also above it, where a separate text is attached as well. For further discussion, see O. A. W. Dilke, *Greek and Roman Maps* (London: Thames & Hudson, 1985), pp. 13–14.

33. Ibid., p. 112.

34. The roads on the Peutinger Table are not drawn to scale. Dilke cautions, however, that "it should not be thought that the Peutinger Table was typical of Roman maps, or that the Romans were incapable of drawing a map to scale. General maps, land surveyors' maps and town plans were drawn to scale, with reasonable approximations to cartography" (ibid., p. 120).

35. As John Noble Wilford writes, "in purpose, if not exactly in presentation, the automobile road map—the map most familiar to the twentieth-century traveler—is the lineal descendent of the Peutinger Table" (John Noble Wilford, *The Mapmakers* [New York: Knopf, 1981], p. 50). Wilford notes that the modern road map did not take on its present form until relatively recently—the 1920s—when the enormous increase in public highway systems made the older, primarily verbal, "route books" too cumbersome to be useful to travelers. Once more, the written word required the addition of the pictorial image—which, far from being a merely secondary feature, shows itself to be quite essential to mapping.

36. I say "symbols proper," for if *symbol* is taken in a sufficiently large sense it can be maintained that "almost every feature on a map is symbolized" (Erwin Raisz, *General Cartography* [New York: McGraw-Hill, 1948], p. 97). But if so, various features are symbolized in importantly different ways—two of which I distinguish under the headings of "pictograms" and "symbols proper."

37. Ibid.; my italics. Raisz italicizes the entire sentence. His use of *symbol* includes pictograms as well as conventional signs, but he recognizes the difference when he writes that a map symbol "should either be reminiscent of the feature it represents or be sanctioned by centuries of use" (ibid.).

38. Until this point in the book, the term *topographic* has been used to stand for any concerted pictorial representation of the landscape—for example, pencil or ink sketches intended to be precise resemblances. But another usage obtains in the case of maps, where *topography* implies a representation of the layout of the land, its protuberances as well as its declivities, for example, by means of contour lines that designate comparative height or depth. (Such representation will be discussed later in this chapter.) Even though I shall occasionally advert to this more technical sense in this and successive chapters (see especially chapter 8, opening section), I shall also draw

upon the more extensive meaning on occasion. Sometimes, "topographic map" refers to a map of a particular place, e.g., a town. *Pictographic* does not share this complexity; it refers to any graphically inscribed pictorial element that represents a discrete object situated in the landscape.

39. Lloyd A. Brown, *Map Making: The Art That Became a Science* (Boston: Little, Brown, 1960), p. 197.

40. This map is drawn from *The Liverpool Directory* of 1776; but maps just like it, with comparable freedom of symbol formation, are still being constructed today. The Liverpool map is reproduced in F. J. Monkhouse and H. R. Wilkinson, *Maps and Diagrams: Their Compilation and Construction* (London: Methuen, 1963), p. 351.

41. If we include color per se, there would be ten such signs, because green represents vegetated areas. Conspicuous by their omission are any signs for rivers as well as any signs for a mixture of landscape and the humanly constructed, for example, signs for bridges over rivers. (The only candidate for such a hybrid sign is that for a park with camping facilities; it features a green mountain with a tree on top and a white area across the mountain's face; but the white area is ambiguous as to its status: is it an icon or a symbol?)

42. "Since the early walled cities were usually round, on small-scale maps their representation was either reduced to the more or less circular layout of the wall or was symbolized by a circle. The circle as a city symbol survives up to the present time, although there is now little analogy between a small circle and a city" (Raisz, *General Cartography*, p. 97). In other words, what is perceptually the same figure has changed its status: from being a pictogram, it has become a symbol. Raisz adds that "it is possible that the origin of the city circle was different. It was customary in early Renaissance maps to designate a city by a small pictorial group of houses. But since this group was very much larger than the size of the city, the exact location of the latter was shown by a small circle within the group of houses" (p. 98). This, however, overlooks the use of circles to designate cities on the clay maps of Babylonia; if anything, the Renaissance practice represented a return to the much earlier sign.

43. Monkhouse and Wilkinson, *Maps and Diagrams*, p. 62.

44. The color map is effective insofar as it includes both the exactitude of linear design and the natural associations of colors. It should be noted that the colors themselves operate by a spontaneously adopted principle of resemblance: "the greenish tints remind us of fertile valleys, the browns [of] the bare rock, the white [of] the snow" (Raisz, *General Cartography*, p. 115).

45. David Greenhood, *Mapping* (Chicago: University of Chicago Press, 1965), p. 77.

46. In fact, this virtue, carried to an extreme, becomes a distinct vice: some maps are "so loaded with hachures that people could not read the map for the data" (ibid.).

47. Raisz, *General Cartography*, p. 109. Raisz adds that "the solution of almost every practical problem connected with contour lines is accomplished with the help of drawing profiles or vertical sections" (ibid.). See the examples given by Raisz on pp. 110–11 (Figures 97–99).

48. Which feature is to be chosen "depends on the topographic sense of the individual surveyor" (ibid., p. 109). On the question of choice of contour, see G. D. Whitmore, "Contour Interval Problems," *Surveying and Mapping* (Washington, D.C.: Quarterly Publication of American Congress on Surveying and Mapping, 1953), vol. 12, pp. 174–77.

49. Contour-line maps "give a somewhat smooth, rounded and rolling effect to the relief, while sudden changes of slope, sharp breaks or edges, and any interruption may be obscured" (Monkhouse and Wilkinson, *Maps and Diagrams*, p. 67). Tanaka Kitro's methods of orthogonal relief are designed to overcome this particular defect: see ibid., pp. 68–70.

50. "This [cartographic] work has much of the nature and quality of landscape drawing and should preferably be carried out in the field" (ibid., p. 77).

51. Erwin Raisz, "The Physiographic Method of Representing Scenery on Maps," *Geographical Review* 21 (1931): 297–304.

52. "In principle, the method goes back to the primitive concepts of early maps, whereby relief features were shown obliquely and in some degree of perspective, instead of by vertical [i.e., planigraphic] conventions" (Monkhouse and Wilkinson, *Maps and Diagrams,* p. 75).

53. For a complete presentation, see Raisz, *General Cartography,* pp. 120–21.

54. Raisz himself regards these units as "symbols": "to show the various surface types [of landscape] a certain symbol system was adopted" (ibid., p. 119).

55. Monkhouse and Wilkinson, *Maps and Diagrams,* p. 77.

56. Raisz, *General Cartography,* p. 122. Thus this method "makes mountains look like mountains" (ibid.).

57. This is so despite Raisz's claim that "in a morphographic map the existing conventional symbols for roads, rivers, and cities can be used without change" (ibid., p. 119). But symbols for detailed topographic features (e.g., numbers for the absolute height of mountains) are nonetheless downplayed in the interest of direct representation. It should also be noted that the obliquity of the angle assumed by a physiographic map brings with it a distortion in the representation of those slopes of a mountain (typically, the northern slopes) that are not straightforwardly in view.

58. On these two examples, see J. B. Harley and David Woodward, eds., *The History of Cartography,* vol. 1, Figure 4.38 (p. 89) and Plate 2 (p. 107), respectively. The cartography of ancient Egypt, like that of ancient China, is characterized by an exquisite aesthetic sensibility that is much less evident in the surviving maps of Mesopotamia or in those of the late Bronze Age in the Valcamonica and Mont Bégo area.

59. Of the colorful and figure-filled *mappae mundi* it has been said that "the great majority of these [maps] are to be regarded as works of art and not of information" (George H. T. Kimble, *Geography in the Middle Ages,* cited in Wilford, *The Map Makers,* pp. 45–46). Portolan charts, to which we shall return, are an even more striking case in point: "the finest examples of the sixteenth century were brilliant works of art" (R. V. Tooley, *Maps and Map-Makers* [New York: Crown Publishers, 1978], p. 16). Dutch maps of the seventeenth century, as we have seen, are intimately linked to the art of that century; and it is an understatement to say that "early Dutch maps [i.e., of the seventeenth century] were among the best for artistic expression, composition, and rendering" (*Encyclopedia Britannica,* 13th ed., vol. 18, article "Maps," p. 473).

60. Claudius Ptolemy, *Geography,* trans. E. L. Stevenson (New York: New York Public Library, 1932), vol. 1, p. 1.

61. Leo Bagrow, *The History of Cartography,* rev. ed. by R. A. Skelton (Chicago: Precedent Publishing, 1985), p. 20.

62. David Turnbull asks the reader to determine whether several examples—ranging from a plan of Leonardo to regulate the Arno (1502) to a picture of the Siege of Enniskillen (1594)—are maps or paintings, with the clear implication that this cannot be easily decided (David Turnbull, *Maps Are Territories: Science Is an Atlas* [Chicago: University of Chicago Press, 1993], pp. 16–17).

63. Dicaerchus, who lived in the third century B.C., was thought to be the first person to draw a parallel and a meridian line on a map. Eratosthenes, his contemporary and better known as the man who gave the first approximately accurate measurement of the earth's circumference, also imposed a grid on his map of the earth: he first "divided the inhabited world by a line going from the Pillars of Hercules to the Taurus mountains and beyond, then subdivided each of these two sections into a number of irregular shapes" (Dilke, *Greek and Roman Maps,* p. 35). These shapes or *sphragides* (literally, "seals"), once given a geometrical regularity thanks to lines of latitude and longitude, became the graticules or grid units of modern maps.

64. For an account of the grid in urban design, see Spiro Kostoff, *City Shaped* (Boston: Little, Brown, 1993), and John Reps, *The Making of Urban America* (Berkeley: University of California Press, 1997).

65. Raisz, *General Cartography,* p. 64; in italics in Raisz's text.

66. Monkhouse and Wilkinson, *Maps and Diagrams,* p. 63.

67. Cortesào, *History of Portuguese Cartography,* vol. 1, p. 3.

68. See Norman J. W. Thrower, ed., *The Compleat Plattmaker: Essays on Chart, Map and Globe Making in England in the Seventeenth and Eighteenth Centuries* (Berkeley and Los Angeles: University of California Press, 1978). The Old English word *platt* meant "flat." *Plat* is still used today to signify a plan or map of a particular piece of land: that is, a "plot" (etymologically affine with *plat*). It is of interest that *plat* also once meant place, spot, or locality.

69. Greenhood, *Mapping,* p. 122.

70. Ibid., p. 113. Greenhood continues: "no other kind of map can represent the true forms of continents in the full as they fit into, and indeed are part of, the rondure of the earth. It is the most nearly true to scale of all maps in each and all of its areas and in its distances from any one point to every other point" (ibid.). Cf. also this statement: "a globe is an accurate model of the earth and is the only possible medium of showing all geographical relationships truly" (David Greenhood, "The Round Earth on Flat Paper," *National Geographic Society* [Washington, D.C., 1950], p. 53).

71. Ibid., p. 114. Buckminster Fuller's Dymaxion Sky-Ocean World Map can be construed as an attempt to refute Greenhood's contention. It succeeds in flattening out a global surface while retaining essential geographical information, including the "correct" shape of continents. The price paid, however, is that the outer edges of this map form a set of discontinuous polygons. See the model in R. Buckminster Fuller, *Critical Path* (New York: St. Martin's Press, 1981), p. 169 (Figure 31).

72. Raisz, *General Cartography,* p. 64. *Orthomorphic* means retaining the right shape. Cf. also Greenhood's statement: "In expressing the earth's spherical surface on a flat-map surface, we cannot keep both the equivalence of areas and the similarity of shapes. We cannot eat our cake and have it" (*Mapping,* p. 117). On the other hand, it would be wrong to regard map projections as merely second-best: "because the selection from among map projections necessarily involves evaluation of deformation, there is a tendency to think of map projections as necessary but poor substitutes of the globe surface. On the contrary, in most instances projections are advantageous for reasons in addition to the fact that it is cheaper to make a flat map than a globe map. Map projections enable us to map distributions and derive and convey concepts that would be either impossible or at least undesirable on a globe" (Arthur H. Robinson, Randall Sale, and Joel Morrison, *Elements of Cartography* [New York: Wiley, 1978], p. 58).

73. For a detailed discussion of these four forms of deformation, see Arthur H. Robinson, *Elements of Cartography* (New York: Wiley, 1953), pp. 29–44.

74. Greenhood, *Mapping,* p. 122; Greenhood's italics. In other words, "we must keep each location at its exact crossing of co-ordinates" (ibid., p. 123). This last sentence could be taken as a perspicuous example of what Whitehead had in mind when he spoke of "simple location." It is a striking fact that the first complete and convincing representation of the earth via a systematic layout of coordinates—a layout that preserved both latitude and longitude as parallel lines, thus creating a grid of rectangles—was achieved by Mercator in 1569, that is to say, less than fifty years before the "century of genius" in which the idea of simple location was to become fully canonized in philosophy and physics.

75. Greenhood, "The Round Earth on Flat Paper," p. 53. Cf. Greenhood's similar claim

that "the importance of globes is that they are the basis of flat maps . . . just as we speak of Mother Earth, we may respect globes as the mothers of maps" (*Mapping,* p. 113).

76. Greenhood, *Mapping,* p. 124.

77. "There are an unlimited number of possibilities for representing the earth grid systematically on a plane" (Robinson, *Elements of Cartography,* p. 27). The reason for the existence of so many possibilities is not purely a matter of projective geometry. It also reflects the multiplicity of aims that determine specific choices of projections: "the number of diverse factors that may influence the choice of a map projection is surprising. The geographer, historian, and ecologist are likely to be concerned with the sizes of areas. The navigator, meteorologist, astronaut, and engineer are generally concerned with angles and distances. The atlas mapmaker often wants a compromise" (Robinson, Sale, and Morrison, *Elements of Cartography,* p. 58).

8. CARTOGRAPHY AND CHOROGRAPHY

1. Thus Barbara Novak remarks that "the botanical veracity of Church's plant studies, the astute observation of his erupting volcanoes, show the accuracy of the best descriptive science" (*Nature and Culture: American Landscape and Painting, 1825–1875* [Oxford: Oxford University Press, 1980], p. 68; with reference to Figures 34 and 35 in the same text).

2. Michael Rosenthal, *Constable: The Painter and His Landscape* (New Haven: Yale University Press, 1983), p. 10.

3. Ibid., p. 9 (Figures 2 and 3). Such checking against maps is a step that is difficult to imagine being taken in the case of the Sung masters of landscape, who, despite their much earlier position in history, were already trans-topographic in their highly imaginative and metaphysical manner of being "true to the motif." Not until the later Turner and the Postimpressionists—for example, Braque's and Picasso's proto-Cubist paintings of L'Estaque in 1907–8—will there be a comparable freedom from topographic concerns in the West. Rosenthal reprints a photograph of Dedham Vale taken from approximately the same spot from which Constable's 1812 sketches were composed (ibid., Figure 94). As in the case of John Rewald's photographs of scenes painted by Cézanne around Aix, this particular piece of evidence of topographic exactitude is not altogether convincing—given that both artists were highly selective and ended by transforming rather than transcribing the views from which they started.

4. Stephen Daniels, "Re-visioning Britain: Mapping and Landscape Painting, 1750–1820," in *Glorious Nature: British Landscape Painting, 1750–1850,* ed. K. Baetjer (New York: Hudson Hills Press, 1993), p. 61.

5. See, for example, Richard DeBruin, *100 Topographic Maps* (Northbrook, Ill.: Hubbard, 1970). The term *physiography,* which we have met before in the work of Erwin Raisz, is interchangeable with *geomorphology.* Both terms refer to the graphic representation of landform features of the natural landscape. For a brief history of the usage of the two terms, see James O. Wheeler and Francis M. Sibley, eds., *Dictionary of Quotations in Geography* (Westport, Conn.: Greenwood Press, 1986), pp. 1–7.

6. Physiographic lines, even if they are not strictly chirographic or typographic in status, not only describe a given locality but edge toward a *geonarrative*: "Monadnock takes its rise just *here* (i.e., where a group of contour lines begins to group together into the shape of its base) and extends upwards toward its peak *there* (i.e., where the last complete contour is traced)" (DeBruin, *100 Topographic Maps,* p. 62). So too a painter's contour lines may describe a scene that is at once unrepeatable and quasi-narrative. In Constable's *Dedham Vale, Morning* (1811), for example, a moment's inspection reveals that the view is of a town in a particular location across a river as seen

from the elevation of a hill on the near side of the pictorial space: where town, river, and hill are each uniquely placed and nonexchangeable with other towns, rivers, or hills. Moreover, an implicit story is being told, a story to the effect that "in the early morning, when shadows are long at the sun's rising, peasants are already at work herding cows and walking horses, etc." (Later, such implicit narrativity will be seen as circumscribed by the deeply descriptive status of most landscape painting.) No such story is even suggested by looking at the Ordnance Survey Map of 1805 cited earlier. Even though this map *refers to* the very same entities mentioned in my description of Constable's painting—"Dedham," "Stour River," and "Langham" (i.e., the locality from which Constable viewed Dedham)—it does not describe them, much less weave them into a proto-narrative. Toponymy should not be confused with topography. Place-names do not constitute the kind of descriptive space that is at once topographically sensitive and narratively suggestive. The map in question is aptly named a "survey," for it gives us what the French call "the point of view of Sirius," a bird's-eye view that suspends the representation of specific features of the landscape. These very features, however, constitute the subject matter of a topographic map—and of Constable's paintings of the region that is thus mapped.

7. I take these designations from DeBruin, *100 Topographic Maps,* p. 10.

8. I am here referring to Gombrich's sense of schema as the organizing principle of visual works of art—a principle that is often conventional and historically specific even though it has been internalized by the artist to become an immanent feature of his or her representation of pictorial space. The schema, in other words, is the schema *of the pictorial image* that, together with other images, constitutes the identifiable subject matter of the sketch or painting. (See E. H. Gombrich, *Art and Illusion: A Study in the Psychology of Pictorial Representation* [Princeton, N. J.: Princeton University Press, 1960], chapter 1.) For a more general treatment of schemata in perception and memory at large, see Ulrich Neisser, *Cognition and Reality* (San Francisco: Freeman, 1976), chapter 4.

9. On the distinction between "geographia" and "chorographia" in Ptolemy's *Geography,* see Svetlana Alpers, *The Art of Describing: Dutch Art in the Seventeenth Century* (Chicago: University of Chicago Press, 1983), p. 134. In what follows, I am indebted to Alpers's insightful discussions in chapter 4 ("The Mapping Impulse in Dutch Art") of her remarkable book.

10. R. V. Tooley, *Maps and Map-Makers* (New York: Crown Publishers, 1978), p. 29.

11. For an account of these figures and others, see ibid., chapter 5 ("Holland and Belgium").

12. See *Catalogue Raisonné of the Works by Pieter Janszoon Saenredam* (Utrecht: Centraal Museum, 1961). I owe this reference to Alpers, who remarks that "In the seventeenth century Netherlands, from Pieter Saenredam early in the century to Gaspar van Wittel toward the end, artists were employed in executing maps and plans of all kinds. These have tended to be included (though mostly passed over lightly) in the monographs of art historians rather than in the studies of cartographers" (*The Art of Describing,* p. 128).

13. Thus Kenneth Clark calls Vermeer's painting a "unique work" in which "the rendering of atmosphere reached a pitch of perfection that, for sheer accuracy, has never been surpassed" (Kenneth Clark, *Landscape into Art,* new ed. [New York: Harper & Row, 1976], p. 65).

14. On this point, see Alpers, *The Art of Describing,* pp. 152–56, especially her statement that "Vermeer's *View of Delft,* then, is dependent on a tradition of topographical prints and is not the first painted view of a Dutch town to take over this design. In the midst of the renewed interest in many kinds of depictions of cities at mid-century, Vermeer's seems essentially to be a traditional, even conservative way to view his hometown" (pp. 154–55).

15. Ibid., p. 158. Alpers affirms that this painting transcends its own topographic origins in printmaking: "The accuracy of Vermeer's view has often been remarked . . . But still it seems mis-

taken to call this picture a topographical view. It is of a different order of rendering from the printed or even the other painted views [i.e., of contemporary mapmakers and painters]. It is endowed with an uncommonly seen and felt presence" (p. 156). It is this presence that I have just called a "clearing" in order to indicate how landscape painting can surpass its own topographic sources.

16. As does Clark in *Landscape into Art,* pp. 59–60: "This has been called the Age of Observation. It could almost be called the age of lenses."

17. Alpers, *The Art of Describing,* p. 136.

18. On the semantic equivalence of the Greek *graphō* and the Latin *scribo*—both of which mean "to write, to trace"—see ibid.

19. Ibid.

20. Ibid., p. 137.

21. "The word *landschap* was used to refer to both what the surveyor was to measure and the artist to render" (ibid., p. 136). Alpers refers to the Dutch title of a 1609 translation of a text by Frisius on triangulation: *Die maniere om te beschrijven de plaetsen ende Landtschappen* (The manner of describing places and landscapes).

22. The consequence for painting is that "like a surveyor, the painter is within the very world he represents" (ibid., p. 168). This is as true for interiors as it is for landscapes: in both cases, there is inclusion of the viewer in the space viewed.

23. Thus Hendrick Goltzius's landscape sketch of 1603—thought to be the very first to embody the conception of space here at stake—"makes us feel that we are situated apart from the land, but with a privileged view. It [exhibits] precisely the curious mixture of distance preserved and access gained" (ibid., p. 141).

24. Ibid., p. 139.

25. Gainesborough, who influenced Constable so greatly in his early efforts at landscape, had himself been decisively impressed by the Dutch landscapists of the seventeenth century. On this last point, see Gombrich, *Art and Illusion,* p. 317.

26. On the influence of Dutch landscape painting of the seventeenth century on the American Luminists, see Novak, *Nature and Culture,* pp. 232–36.

27. Ultimately, the "distance point" conception is Ptolemaic. See Alpers, *The Art of Describing,* pp. 53–57, 134ff.

28. Ibid., p. 147.

29. It is of interest that Kenneth Clark terms Dürer's studies "topographical": "the curiosity about the precise character of a particular spot, which was a part of the general curiosity of the fifteenth century, culminated in the topographical water-colours of Dürer" (*Landscape into Art,* p. 41).

30. For a discerning discussion of early modern mapping as parallel to the psychoanalytic exploration of the unconscious, see Tom Conley, *The Self-Made Map: Cartographic Writing in Early Modern France* (Minneapolis: University of Minnesota Press, 1996), pp. 7–13.

31. Alpers, *The Art of Describing,* p. 166. In this regard, the cartographic image was only an especially prominent case of a more general passion for learning about the world through images: "no other culture assembled knowledge through images as did the Dutch" (p. 165).

32. Ibid., p. 134. The same goes for late Renaissance and early modern literature—from Rabelais through Montaigne to Descartes—as Conley shows so tellingly (see his *Self-Made Map,* esp. chapters 4–8).

33. Immanuel Kant, *The Critique of Judgment,* trans. J. C. Meredith (Oxford: Clarendon Press, 1952), p. 107.

34. Alpers, *The Art of Describing,* p. 127.

35. It was thanks to this space that "the reach of mapping was extended along with the role of pictures, and time and again the distinctions between measuring, recording, and picturing were blurred" (ibid., pp. 134–35). For further discussion of the pictoriality of maps—especially in the explicit form of the pictogram—see Conley, *The Self-Made Map*, pp. 17–20. See also the excellent treatment of the pictoriality in maps and paintings in Malcolm Andrews, *Landscape and Western Art* (Oxford: Oxford University Press, 1999), chapter 4.

36. Alpers, *The Art of Describing*, p. 162.

37. This is Alpers's paraphrase of a remark found in chapter 19 of Apianus's text.

38. Cited by Alpers, *The Art of Describing*, p. 159, from Joan Blaeu, *Le Grand Atlas* (Amsterdam, 1663), pp. 1, 3. Joan Blaeu was the son of Willem Janszoon. We shall return to Blaeu's remark in the Epilogue.

39. Cited from Edward Norgate, *Miniatura*, p. 51, by Alpers, *The Art of Describing*, p. 157.

40. It is at this point that I begin to part company with Alpers, who fails to focus on place per se in her otherwise remarkable attempt to assimilate maps and paintings in seventeenth-century Dutch culture.

41. Martin Heidegger, *Being and Time*, trans. John Macquarrie and Edward Robinson (New York: Harper & Row, 1962), p. 138.

42. Ibid., p. 137.

43. Clark, *Landscape into Art*, pp. 60–61.

44. This overall lack of follow-through on Rembrandt's part stands in contrast with Constable, for whom (as we have seen) this was an exceptional circumstance.

45. Clark, *Landscape into Art*, p. 60.

46. Alpers, *The Art of Describing*, p. 139; my italics.

47. Ibid., p. 124; my italics.

48. Ibid., p. 160.

49. Ibid., p. 161. It is altogether revealing that this phrase of Alpers's is meant to describe what *maps* do. In my view, this is precisely what they *cannot do*. I certainly agree, however, with Alpers that neither Dutch maps nor Dutch landscapes represent actions or events. In Rembrandt's case, "the drama of human events" is the proper subject of his allegorical and historical paintings (e.g., *Aristotle Contemplating the Bust of Homer, The Night Watch*).

50. Ibid., p. 122.

51. Ibid., p. 141.

9. DISCURSIVE AND PRESENTATIONAL SYMBOLISM IN MAPS

1. See Edmund Husserl, "The Origin of Geometry," appendix to *The Crisis of European Sciences and Transcendental Phenomenology*, trans. David Carr (Evanston, Ill.: Northwestern University Press, 1970), pp. 353–78, as well as Jacques Derrida's Introduction to this essay in Jacques Derrida, *Edmund Husserl's Origin of Geometry: An Introduction*, trans. J. Leavy (New York: Nicolas Hayes, 1978).

2. On the idea of proto-geometric "limit-shapes" as constraints on the purely qualitative aspect of perceptual experience, see Husserl, *The Crisis of European Sciences*, pp. 26, 35.

3. A. R. Luriya has argued that such nonabstractive geometry is tied to an oral culture. His argument is summarized by C. D. Smith: "oral persons tend not to recognize or to have a discrete category for abstract shapes. They see a circle, for instance, as the object they know it represents, so that one circle is described as a plate, another as the moon, and so on" (C. D. Smith, "Cartography in the Prehistoric Period in the Old World: Europe, the Middle East, and North Africa," chapter 4 in J. B. Harley and David Woodward, eds., *The History of Cartography: Car-*

tography in Prehistoric, Ancient, and Medieval Europe and the Mediterranean [Chicago: University of Chicago Press, 1987], vol. 1, p. 58, with reference to Luriya's *Cognitive Development: Its Cultural and Social Foundations*).

4. Ibid., p. 88.

5. Strabo, *Geography,* book 2, chapter 5, section 16 (in the English translation of J. R. S. Sterrett [New York: Putnam's, 1917], vol. 1, p. 463). This book was written in the last years of the first century B.C.

6. A graticule is definable as "an imaginary network of meridians and parallels on the surface of the Earth or other celestial body" or as "a network of lines, on the face of a map which represents meridians and parallels" (H. M. Wallis and A. H. Robinson, eds., *Cartographical Innovations* [London: Map Collector Publications, 1987], p. 172). A grid, strictly speaking, is applicable to *any* flat or curved surface and is not confined to that of the earth alone.

7. John Noble Wilford, *The Mapmakers* (New York: Knopf, 1981), p. 36.

8. Arthur H. Robinson and Barbara B. Petchenik, *The Nature of Maps* (Chicago: University of Chicago Press, 1976), p. 49. Robinson and Petchenik are here restating Langer's position for their own purposes.

9. Susanne Langer, *Philosophy in a New Key* (Cambridge: Harvard University Press, 1951), p. 75; Langer's italics.

10. Ibid., p. 89.

11. Robinson and Petchenik, *The Nature of Maps,* p. 45.

12. For this conception of indexical signs, see Edmund Husserl, *Logical Investigations,* trans. J. N. Findlay (New York: Humanities Press, 1970), vol. 1, sections 1–15.

13. Robinson and Petchenik, *The Nature of Maps,* p. 52.

14. Ibid., p. 48. Despite providing information about places and more generally referring to them, maps still cannot be said to *represent* places (or regions) in any strong sense, that is, in their very particularity. This is a point I stressed in chapter 8.

15. Ibid., p. 53.

16. Ibid., p. 52. The authors add: "certainly maps are fundamentally much more nearly presentational than discursive" (ibid.). I believe that this is true, but the claim needs to be supported more thoroughly than is done in the brief compass of *The Nature of Maps.*

17. "The two systems, maps and language, are essentially incompatible" (ibid., p. 43). The authors also speak of the "nonconformity" of the two systems.

18. The close affinity of maps and certain literary forms is, however, undeniable. For a convincing illustration of this affinity in the case of the Renaissance and the early modern period, see Tom Conley, *The Self-Made Map: Cartographic Writing in Early Modern France* (Minneapolis: University of Minnesota Press, 1996), esp. chapters 1–3.

19. It has been estimated that the ordinary map conveys to its reader between 100 and 200 million "bits" of information.

20. Robinson and Petchenik, *The Nature of Maps,* p. 44.

21. The term *picto-ideogram* is found in Conley, *The Self-Made Map,* p. 19. What is for Conley an outstanding feature of early modern mapping (i.e., from c. A.D. 1400 to 1650) and is evidenced in such special formats as ornately decorated letters, condensed images, and signatures, I take to be characteristic of mapping of many periods, from Babylonian times to the present. Freud's analogy of the dream to the rebus—both being forms of *Bilderschrift*—is found in the opening paragraphs of chapter 6 of his *Interpretation of Dreams.*

22. Tony Campbell, "Portolan Charts from the Late Thirteenth Century to 1500," in Harley and Woodward, *The History of Cartography,* vol. 1, p. 446.

23. Amando Cortesào, *History of Portuguese Cartography* (Lisbon: Coimbra, 1969), vol. 1,

pp. 215–16. The portolan chartists were the first "to pursue mapmaking as a full-time commercial craft" (Wilford, *The Mapmakers,* p. 50).

24. Campbell, "Portolan Charts from the Late Thirteenth Century to 1500," p. 445.

25. "From the earliest extant copies, probably a little before 1300, the outline they gave for the Mediterranean was amazingly accurate" (ibid., p. 371).

26. See, inter alia, the chapter on medieval cartography in J. K. Wright, *The Geographical Lore of the Time of the Crusades* (New York: American Geographical Society, 1925), pp. 247ff. Wright comments, for example, on the "extreme of confusion and disregard for reality" that is found in one representative (Beatus type) of map: "here it is difficult to make out which continent is which" (p. 251).

27. This is not to say, however, that mythical regions were altogether absent from portolan charts. On the celebrated "Cantino" planisphere of 1502, for example, an inscription attached to an icon of the Atlas mountains of North Africa reads: "Land of the King of Nubia, who is continually at war with Prester John, and whose King is a Moor and a great enemy of Christians."

28. Campbell, "Portolan Charts from the Late Thirteenth Century to 1500," p. 429. A "rubricator" is someone who inscribes in red and, more generally, someone who colors.

29. "It became accepted practice for Portuguese overseas expeditions to carry a 'painter' to depict coastal scenes, which were later incorporated into the finished charts . . . The hydrographers of Dieppe, who learnt much from the Portuguese, also employed artist/mapmakers ('le peintre')" (Wallis and Robinson, *Cartographical Innovations,* p. 44).

30. In fact, we may trace a line of influence between the early Flemish landscapists and portolan chart makers in Portugal: "Flemish miniaturists' influence was predominant in Portugal from the second half of the fifteenth century [i.e., at the height of portolan chart creation]." The influence in question is especially evident in the Cantino planisphere. Of it and other like creations, we can say that "the style of the draftsmanship and decoration can be attributed to the influence of the Flemish miniaturists" (Michel Mollat du Jourdin and Monique de La Roncière, *Sea Charts of the Early Explorers,* trans. L. Dethan [London: Thames & Hudson, 1984], p. 215). Campbell points out that the production of portolan charts was almost exactly contemporary with the creation of the Book of Hours in Holland and in Germany—texts in which miniature landscapes first came into their own.

31. Lloyd A. Brown, *The Story of Maps* (Boston: Little, Brown, 1949), p. 163. Brown describes Ortelius's masterwork as "the first modern geographical atlas" and "a new kind of publication" (pp. 162, 163).

32. As a result, few portolan charts have survived. "The approximately 180 charts and atlases that can now be assigned to the fourteenth and fifteenth centuries must be a minute fraction of what was originally produced, and they are not necessarily representative" (Campbell, "Portolan Charts from the Late Thirteenth Century to 1500," p. 373).

33. See ibid., pp. 415–27.

34. Ibid., p. 371.

35. The rhumb lines, also called "loxodromes," are themselves aesthetically appealing parts of portolan charts, lending to them a special sense of order and scale while also interconnecting the various regions represented.

36. On this subject, see O. A. W. Dilke, *Greek and Roman Maps* (London: Thames & Hudson, 1985), chapter 6, "Land Surveying."

37. Campbell, "Portolan Charts from the Late Thirteenth Century to 1500," p. 372.

38. Ibid.

39. Ibid., p. 377. Campbell adds that "the headlands themselves frequently conform to one of a number of repeated types: pointed, rounded, or wedge-shaped" (ibid.).

40. The schematization is not always geometric. The Atlas Mountains are sometimes represented in the form of a bird's leg complete with claws and a spur: here we might speak of a spontaneous "animalization" of the landscape, akin to Gaston Bachelard's idea of the irrepressible "animalizing function" of the human imagination.

41. An inscription on this map reads: "Before and after *Taprobana* [Sumatra] there is a multitude of islands . . . 1,378 in all." As Mollat du Jourdin and La Roncière remark, these islands "gleam on this chart like many-hued enamels" (*Sea Charts of the Early Explorers,* p. 221). The anonymous author of the map has decided to create a cartographic presentation whose artistry and geometricity more than compensate for lack of specific detail.

42. "The Miller Atlas is too often thought of simply as a work of art, but its scientific value, which is of the first order, should not be underestimated" (ibid., p. 219).

43. The reason for this development is usually held to be the increasing use of the compass—which had been employed by the Chinese and the Arabs since at least A.D. 1100—with its orientation to magnetic north. On this question, see the article "Magnetic North" in Wallis and Robinson, *Cartographical Innovations,* section 4.133.

44. Campbell, "Portolan Charts from the Late Thirteenth Century to 1500," p. 378.

45. It is notable that the network of rhumb lines also "shares a common orientation with the coastal outlines [even if it] is otherwise unrelated to them" (ibid., p. 377).

46. I here allude to Lacan's theory of the "mirror stage" in the early formation of the human ego. Cf. Jacques Lacan, "The Mirror Phase as Formative of the I," in *Écrits,* trans. Alan Sheridan (New York: Norton, 1977), pp. 1–9.

47. Conley's entire book *The Self-Made Map* considers the consequences of mapping as made for selves by other selves—rather than as objective records of homogeneous space. In making his case, he invokes psychoanalytic categories that possess cartographic implications—for example, the unconscious, object of desire, and transitional object. The result is a provocative psychocartographic reading of the literary culture of early modernity. See especially the Introduction to his book, as well as the Conclusion: commenting on Jan van der Straet's copperplate engraving titled *Vespucci Discovering America* (1624), in which Vespucci gazes at an unclad female America rising from a hammock, Conley writes that "where [Vespucci] finds his origin in the virtual feminization of himself, in the female he believes [to be] arising from his own cartographic hammock, his map is surely self-made" (*The Self-Made Map,* p. 309).

48. Wallis and Robinson, *Cartographical Innovations,* p. 73 (section 1.203).

49. For a full treatment of such maps, see P. D. A. Harvey, "Local and Regional Cartography in Medieval Europe," in Harley and Woodward, *History of Cartography,* vol. 1, pp. 464–501.

50. See Susanne Langer, *Feeling and Form* (New York: Scribner's, 1953), chapter 1, on the special dynamics of the pure decorative line, above all the arabesque.

51. "The usual name for the system of [Roman] land division was *limitatio,* but as 'limitation' in English has a different sense, it is usual in English writing to have recourse to the alternative word *centuratio,* which is anglicized as 'centuration'. *Limitatio* was named from *limites:* a *limes* was literally a balk separating two ploughed fields, but in centuriation it normally became a road" (Dilke, *Greek and Roman Maps,* pp. 89–90).

52. The only significant exception to this rule occurs when the coastline itself is unknown, in which case the painting frames itself, often in a traditional rectangular form. This is just what occurs, for example, in the Miller Atlas miniature of the unexplored "Terra Bimene" above the south Florida coast. (Cf. Plate 34 in Mollat du Jourdin and La Roncière, *Sea Charts of the Early Explorers.*) The vignettes of life in Brazil and in Africa, in contrast, are carefully contained by the coastal outlines of these better-known regions.

53. Definitive and delimiting as a coastline is, in its actuality it possesses a complexity and ambiguity that cannot be captured by linear representation. It is complex in its fractal geometry because it is unending in its recursive substructures. It is ambiguous insofar as we cannot say precisely where land ends and water begins, given that the land usually extends under the water, which changes its own level continually in accordance with the changing tide. Maps assume fixed edges where there are none! (I owe this observation to Ann Cahill.)

54. Campbell, "Portolan Charts from the Late Thirteenth Century to 1500," p. 373.

55. Claudius Ptolemy, *Geography*, trans. E. L. Stevenson (New York: New York Public Library, 1932), p. 25.

56. Mollat du Jourdin and La Roncière, *Sea Charts of the Early Explorers*, p. 16.

57. Fernand Braudel, *The Mediterranean and the Mediterranean World in the Age of Philip II*, trans. Siân Reynolds (New York: Harper & Row, 1972), vol. 1, p. 106.

58. Ibid., p. 108.

59. Concerning the dread of the high sea, and above all the ocean—which circumscribes the known world and yet is itself boundless (*apeiron*: Anaximander's term)—see James S. Romm, *The Edges of the Earth in Ancient Thought: Geography, Exploration, and Fiction* (Princeton, N. J.: Princeton University Press, 1994), pp. 11–31. This book also contains an excellent treatment of circumnavigation or *periplous* (pp. 27–28, 31, 122, 164).

60. On these two terms, see Braudel, *The Mediterranean and the Mediterranean World in the Age of Philip II,* vol. 1, p. 103 n. 3.

61. "In the separate Mediterranean basins the coast would be seen everyday" (Campbell, "Portolan Charts from the Late Thirteenth Century to 1500," p. 441). Indeed, until the eighteenth century, sailing in a *costeggiare* fashion was accomplished by the use of portolan charts in simple techniques of dead reckoning (ibid.). For an informative discussion of *periploi,* see Dilke, *Greek and Roman Maps,* pp. 130–44.

62. I say "almost certainly," because an unresolved controversy concerns the question of the exact parentage of portolan charts: see Campbell's skeptical remarks ("Portolan Charts from the Late Thirteenth Century to 1500," pp. 386–89).

63. "Homer is said to have lived either on the west coast of Asia Minor or on an adjacent island. Since Miletus was the birth-place of Greek map-making, and since Homer not only showed a keen awareness of geography but evidently lived not far from there, it is appropriate that many later Greeks [e.g., Strabo] should have thought of him as the father of geography" (Dilke, *Greek and Roman Maps,* p. 20).

64. For a detailed discussion of this shield and its significance, see "The Foundations of Theoretical Cartography in Archaic and Classical Greece," edited from materials supplied by German Aujac in Harley and Woodward, *History of Cartography,* vol. 1, pp. 131–32. It is notable that Achilles' shield is present as both a map *and* a work of art.

65. See, for example, the reconstruction of Hecataeus's world map (c. 500 B.C.) by E. H. Bunbury, *A History of Ancient Geography among the Greeks and Romans from the Earliest Ages till the Fall of the Roman Empire* (New York: Dover, 1959), vol. 1, p. 148. Dilke remarks that "the river Oceanus surrounding the inhabited earth was a permanent concept in Graeco-Roman antiquity" (*Greek and Roman Maps,* p. 56). For more detail, see Romm, *The Edges of the Earth in Ancient Thought,* pp. 20–26, 148, 177–79. Romm interprets *oikumēnē* as "familiar world," that is, "a region made coherent by the intercommunication of its inhabitants, such that, within the radius of this region, no tribe or race is completely cut off from the peoples beyond it" (p. 37).

66. For an account of these surviving *periploi,* see Dilke, *Greek and Roman Maps,* pp. 130–44.

67. From the second half of the fifteenth century onward, sailing manuals often contained

such profiles along with "views of landmarks as seen from the sea" (Leo Bagrow, *History of Cartography*, rev. ed. by R. A. Skelton [Chicago: Precedent Publishing, 1985], p. 119). An example is Cornelius Anthontsz's printed *Caerte van die Oostersche See*. Anthontsz was a pilot, a painter, and a hydrographer.

68. Braudel, *The Mediterranean and the Mediterranean World in the Age of Philip II*, p. 108.

69. David Woodward, addition to the article by Tony Campbell in Harley and Woodward, *History of Cartography*, vol. 1, p. 388. "Traverse" here means route; more exactly, it is the zigzag course taken by a boat that is tacking against contrary winds.

70. Brown, *The Story of Maps*, p. 112; my italics.

71. E. L. Stevenson, *Portolan Charts: Their Origin and Characteristics* (New York: Hispanic Society of America, 1911), p. 4. Stevenson's book was the first in English to treat portolan charts as a distinctive genre of maps.

72. Cited from Scylax, in ibid., p. 6.

73. Ibid., p. 7. A "stade," from *stadion*, the Greek word for distance, was originally the distance covered by a plough before it had to be turned over; it came to signify about 607 feet—which would mean that there are approximately 8⅓ stades to a mile. See the discussion by Dilke, *Greek and Roman Maps*, pp. 32–33.

74. Cited from *M. Blundeville His Exercises* (London, 1622) by Brown, *The Story of Maps*, p. 113.

75. Cited by Stevenson, *Portolan Charts*, p. 11.

76. Cited by Dilke, *Greek and Roman Maps*, pp. 142–43.

77. Ibid., p. 143. By mentioning the arc of a smaller circle, Marcian refers to the situation in which walking around the arc (as one does on a beach in a cove) means more stades traversed than moving in a straight line across the arc—as happens with a ship that cuts straight across a cove. "Dead reckoning" is navigation without recourse to astronomy; typically, it takes into account only the course and the distance from a former position at sea. In "triangulation," an unknown point is located by constructing a triangle whose vertices are two known points and the unknown point itself.

78. In this instance—Brazil in the Miller Atlas—the inscription reads as follows: "The True Cross †, called by this name, which was found by Pedro Alvares Cabral, a nobleman of the house of the King of Portugal, and he discovered it when he went as captain-major of fourteen ships which the said King was sending to Calicut and going this way he met with this land here, which is believed to be a continent, in which there are many people who go about naked as their mothers delivered them: they are more white than brown and have very lanky hair. This said land was discovered in the year 500."

79. Francesco Beccari, "Address to the Reader," affixed to his chart of 1403 (Beinecke Library, Yale University); my translation.

80. Another instance of representational compounding is the inclusion of an image of the whole world, a cosmogram, as an inset within a map of a particular region: as occurs, for example, in Joan Riezo's 1590 portolan chart (in which a miniaturized hemisphere is inserted as an icon in an empty part of Africa) or in Giovannia Misima's chart of 1643, in which two hemispheres are cozily placed at the lower borders of a map of the Western Hemisphere. I have consulted both of these charts in the Yale University Map Collection.

81. I follow Brown's account of this development (see *The Story of Maps*, p. 121).

82. This brief review of the advantages of imagery in portolan charts leaves us with yet another pair of contrasting terms, simplicity-*cum*-complexity. On the one hand, such imagery assures clarity and convincingness in the representation of particular landscape features; it acts to simplify what often remains confused and cluttered in the sheerly verbal descriptions of *periploi*

and *portolani*. If a portolan chart can be defined as "a catalog of directions to follow between notable points," the directions are made graphically vivid by rhumb lines, and the notable points are brought lucidly to our attention by various cartographic images, iconic and symbolic alike. In both instances, imagery simplifies by intensifying. Yet, on the other hand, Mollat du Jourdin and La Roncière observe that "in place of the [medieval] T.O. ideogram, enclosed in its conventional frame, there is substituted [in portolan charts] the possibility of infinite representations of reality, based on the calculation of the positions of places and the distances separating them, and the construction of a system of compass roses proliferating from one to another" (this definition of L. Denoix is cited by Mollat du Jourdin and La Roncière, *Sea Charts of the Early Explorers,* p. 15). The five aspects of imagistic representation just reviewed all contribute to such "infinite representations," and thereby to an increasing complexity. The result is almost precisely the converse of written accounts of the same landscape: in these accounts, complexity of the inscription (a complexity that is at once discursive and syntactical) is paired with restriction in the range of representations of the geographically real. Both the verbal and the imagistic representation of a given coastline offer "a reflection of an observed reality" (ibid.)—that is to say, an at least *adequate* representation of it—but the reflection is ultimately much more delimited in the former case than in the latter.

83. This is not to deny that pictographic elements are often closely attached to the places they designate: in many early modern maps "the pictogram stands attached to the place that it both represents and remotivates" (Conley, *The Self-Made Map,* p. 19). But the place-name is free from the ambiguity that often afflicts a pictorial image: "Venice" designates one city alone in Italy, whereas a miniature icon of this city might also stand for another city with canals. Conversely, a verbal account of Venice might be long-winded in a way that no single emblem could ever be: Conley contrasts "a silent, spatial, schematic rendering of an area (in visual form) and a voluble, copious, emphatic discourse that strives to tell of the invisible history that the image cannot put into words" (p. 16).

84. I take the expression "verbal map" from Dilke, who employs it as a description of ancient *periploi* (*Greek and Roman Maps,* p. 134).

85. Mollat du Jourdin and La Roncière, *Sea Charts of the Early Explorers,* p. 16.

86. Cited in Stevenson, *Portolan Charts,* pp. 13–14.

87. This phrase is discussed in P. D. A. Harvey's *The History of Topographical Maps: Symbols, Pictures, and Surveys* (London: Thames & Hudson, 1980).

88. Jonathan T. Lanman, *On the Origin of Portolan Charts* (Chicago: Newberry Library, 1987), p. 6. For a critical assessment of Lanman's work, see Campbell, "Portolan Charts from the Late Thirteenth Century to 1500," p. 383.

89. As an instructive case of such noncontingent development, I would cite the evolution of portolan charts—with their exemplary combination of images and words—out of ancient *periploi*. The unremittingly verbal character of the nautical guides (whose origins are to be found in Phoenician sailors' manuals) is deftly and effectively supplemented by the highly imagistic nature of portolan charts, which at the same time retain an important verbal component. The evolution from one to the other is from a less to a more complete representation, and thus exhibits an internal necessity. This is so even if there is a lack of definitive empirical evidence for this evolution: "despite careful investigations, no direct line of descent has been able to be established between ancient *periploi* and these portolans" (Dilke, *Greek and Roman Maps,* p. 143). In spite of his skepticism, Dilke goes on to single out five specific features of *periploi* that could well have been taken over by portolan charts; forms of spatiotemporal measurement, compass directions, knowledge of world cartography, information about harbors and shores, and trade information.

90. Kenneth Clark, *Landscape into Art,* new ed. (New York: Harper & Row, 1976), p. 35.

91. A striking juxtaposition of these two kinds of map is found in Plates 1 and 2 in Dilke, *Greek and Roman Maps:* Plate 1 shows a detailed fragmentary plan of Nippur; Plate 2 is a world map that represents a transoceanic cosmic space.

92. Wallis and Robinson, *Cartographical Innovations,* p. 55. The authors add: "originally a stereographic projection of the celestial sphere, later a name associated with the astrolobe and, more recently, used to denote a number of projections of the sphere on the plane" (pp. 55–56).

93. Another alternative is presented by the inclusion of a flattened *mappa mundi* alongside a portolan chart—as occurs on the "Christopher Columbus Chart" of 1492. In this particular instance, a set of nine concentric circles around an image of the earth is placed on the neck of the chart with an inscription that reads (in part): "that the *mappa mundi,* although drawn on a plane, should be considered to be spherical." Here the chart maker appeals to the chart reader to *imagine* the three-dimensionality of the original on which the cosmogram is based and from which it is projected. (See Plate 21 in Mollat du Jourdin and La Roncière, *Sea Charts of the Early Explorers,* and the commentary on pp. 211–12.)

94. On the nature of cylindrical projections, first invented by Marinus of Tyre (c. A.D. 100), see Arthur H. Robinson, *Elements of Cartography* (New York: Wiley, 1953), pp. 27–28; and David Greenhood, *Mapping* (Chicago: University of Chicago Press, 1965), pp. 124–25. As Greenhood makes clear, the procedure of cylindrical projection is twofold: first, a projection of the spherical global surface onto a cylinder, then the imprinting of the cylindrical image onto a flat, two-dimensional receptive surface.

95. For an insightful and subtle discussion of mapping islands, real and imaginary, see the chapter titled "An Insular Moment: From Cosmography to Ethnography" in Conley's *The Self-Made Map,* pp. 167–201.

96. Cited in Mollat du Jourdin and La Roncière, *Sea Charts of the Early Explorers,* p. 206, with reference to Plates 13 and 14.

97. See Plate 76 in ibid. and the comment on p. 254 that "the relief is painted in elevation, accurately so in the case of Crete."

98. On this question, opinion remains divided. On the one hand, it has been assumed for a long time that portolan charts are altogether projectionless—that, as the naive, rough-and-ready creations of seafarers, they are unlikely to have been constructed on the basis of classical conic projections, which require an awareness of fairly advanced geometric conceptions. On the other hand, it is arguable that, nevertheless, an at least *implicit* projective geometry is at work: "to start with, the expression 'projectionless chart' contains an internal contradiction. Any drawing that aims at delineating a portion of the earth's surface must involve a certain connection between that portion and the points of the chart, that is to say, a certain type of projection, even if the connection is not made explicit" (ibid., p. 15). In this latter spirit, it is even possible to speak of portolan charts dating from the fourteenth century onwards as involving "a kind of implicit Mercator's projection" (ibid.). For a thorough discussion of this debate, and a despairing sense that it will not be soon (if ever) resolved, see Campbell, "Portolan Charts from the Late Thirteenth Century to 1500," pp. 385–86. Reinforcing the argument for the comparative sophistication of portolan charts is the fact that most of them made use of magnetic north and that sailing by alignment with rhumb lines—which stem from the wind roses that incorporate information regarding magnetic north—was an effective, if not infallible, way of sailing across open seas. On this last point, see Norman J. W. Thrower, *Maps and Civilization: Cartography in Culture and Society* (Chicago: University of Chicago Press, 1996), pp. 51–57.

99. On this influence, especially that of Ortelius, on Gramolin's map, see Mollat du

Jourdin and La Roncière, *Sea Charts of the Early Explorers,* p. 254. Still other Dutch mapmakers also created planispheric maps: for example, William Barentsz in his *Caertboeck van de Midlandtsche Zee* of 1595. (For an example of Barentsz's work, see Plate 4 in R. Putnam, *Early Sea Charts* [New York: Abbeville Press, 1983].)

100. For further details of this connivance, see Svetlana Alpers, "The Mapping Impulse in Dutch Art," chapter 4 of Alpers's *The Art of Describing: Dutch Art in the Seventeenth Century* (Chicago: University of Chicago Press, 1983), pp. 119–68.

101. Concerning the landscape of fact, see Clark, *Landscape into Art,* chapter 2.

102. *La Cosmographie universelle* is the title of a book written by André Thevet in 1575 at the height of the French Renaissance. It is discussed by Conley in *The Self-Made Map,* pp. 187–201.

103. See, for example, Plate 36 in Putnam, *Early Sea Charts,* p. 69. The only vestige of depth in a chart such as this is the paradoxical visual effect of rhumb-line centers, which *seem* to recede into (and sometimes, conversely, to rise out of) the flat surface of the map. But the depth at stake here is a pseudodepth: wholly abstract, it does not *represent* any depth on the earth's surface itself.

10. FAR-OUT MAPPING

1. Cited from the *Tso Chuan* in Joseph Needham, *Science and Civilization in China* (Cambridge: Cambridge University Press, 1959), vol. 3, p. 503. I am deeply indebted to Needham's magisterial study of Chinese geography and cartography. It is the only comprehensive single study in any Western language. But to be consulted as well are the seven chapters on the topic "The Cartography of China" in J. B. Harley and David Woodward, eds., *The History of Cartography,* vol. 2, book 2, *Cartography in the Traditional East and Southeast Asian Societies* (Chicago: University of Chicago Press, 1994).

2. Needham remarks that the "numina of localities" on the Cauldrons are instances of "*pai wu,* the 'hundred beings', though the less familiar wild animals would also be included. Travelling officials or envoys would know what god should be honored with sacrifice *in what place*" (*Science and Civilization in China,* vol. 3, p. 504; my italics). Here the *genii loci* not only inhabit local places but establish these places themselves as culturally and religiously significant.

3. "From the first century A.D. onwards, each of the official histories contains a geographical section (Ti Li Chih), the whole forming an immense compilation concerning the changes in place-names and local administrative divisions controlled by the dynasty, descriptions of mountain ranges, river systems, etc." (ibid., p. 508). Needham also comments that local topographical writings in China "are probably unrivalled by any nation for extent and systematic comprehensiveness . . . there is hardly a town, however small, which does not have its own historical geography" (pp. 517–19).

4. On this ceremony, see ibid., p. 537.

5. See Needham's discussion at ibid., pp. 547–49, in reference to Figures 225 and 226. We shall return to these maps, together called *yujitu,* later (see nn. 22, 59).

6. The use of silk for maps "would invite the suggestive idea that the position of a place could be fixed by following a warp and a weft thread to their meeting place" (ibid., p. 538). It is to be noted that the employment of silk goes back to an extremely early date. Two silk maps have been discovered in a Han tomb of the second century B.C. This means that they preceded the two stone stele maps—previously the earliest known surviving maps—by thirteen hundred years and that on the earth as a whole they are preceded only by the Mont Bégo and Valcamonica petroglyphs, Babylonian clay tablet maps, and the wall map at Çatal Hüyük. Mei-Ling Hsu comments

that "these Han maps, which are of high quality in terms of scale consistency, information content, and use of symbols, are much more advanced than the Babylonian tablets . . . These maps provide evidence of a much higher level of cartographic achievement in ancient China than had previously been realized" (Mei-Ling Hsu, "The Han Maps and Early Chinese Cartography," *Annals of the Association of American Geographers* 68 [1978]: 45).

7. The Inuit of Greenland carve relief maps out of twisted lengths of wood so as to represent groups of islands in relation to a landed coast. These three-dimensional, nonverbal *periploi* are carried in kayaks as navigational aids. In the West, the earliest known sculpted relief map is that reported by Ibn Battuth, an Arab mapmaker of the fourteenth century who saw such a map at Gibraltar.

8. "Eight pieces of wood were used, with hinges to connect them together. The map could be folded up and one person could carry it. Whenever he travelled he took this along with him" (cited from the *Ho Lin Yü Lu* [twelfth century A.D.] by Needham, *Science and Civilization in China,* vol. 3, p. 580).

9. Another, even earlier relief map bears mentioning: a tomb chamber of the third century B.C. in which the Yangtze and the Yellow rivers, the Eastern Sea, the nighttime sky, and the geography of the earth were all carved. Mercury flowed through special channels to imitate the waterways. (Cf. ibid., p. 582.) This was in effect an entire world-model in three dimensions.

10. Even this single case is that of a globe imported into China by the Arab cartographer Jamal al-Din in A.D. 1267. This is discussed in ibid., pp. 374 and 583.

11. For this interpretation of Caravaggio and of Picasso's paintings of the "monumental" period of the 1920s, see Frank Stella, *Working Space* (Cambridge: Harvard University Press, 1987). I owe this reference to Eve Ingalls.

12. I have in mind Eugen Fink's notion of *"Träger"* as discussed in his 1927 monograph "Vergegenwärtigung und Bild," reprinted in Eugen Fink's *Studien zur Phänomenologie* (The Hague: Nijhoff, 1966), pp. 1–78.

13. Cited by Needham from the *ShihI Chi* of Wang Chia (late third century A.D.) in *Science and Civilization in China,* vol. 3, p. 538; my italics. It is of interest to note that the woman who was thus called upon refused to paint a map and instead "suggested that as the colors of a drawing [i.e., painting] would fade it would be better to make the map in embroidery, and this was accordingly done" (ibid.). In this way, silk came to be preferred to paper as a cartographic medium.

14. Cited by Needham from the *Ho Lin Yü Lu* (ibid., p. 580).

15. Compare the two examples reproduced side by side in Figure 224 in ibid., p. 546. The first of these comes from the seventeenth-century edition of the *Wu Yo Chen Hsing Thu* (whose printed text is of a still earlier origin); the second comes from a modern textbook of mapping.

16. Only a few authentically Chinese cosmographic maps have survived—for example, that of Jen Chao of 1607, which is based on much earlier examples. Almost all of these are orbocentric with regard to Mount Khun-Lun (the equivalent of Mount Meru), reminding one of the *mappae mundi* that placed Jerusalem at their center. But the main interest in this kind of map appears to have arisen in Korea. As Needham remarks, "there can be no reason for doubting that the Koreans received this tradition from China, though it seems never to have been so popular there" (ibid., p. 565).

17. This is recorded in the *Chou Li* as cited by Chen Cheng-siang, "The Historical Development of Cartography in China," *Progress in Human Geography* 2 (1978): 102. The Chinese shared with the Romans a passion for both land surveying and the military use of cartography; but it is notable that few major advances in cartography itself were made in the period of Roman supremacy in the West. On this question, see O. A. W. Dilke, *Greek and Roman Maps* (London:

Thames & Hudson, 1985), chapters 6–8. On Chinese land survey methods, see Needham, *Science and Civilization in China,* vol. 3, pp. 569–79. Another motive for precise mapping was provided by the sea expeditions undertaken by the Chinese, who reached Africa long before the Portuguese circumnavigated that continent.

18. As Needham says, "just as the scientific cartography of the Greeks was disappearing from the European scene, the same science in different forms began to be cultivated among the Chinese" (*Science and Civilization in China,* vol. 3, p. 533).

19. Needham's considered judgment is that Western portolan charts were "at least partly dependent upon the transmission of the magnetic compass from China" (ibid., p. 587). He speculates that Arab traders carried the discovery to the West; an Arab community was established in Canton from the eighth century A.D. onward. (Cf. ibid., pp. 561–65.) On the other hand, it must be observed that the Chinese developed their *own* version of portolan charts before any European influence had occurred: for example, the sea charts used by the eunuch general Chêng Ho in his fifteenth-century expeditions. (An example of such a chart is reproduced as Figure 236 in ibid., p. 560.) These charts are notable for their lack of grid combined with attention to topographic detail. "Indeed," comments Needham, "they correspond, not only in nature, but also in date (early fifteenth century A.D.) with the portolan charts of Europe, the only difference being that they give their compass bearings in words instead of drawing rhumb-lines from arbitrarily chosen centers" (ibid.). The specific influence of the Europeans manifests itself in the addition of rhumb lines to such charts after the early seventeenth century. For a skeptical assessment of the extent of Western influence on Chinese cartography, see Cordell D. K. Yee, "Traditional Chinese Cartography and the Myth of Westernization," in Harley and Woodward, eds., *The History of Cartography,* vol. 2, book 2, pp. 170–72, especially the conclusion that, despite interest in Ptolemaic geography at the imperial court, "traditional Chinese cartographic practices continued unabated" (p. 187).

20. On these technical questions, see the articles titled "Magnetic North," "Loxodromes," and "Portolan Charts" in H. M. Wallis and J. H. Robinson, eds., *Cartographical Innovations* (London: Map Collector Publications, 1987).

21. This is the definition of "grid" given in section 4.073 of ibid., p. 174. It is important here to distinguish once again between grid and "graticule," which is defined as "a network of lines, on the face of a map, which represents meridians and parallels" (p. 172). In contrast with the system that is at play in the grid, a coordinate system composed strictly of latitude and longitude lines constitutes the graticule, which involves the two-dimensional projection of the terrestrial sphere. Such a system was first introduced by Hipparchus of Rhodes in the second century B.C. and was refined by Peter Apian in his appendix to the 1486 edition of Ptolemy's *Geography.* It was not employed by the Chinese or Japanese until well into the eighteenth century.

22. The earliest extant map to employ an unmistakable grid system is known as the *yujitu,* a stone stele on which the map is carved. It dates from 1136. Another version of the same map dates from 1142. It has been speculated that the grid unit, the square, derives from a Han graph for "field," *tian,* or from the graph *jing,* "well."

23. None of these roots is sufficient by itself alone, and each is problematic. Thus, Wallis and Robinson warn that a cartographically significant grid system "is not to be confused with the grid-like appearance on a map of the lines established by a rectangular land division system" (*Cartographical Innovations,* p. 174). Even if it can be claimed that the earliest known Chinese map was a set of encircling rectangles—see Figure 204 in Needham, *Science and Civilization in China,* vol. 3, p. 502—the set does not constitute a grid in any strict sense, any more than does an abacus. At the most, we might speak of adumbrations of the grid system in these suggestive phenomena.

24. Scale is shown by the inclusion of a written statement indicating the linear distance that

each side of a square on the map represents. This is presented as a ratio. Thus, on the *yujitu* a note engraved on the side of the map gives the working ratio as 1:100, that is, each side of a square equals one hundred *li*. This allows for the determination of distance and area, in contrast with the graticule, which only permits the determination of relative location on the map as a whole. Furthermore, the Chinese also used the grid to alter map scales by means of the method of "similar squares," for example, in Lo Hung-Hsien's world atlas *Kuang Yu Thu* of 1555: here the sizes of the map squares remain similar, but they represent different linear distances. Finally, the square grid may be used as a referencing system: by counting off squares, one can locate any one point in relation to another point. (On all three functions, see ibid., p. 203 [article titled "Square Grid"].)

25. Cited from Fan Ye, *Hou Han Shu,* chapter 59, in Needham, *Science and Civilization in China,* vol. 3, p. 538.

26. Cordell D. K. Yee, "Taking the World's Measure: Chinese Maps between Observation and Text," in Harley and Woodward, *The History of Cartography,* vol. 2, book 2, p. 112.

27. Cited from the *Chin Shu* in Needham, *Science and Civilization in China,* vol. 3, p. 539; my italics. It is to be noted that the interpretation of *chun wang* as "grid" is controversial. Needham takes it for granted, but Yee questions it in "Taking the World's Measure," p. 110 n. 45 and pp. 125–27. According to Yee, *chun wang* is rather the use of a single reference point for measurements of distance and direction. Thus, according to Yee, *chun wang* should be translated not as "rectangular grid" but as "regulated sighting." I shall, however, retain the traditional translation.

28. I adapt these principles from Needham's translation, *Science and Civilization in China,* vol. 3, pp. 539–40.

29. Chen Cheng-siang, "The Historical Development of Cartography in China," p. 104.

30. Cited in Needham, *Science and Civilization in China,* vol. 3, p. 540; my italics.

31. See Martin Heidegger, "The Age of the World Picture," in *The Question concerning Technology and Other Essays,* trans. W. Lovitt (New York: Harper, 1977), pp. 115-54. See also chapter 12 in this volume.

32. Phei Hsiu cited in Needham, *Science and Civilization in China,* vol. 3, p. 540.

33. These three methods correspond, respectively, to the fourth, fifth, and sixth of Phei Hsiu's six principles. For further discussion of each, see Yee, "Taking the World's Measure," pp. 110–11. All three methods "deal with the problem of converting actual ground distances, which may be curved in both horizontal and vertical dimensions, to straight-line distances and depicting them on a flat map" (ibid., p. 111).

34. For a discussion of this iconomorphic symbolism, see Chen Cheng-siang, "The Historical Development of Cartography in China," p. 102. Such symbolism is already evident in the maps recovered from the Han tomb of the second century B.C.; see Mei-Ling Hsu, "The Han Maps and Early Chinese Cartography," pp. 47–54, where the extreme ingenuity of this earliest known system of symbols is strikingly apparent.

35. On the naturalistic-pictorial origin of such Chinese characters, see Needham, *Science and Civilization in China,* vol. 3, pp. 497–98.

36. Chen Cheng-siang, "The Historical Development of Cartography in China," p. 118.

37. Yee, "Taking the World's Measure," p. 127. This is reminiscent of the "practical complementarity" that we have observed in the case of Western maps that combine presentational with discursive symbolism. Yee notes, however, that "the ratio of image to text in traditional Chinese geographic words varies, but it is not uncommon to find a disproportionate amount of text" (p. 59). In general, "maps were often placed in contexts where they complemented verbal representations of geographic knowledge. As a means of storing information, both verbal and graphic modes of representation were held to have their uses" (p. 128).

38. Cited in Needham, *Science and Civilization in China,* vol. 3, p. 540.

39. For other examples of surviving Chinese square grid maps, see Wallis and Robinson, *Cartographical Innovations,* pp. 202–3. I say "appear to be" in view of Yee's skepticism cited in earlier notes.

40. See Marcel Destombes, "A Venetian Nautical Atlas of the Late 15th Century," *Imago Mundi* 12 (1955): 14–33. Wallis and Robinson remark on the basis of this example that "it has been suggested that the Chinese square grid stimulated 13th and 14th century Mediterranean navigators and cartographers to create portolan charts" (*Cartographical Innovations,* p. 203). The authors themselves, however, prefer the view that the grid system in question may derive from the classical period, perhaps even from a lost sea chart of Marinus of Tyre. But in favor of a Chinese origin is the fact that the placement of south at the top of maps was an unusual feature that is found, with rare exceptions, in Chinese military and topographic maps from the second century B.C. onward and in a significant number of portolan charts (which were either omnidirectional or south-oriented).

41. For an account of the intimate relationship between landscape painting and maps in China, see Cordell D. K. Yee, "Chinese Cartography among the Arts: Objectivity, Subjectivity, Representation," in Harley and Woodward, *The History of Cartography,* vol. 2, book 2, pp. 147–53. Yee establishes that at significant moments (e.g., the Tang) maps were considered a form of painting. More generally, "cartography in China did not emerge as a representational practice fully independent from the visual and literary arts until late in the imperial period, under the influence of Western examples" (p. 128).

42. Nanba Matsutaro, "The Pleasures of Collecting Old Maps," in *Old Maps in Japan,* ed. Nanba Matsutaro, Muroga Nobuo, and U. Kazutaka, trans. P. Murray (Osaka: Sogensha, 1973), p. 146. Mapmaking, adds Matsutaro, "penetrated deeply into the lives of the Japanese people" (ibid.).

43. "In order to accomplish the division of farm land and its uniform distribution to the cultivators, detailed cadastral maps and registries were constructed" (Muroga Nobuo, "The Development of Cartography in Japan," in ibid., p. 158).

44. Matsutaro, "The Pleasures of Collecting Old Maps," p. 147.

45. For examples, see Plate 23 and Figure 2 in Matsutaro, Nobuo, and Kazutaka, *Old Maps in Japan.* The dragon, a Buddhist symbol, was originally thought to protect Japan from Mongolian invaders. At a later point, Japan was sometimes shown in the form of a sheatfish (e.g., as in ibid., Figure 45).

46. Nobuo, "The Development of Cartography in Japan," p. 159. It should be mentioned that there was also a tradition of cosmographic maps—surprisingly akin to *mappae mundi* in the West during this same long period. Such maps depicted "the Land of the Buddha," that is to say, India, with Mount Meru at its center. See, for example, Gotenziku Zu's "Map of the Five Indies" of 1364 (Plate 1 in Matsutaro, Nobuo, and Kazutaka, *Old Maps in Japan*), which is strikingly reminiscent of the Hereford and Ebstorf maps of approximately the same period in the West. However, no direct lines of influence, from East to West or vice versa, have been detected for this case of cultural parallelism.

47. "Instead of being based on an out-of-date map on an oval projection such as Matteo Ricci's [i.e., stemming from his mission in China], these [maps of the world] were based on Dutch hemispheric world maps or stereographic projections" (Nobuo, "The Development of Cartography in Japan," p. 164, with reference to Plates 13 and 14 in Matsutaro, Nobuo, and Kazutaka, *Old Maps in Japan*). For splendid examples of early Edo screen maps of the world, see Plates 2, 3, and 5 in *Old Maps in Japan.* For an account of the effect of the Dutch on Japanese cartography, see Katzataka Unno, "Cartography in Japan," in Harley and Woodward, *The History of Cartography,* vol. 2, book 2, pp. 432–43.

48. For Senseki's map, see Figure 20 in Matsutaro, Nobuo, and Kazutaka, *Old Maps in Japan.*

49. Two excellent examples are the map of Sunpu of 1616 and the map of Kyoto of 1641, reproduced as Plates 91 and 93, respectively, in Takejiro Akioka, ed., *Collection of Old Maps of Japan,* (Tokyo: Kajima Institute, 1971).

50. The Japanese remain indebted, for example, to the Dutch employment of copperplate engraving, which foreshadowed their own multicolored wood-block prints of the late eighteenth century.

51. Matsutaro, "The Pleasures of Collecting Old Maps," p. 154.

52. Ibid.

53. By saying "fully at one," I mean a mode of fusion that cannot even be described as hybrid in character. Thus I disagree with Nobuo's way of putting the revolution in question: "these beautiful works, produced with no concern for practicality, are hybrids that lie between maps and *ukiyoe*-style landscapes that appeared toward the close of the eighteenth century" (Nobuo, "The Development of Cartography in Japan," p. 160). Hybrids (e.g., centaurs or donkeys) keep visible at least remnants of their disparate origins; in full fusion, these remnants are dissolved in each other's presence.

54. See, for example, Plate 35 in Matsutaro, Nobuo, and Kazutaka, *Old Maps in Japan*: "Picture-Scroll of the Sea and Land Routes from Edo to Nagasaki" by Dochu Emakimono. Such a strip map bears comparison with pilgrimage maps made in nineteenth-century Tibet. (On the latter, see Barbara Nimri Aziz, "Tibetan Manuscript Maps," *Canadian Cartographer* 12:1 [1975]: 71–78; and the same author's "Maps and the Mind," *Human Nature* [1978]: 50–59.)

55. In this way, we may say that Japanese effected its *second* triumph: having outdistanced the Dutch in their own enterprise, Japanese mapmakers took a step that their own earlier dependency on Chinese cartography would not have led one to predict. Both of these cases are to be contrasted with that of portolan charts, which the Japanese took over without significantly modifying (e.g., they simplified the wind rose from sixteen to twelve points). Similarly, the Japanese paid only lip service to the graticule employed on most Western maps made after Mercator, using latitude and longitude lines mainly for decorative purposes.

56. *Ukiyoe* techniques were also employed in the creation of rebus maps in which the acoustic properties of words signified by lively pictograms are exploited. These maps, made originally for illiterates, were constructed such that "the place names are not given in letter but in pictures . . . the sound of the object shown makes up a syllable of the place name" (Matsutaro, "The Pleasures of Collecting Old Maps," p. 150; with reference to Plate 89 of Matsutaro, Nobuo, and Kazutaka, *Old Maps in Japan*).

57. See Plate 50 in Takejiro Akioka, *Collection of Old Maps of Japan.*

58. Merleau-Ponty writes that "the enigma [of depth] consists in the fact that I see things, each one in its place, precisely because they eclipse one another, and that they are rivals before my sight precisely because each one is in its own place" (Maurice Merleau-Ponty, "Eye and Mind," trans. C. Vallery, in *The Primacy of Perception,* ed. J. Edie [Evanston, Ill.: Northwestern University Press, 1964], p. 180).

59. In a map of 1596, van Linschoten was "the first to attain correct views on the Chinese coastline" (Needham, *Science and Civilization in China,* vol. 3, p. 586). This is not to deny that the *yujitu* of 1136 and 1142 has an accuracy that, "especially in its representation of the rivers and coastline [of China] is remarkable" (Cordell D. K. Yee, "Reinterpreting Traditional Chinese Geographical Maps," in Harley and Woodward, *The History of Cartography,* vol. 2, book 2, p. 48).

60. Thus, even if it is true that maps by *ukiyoe* artists "cannot be expected to incorporate a high degree of precision by cartographic standards" (Nobuo, "The Development of Cartography

in Japan," p. 172), this is not of great concern. The Japanese (and certainly we today) have the pleasure of savoring a map such as Keisai's without being forced to use it as the only source of cartographic precision. It is striking that just as his map was being produced, the most accurate and reliable cartographic representation of Japan was in the process of completion—namely, Ino Tadataka's monumental *Dai-Nihon Enkai Jissokuroku* (Coastal survey record of Japan), which appeared in fourteen volumes in 1821. With the availability of this latter resource, map connoisseurs and ordinary map lovers could turn unashamedly to the delights of *ukiyoe* landscape maps.

61. An example of a nondirectional *ukiyoe* map is Katsushika Hokusai's panoramic map of Kisoji highway of 1819 (Plate 37 in Matsutaro, Nobuo, and Kazutaka, *Old Maps in Japan*). Maps oriented around particular cities are found in ibid., Plates 62 and 63. Wallis and Robinson remark that "prominent structures such as a castle were often used to orient city maps, as found in some maps of Osaka with eastern orientation. Major topographical features also influenced the arrangement. An example is *Fujima Jusanshu Yochi Zenza* (Complete map of the thirteen provinces commanding a view of Mount Fuji) by Akiyama Einen, Edo period, 1843" (Wallis and Robinson, *Cartographical Innovations,* section 4.151, "Orientation").

62. Wallis and Robinson, *Cartographical Innovations,* section 4.151. I do not mean, however, to minimize the importance of the practical consideration raised in this statement. There is an important correlation—a scale of yet another sort—existing between the space of cartographic presentation (i.e., the actual map surface) and the space of what is represented in that presentation itself. The *ukiyoe* map artists were masters of this correlation, as we can see in the case of Keisai's map.

63. See, for example, Hashimoto Gyokuransai's map of Mikawa province of circa 1870 (Plate 54 in Matsutaro, Nobuo, and Kazutaka, *Old Maps in Japan*).

64. Claudius Ptolemy, *Geography,* trans. E. L. Stevenson (New York: New York Public Library, 1932), p. 26.

11. RECTANGULARITY AND TRUTH

1. Claudius Ptolemy, *Geography,* trans. E. L. Stevenson (New York: New York Public Library, 1932), p. 25.

2. Ibid., p. 26.

3. "Chorography is most concerned with what kind of places those are which it describes, not how large they are in extent. Its concern is to paint a true likeness, and not merely to give exact position and size . . . Chorography does not have need of mathematics, which is an important part of Geography" (ibid.). See also Lucia Nut, "Mapping Places: Chorography and Vision in the Renaissance," in Denis Cosgrove, ed., *Mappings* (London: Reaktion Books, 1999), pp. 90–108.

4. An example of a nested rectangular map—consisting in a series of rectangles encased in each other—is what is now regarded as the earliest known Chinese map, a bronze tablet depicting a projected mausoleum complex in plan from the fourth century B.C. For discussion, see Cordell D. K. Yee, "Reinterpreting Traditional Chinese Geographical Maps," in J. B. Harley and David Woodward, eds., *The History of Cartography* (Chicago: University of Chicago Press, 1994), vol. 2, book 2, pp. 36–37.

5. For further discussion of this map, see McGuire Gibson, "Nippur: New Perspectives," *Archeology* 30:1 (1977): 34–37.

6. O. A.W. Dilke, *Greek and Roman Maps* (Ithaca, N. Y.: Cornell University Press, 1985), p. 87.

7. On the system of centuriation, see ibid., pp. 88–91.

8. Cited from Frontinus in ibid., p. 96.

9. For examples, see Plates 9–15 in ibid.

10. What distinguishes the case of the United States from that of Rome is its attitude toward wilderness—an encounter with the untamed and the unknown that has no exact precedent in Roman imperialism. On this point, see William H. Goetzmann, *Exploration and Empire* (New York: Knopf, 1966).

11. For this debate, in which the two phrases quoted in this sentence figured prominently, see Lowell O. Stewart, *Public Land Surveys: History, Instructions, Methods* (Ames, Iowa: Collegiate Press, 1935), pp. 3ff.

12. Cited from the Ordinance of 1785 as reprinted in C. Albert White, *A History of the Rectangular Survey System* (Washington, D.C.: U.S. Department of Interior, n.d.), p. 12. The north/south lines are meridians, and those crossing them at right angles are parallels. Left out of consideration is the fact that the meridians converge slightly as they proceed northward—thus making any rectangle based on them inherently imperfect.

13. Stewart, *Public Land Surveys,* p. 87. Burt was well known for his invention of the solar meridian compass, which allows the determination of the true meridian of a place.

14. Reported by Norman J. W. Thrower, *Maps and Civilization: Cartography in Culture and Society* (Chicago: University of Chicago Press, 1996), p. 138. Thrower comments that in the county atlases that flourished in the wake of the Rectangular Survey "the property lines and roads commonly extend north-south and east-west without reference to rivers or, for that matter, to the form of the land. As is typical of older cadastral maps, no relief is shown, but the size and ownership of properties and the location of dwellings are emphasized" (ibid.).

15. See, for instance, the surveys of a Louisiana township with waterfront tracts and of a Florida township in White, *A History of the Rectangular Survey System,* Figures 24 and 40.

16. Cited in Stewart, *Public Land Surveys,* p. 32.

17. Cited in ibid., p. 15. Cf. also the statement from the 1856 Report of the Surveyor General of Oregon: "scarcely an unsurveyed township can be found without canyons, ravines, or precipitous hills" (cited in ibid., p. 86). As if to anticipate such a case, in an amended ordinance of March 11, 1811, Congress expressly authorized deviations from the standard rectangular system.

18. Reprinted in White, *A History of the Rectangular Survey System,* p. 12.

19. It is noteworthy that even in the absence of such obstacles many survey maps included significant sinuosities of boundary lines: for example, in the Michigan Resurvey Map of 1845 that is given as Figure 38 in ibid.

20. "The act of 1785 was defective . . . in that it made no provision for the convergence of the meridians . . . Inaccurate surveys [thus] make some of the sections east of the Scioto River in southern Ohio very uncertain in size" (ibid., p. 36). This consequence makes it clear that the grid system of mapping—whether employed in China or the United States—presumes a flat earth, in contrast with the graticule, which expressly takes the earth's curvature into account.

21. Stewart, *Public Land Surveys,* p. 11; my italics.

22. Cited in ibid., p. 50.

23. Ibid., p. 46.

24. Cited from the "General Instructions" of 1833 in White, *A History of the Rectangular Survey System,* p. 294; my italics.

25. Cited in Stewart, *Public Land Surveys,* p. 142.

26. "At all those points where the township or section lines intersect the banks of such rivers, bayous, lakes or islands, posts are to be established" (cited in White, *A History of the Rectangular Survey System,* p. 298).

27. "Of Field Notes," cited in ibid., p. 299.

28. An example of an entry in such a book is as follows: "North boundary T. 26, R. 20 E.,

3d mile due East. Through open woods, hilly poor land, pine and oak . . . touch a bend of the river thus high bluff bank" ("Specimen Field Notes for the Use of Deputy Surveyors in the State of Mississippi," prepared by Elijah Hayward, Esq., July 28, 1831; cited in ibid., p. 276).

29. As reported in Cordell D. K. Yee, "Chinese Cartography among the Arts: Objectivity, Subjectivity, Representation," in Harley and Woodward, *The History of Cartography*, vol. 2, book 2, p. 154 n. 75.

30. Cited in Seymour I. Schwartz and Ralph E. Ehrenberg, *The Mapping of America* (New York: Abrams, 1980), p. 263.

31. See Plate 192 in ibid., p. 304. Schwartz and Ehrenberg comment that "awe-inspiring views such as this one of a portion of the Kaibab division of the Grand Canyon were drawn by topographic artists accompanying many of the Federally sponsored western expeditions from the 1850's onwards" (p. 305). Examples of aesthetically pleasing and topographically sensitive survey maps include William H. Jackson's *Map of the Mining District of California* of 1851 (Plate 172 in ibid.) and a U.S. Coast survey map of California of 1854 (Plate 174). The latter displays as well a series of profile views of the coastline as seen from sea, reminding us of the "topographic views" at the edges of Dutch maps of the seventeenth century.

32. I allude to the title of a book by Thomas Nagel: *The View from Nowhere* (New York: Oxford University Press, 1986).

33. For a more extensive discussion of the *Timaeus* from the standpoint of place and space, as well as a treatment of these topics in the medieval and Newtonian era, see Edward S. Casey, *The Fate of Place: A Philosophical History* (Berkeley: University of California Press, 1997), chapters 2–6.

34. See Figure 227 in Joseph Needham, *Science and Civilization in China* (Cambridge: Cambridge University Press, 1959), vol. 3, p. 548. As we have seen, the dialectic between verbal descriptions and mapped space is intense for the Chinese from time immemorial to the present: "text together with measurement serves as the final authority" (Cordell D. K. Yee, "Taking the World's Measure: Chinese Maps between Observation and Text," in Harley and Woodward, *The History of Cartography*, vol. 2, book 2, p. 126).

35. Even a minimal or tacit projection such as is found in portolan charts still involves distortion of exact representation.

36. C. Board, "Maps as Models," in R. J. Chorley and Peter Haggett, eds., *Models in Geography* (London: Methuen, 1967), p. 676.

37. See Figure 224 in ibid., p. 546.

38. On this jar, see A. J. Tobler, *Excavations at Tepe Gawra: Joint Expedition of the Baghdad School and the University Museum to Mesopotamia* (Philadelphia: University of Pennsylvania Press, 1950), vol. 2, pp. 150–51.

39. See the extraordinary instance of triple perspective in the map of Ningcheng discovered in a tomb at Horingen, Inner Mongolia. It is reproduced and discussed in Yee, "Chinese Cartography among the Arts," pp. 147–48. It will be recalled from the discussion of chapter 5 that Northern Sung landscape painters were also attuned to three different forms of representing depth that were perfectly compatible with each other.

40. "North, whilst being one end of the Earth's axis of rotation, is not a privileged direction in space, which after all has no 'up' or 'down.' That North is traditionally 'up' on maps is the result of a historical process, closely connected with the global rise and economic dominance of Northern Europe" (D. Turnbull, *Maps Are Territories: Science Is an Atlas* [Chicago: University of Chicago Press, 1989], p. 8).

41. "Gridding" is here employed in a general sense that includes both the grid system prop-

er (i.e., as a nonprojected representation) and the graticule (which depends on projection because it involves lines of latitude and longitude).

42. This last-named map, created in A.D. 1315, is reproduced in Joseph Needham, *Science and Civilization in China* (Cambridge: Cambridge University Press, 1959), vol. 3, Figure 231.

43. On mental maps, see Peter Gould and Rodney White, *Mental Maps* (Boston: Allen & Unwin, 1986).

44. J. Wreford Watson, *Mental Images and Geographical Reality in the Settlement of North America* (Nottingham, England: University of Nottingham Press, 1967), p. 3.

45. See Immanual Kant, "Concerning the Ultimate Ground of the Differentiation of Regions in Space," trans. D. Walford and R. Meerbote, in *Kant: Theoretical Philosophy 1755–1770* (Cambridge: Cambridge University Press, 1988), as well as the discussion of "Directions" and "Built Places" in Edward S. Casey, *Getting Back into Place: Toward A Renewed Understanding of the Place-World* (Bloomington: Indiana University Press, 1992), chapter 3.

46. See Edward S. Casey, "Between Geography and Philosophy: What Does It Mean to Be in the Place-World?" (forthcoming in the *Annals of the American Association of Geographers*).

47. For this distinction, see Maurice Merleau-Ponty, *Phenomenology of Perception,* trans. C. Wilson (New York: Humanities Press, 1962), p. 244.

48. For further discussion of the use of homogeneous versus heterogeneous space in maps, see Harley and Woodward, *The History of Cartography,* vol. 1, pp. 505–6. On the more general significance of homogeneous versus heterogeneous space, see Henri Bergson, *Time and Free Will: An Essay on the Immediate Data of Consciousness,* trans. F. L. Pogson (New York: Harper, 1960), chapter 2; and Gilles Deleuze and Félix Guattari, *A Thousand Plateaus: Capitalism and Schizophrenia,* trans. Brian Massumi (Minneapolis: University of Minnesota Press, 1987), chapters 12 and 14.

49. Ptolemy, *Geography,* p. 25.

50. Ibid; my italics.

51. Ibid.

52. Ibid.

53. On the logic of "merology," that is, parts as parts of wholes, see Edmund Husserl, *Logical Investigations,* trans. J. N. Findlay (New York: Humanities Press, 1970), vol. 2, Third Investigation.

54. This citation, which I have given before, is from Ptolemy, *Geography,* p. 26.

55. All citations are from ibid.

56. Both citations are from ibid., p. 26.

57. Ibid.

58. Concerning "true likeness," see ibid.

12. RE-PRESENTING REPRESENTATION

1. Martin Heidegger, "The Age of the World Picture," in *The Question concerning Technology,* trans. William Lovitt (New York: Harper, 1977), p. 129.

2. "Metaphysics grounds an age, in that through a specific interpretation of what is and through a specific comprehension of truth it gives to that age the basis upon which it is essentially formed. This basis holds complete dominion over all the phenomena that distinguish the age" (ibid., p. 115). Collingwood would term such a metaphysical premise an "absolute presupposition" (see Robin George Collingwood, *An Essay on Metaphysics* [Oxford: Clarendon Press, 1940], chapter 5).

3. "The fundamental event of the modern age is the conquest of the world as picture. The

word 'picture' *[Bild]* now means the structured image *[Gebild]* that is the creature of man's producing which represents and sets before" (Heidegger, "The Age of the World Picture," p. 134). This is nowhere more evident than in maps and in landscape painting.

4. Heidegger's use of *Gestell* is special; it signifies not so much "frame" as the ground of all framing actions. "We now name that challenging claim which gathers man thither to order the self-revealing as standing-reserve: *'Ge-Stell'*" (ibid., p. 19).

5. On the intertwining of *Gestell* as ordering and revealing, see ibid., pp. 32–33. On the theme of "delimitation" *(Entschränkung),* see ibid., p. 151.

6. Maurice Merleau-Ponty, *Phenomenology of Perception,* trans. C. Wilson (New York: Humanities Press, 1962), p. 68; my italics.

7. Martin Heidegger, "Conversation on a Country Path," in *Discourse on Thinking,* trans. J. M. Anderson and H. E. Freund (New York: Harper, 1966), p. 64. He continues: "We determine what is called horizon and transcendence by means of this going beyond and passing beyond . . . which refer back to objects and our re-presenting of objects" (ibid.). Heidegger goes on to argue that the "openness" here referred to in the form of "horizon" can also be thought as "region": "if we now comprehend the horizon through the region, we take the region itself as that which comes to meet us" (p. 65). It would be tempting to compare the Heideggerian sense of region *(Gegend)* with that which we have seen to be at work in the paintings of John Constable and other chorographic painters and mapmakers.

8. Heidegger, "The Age of the World Picture," p. 131.

9. Indeed, this situation is such that there is a positive correlation between the objectivity of the represented object and the subjectivity of the subject who represents it: "the more objectively the object appears, all the more subjectively, i.e., the more importunately, does the *subiectum* rise up" (ibid., p. 133).

10. Ibid., p. 150; my italics. The last sentence reads in German: *"Das Vor-stellen ist vorgehende, meisternde Ver-gegen-ständlichung."*

11. Heidegger, "Conversation on a Country Path," p. 67. Compare also the statement that "every relation to something—willing, taking a point of view, being sensible of [something]—is already representing" ("The Age of the World Picture," p. 150).

12. Merleau-Ponty, *Phenomenology of Perception,* p. 71; Merleau-Ponty's italics.

13. Heidegger, "The Age of the World Picture," p. 150.

14. Immanuel Kant, *The Critique of Pure Reason,* trans. N. K. Smith (New York: St. Martin's Press, 1965), A 34 B 50. Here, quite explicitly, the in-itself (i.e., representations "in themselves") has become for-us (it belongs to "our inner state").

15. Ibid., A 507 B 535.

16. As Heidegger says, "man 'gets into the picture' in precedence over whatever is. But in that man puts himself into the picture in this way, he puts himself into the scene, that is, into the open sphere of that which is publicly represented. Therewith man sets himself up as the setting in which whatever is must henceforth set itself forth, must present itself, i.e., be picture. Man becomes the representative *[der Repräsentant]* of that which is, in the sense of that which has the character of object" (Heidegger, "The Age of the World Picture," pp. 131–32).

17. "The objects of experience, then, are *never* given *in themselves,* but only in experience, and have no existence outside it" (Kant, *Critique of Pure Reason,* A492 B521; Kant's italics).

18. Heidegger, "The Age of the World Picture," p. 134.

19. Ibid., p. 130.

20. R.V. Tooley, *Maps and Map-Makers* (New York: Crown Publishers, 1978), p. 34.

21. Ibid.

22. F. C. Wieder, cited in ibid.

23. Heidegger, "The Age of the World Picture," p. 132.

24. *Theatrum* was a common word employed in the seventeenth century for designating maps of extensive regions of the earth. Beyond its literal root in "place for viewing," it has dramatic connotations that suggest theatrical productions on a grand, even operatic scale.

25. On the cartographer's signature, see Tom Conley, *The Self-Made Map: Cartographic Writing in Early Modern France* (Minneapolis: University of Minnesota Pess, 1996), pp. 20–22.

26. Kant, *Critique of Pure Reason,* A 370. The full statement is: "External objects (bodies), however, are mere appearances, and are therefore nothing but a species of my representations, the objects of which are something only through these representations. Apart from these they are nothing."

27. "We are conscious *a priori* of the complete identity of the self in respect of all representations which can ever belong to our knowledge . . . For in me they can represent something only insofar as they belong with all others to one consciousness" (ibid., A 116).

28. For Kant's own account of the "serial arrangement" of representations, see ibid., A 320 B 377.

29. Ibid., A 108.

30. For an analysis of this example, see ibid., A 69 B 93–94, where Kant says that a "judgment is therefore the mediate knowledge of an object, that is, the representation of a representation of it."

31. See Edward S. Casey, *The Fate of Place: A Philosophical History* (Berkeley: University of California Press, 1997), chapter 7.

32. This is not to say that such idiosyncrasies are simply subsumed into larger units, as often happens in cartographic representation. But even in this latter case, localities and regions are at least implicitly contained in a map, although the chosen scale and manner of representation most often eventuates in a depiction that ignores the peculiarities of a given particular place. This is not to deny that even a map of considerable scope can manage to represent a delimited locale, for example, a city or a county, by various devices such as interpellated tableaux—or else by detailed insets in its margin, as was notably the case in the topographic views of cities that festooned the edges of many early modern Dutch maps. In this way, the cartographers of seventeenth-century Holland no less than the painters of the same country kept place on the agenda; marginal for mapmakers and central for painters, it was a topic of continual, ramified representation.

33. See Kant, *Critique of Pure Reason,* A 23 B 38; A 32 B 48; A 49 B 67; A 272 B 328.

34. Ibid., A 272 B 328.

35. Ibid., A 375.

36. Ibid., A 492 B 520. Cf. also A 378: "space itself, however [i.e., despite its status as the form of 'outer sense'], is nothing but an inner mode of representation." For further discussion of Kant on space and place, see Casey, *The Fate of Place,* chapters 9 and 10.

37. For further treatment of place in Kant's conception of the sublime, see chapters 2–4 in this volume, as well as Edward S. Casey, "The Place of the Sublime," in *Analecta Husserliana,* ed. A. M. Tymienecka (Dordrecht: Kluwer, 1996), pp. 37–49.

38. See Immanuel Kant, "Concerning the Ultimate Ground of the Differentiation of Regions in Space," trans. D. Walford and R. Meerbote, in *Kant: Theoretical Philosophy, 1755–1770* (Cambridge: Cambridge University Press, 1988), p. 367. I discuss this essay and other of Kant's writings on space in *The Fate of Place,* pp. 187–210.

39. "The 18th century witnessed a considerable revival of activity: Anich and Hueber issued an *Atlas Tyrolensis* in Vienna in 1774 with 40 maps; M. Koops published a fine map of the Rhine

336 — Notes to Chapter 12

on ten sheets in 1796 . . . [J. B. Homann] produced general atlases covering the whole world [between 1714 and 1780]" (Tooley, *Maps and Map-Makers,* p. 27). The same enterprising activity was taking place in England and France as well (cf. ibid., chapters 6 and 7).

40. The contemplative sublime of Lane, affiliated as it was with the transcendentalism of Thoreau and Emerson, was mental only in the expanded sense of linking up with an "over-soul" that exceeds the finite boundaries of mind that were laid down by Kant in his doctrine of knowledge as confined to what can be apprehended in sensuous intuition.

41. Kant, *Critique of Pure Reason,* A 374–75.

42. Edmund Husserl, *Ideas: General Introduction to Pure Phenomenology,* trans. W. R. Boyce Gibson (New York: Macmillan, 1931), p. 181.

43. Kant, *Critique of Pure Reason,* A 34 B 50; my italics.

44. Richard Rorty, *Philosophy and the Mirror of Nature* (Princeton, N. J.: Princeton University Press, 1979), p. 46.

45. The complete statement is: "The mind is its own place, and in itself can make a Heaven of Hell, a Hell of Heaven" (John Milton, *Paradise Lost,* book 1, ll. 253–55).

46. Rorty, *Philosophy and the Mirror of Nature,* p. 45.

47. Ibid., p. 3. Rorty continues: "So to understand the possibility and nature of knowledge is to understand the way in which the mind is able to construct such representations. [Modern] philosophy's central concern is to be a general theory of representation" (ibid.). This view accords closely with that of Heidegger, despite the latter's proclivity for diagnosing the circumstance as a metaphysical rather than an epistemological issue. It is thanks to the central notion of representation that the circumstance can be characterized either way.

48. Kant, *Critique of Pure Reason,* A 493 B 521.

49. Ibid.

50. Ibid., A 494 B 522. The full statement is: "we are here speaking only of an appearance in space and time, which are not determinations of things in themselves but only of our sensibility. Accordingly, that which is in space and time is an appearance; *it is not anything in itself but consists merely of representations, which, if not given in us—that is to say, in perception—are nowhere to be met with*" (my italics).

51. Rorty, *Philosophy and the Mirror of Nature,* p. 52.

52. Kant, *Critique of Pure Reason,* A 197 B 242.

53. "If we enquire what new character *relation to an object* confers upon our representations, what dignity they thereby acquire, we find that it results only in subjecting the representations to a rule, and so in necessitating us to connect them in some one specific manner" (ibid.; Kant's italics).

54. Robert Paul Wolff, *Kant's Theory of Mental Activity* (Cambridge: Harvard University Press, 1963), p. 264; Wolff's italics. "Way of organizing" here signifies *rule.*

55. Kant, *Critique of Pure Reason,* A 105. The complete statement is: "it is clear that, since we have to deal only with the manifold of our representations, and since that x (the object) which corresponds to them is nothing to us—being, as it is, something that has to be distinct from all our representations—the unity which the object makes necessary can be nothing else than the formal unity of consciousness in the synthesis of the manifold of representations."

56. Ibid., A 197 B 242: "We have representations in us, and can become conscious of them. But however far this consciousness may extend, and however careful and accurate it may be, they still remain mere representations, that is, *inner determinations of our mind* in this or that relation of time" (my italics).

57. Ibid.; my italics.

58. Ibid., A 198 B 243.

59. For an intriguing account of mapping that emphasizes the way in which (especially in early modern avatars in France) it also encompasses the *unknown* world, both in the form of geographic "terra incognita" and of the psychical unconscious, see Conley, *The Self-Made Map,* Introduction.

60. Heidegger, "The Age of the World Picture," p. 132.

61. Maurice Merleau-Ponty, *The Visible and the Invisible,* trans. Alfonso Lingis (Evanston, Ill.: Northwestern University Press, 1968), p. 220.

62. See Wilfrid Sellars, *Empiricism and the Philosophy of Mind,* ed. Robert Brandom (Cambridge: Harvard University Press, 1997).

63. Hans-Georg Gadamer, *Truth and Method* (New York: Seabury Press, 1975), p. 100.

64. Gadamer, *Truth and Method,* p. 97.

65. Heidegger, "The Age of the World Picture," p. 150.

66. On this notion, see Mikel Dufrenne, *L'inventaire des a priori: recherche de l'originaire* (Paris: Bourgois, 1981), Parts III and IV.

67. "The dependence of aesthetic being on presentation does not mean any deficiency, any lack of autonomous determination of meaning" (Gadamer, *Truth and Method,* p. 114). On the idea of the "augmentation of being" *(Seinszuwachs),* see ibid., pp. 124–25, 132, 134–35.

68. Ibid., p. 104.

69. Ibid., p. 98.

70. Immanuel Kant, *The Critique of Judgment,* trans. J. C. Meredith (Oxford: Clarendon Press, 1952), p. 192.

71. Ibid., p. 175.

72. Ibid., p. 180: "the free harmonizing of the imagination with the understanding's conformity to law."

73. Ibid., p. 177.

74. Ibid., p. 179.

75. See Gadamer, "The Subjectivization of Aesthetics in the Kantian Critique," in *Truth and Method,* pp. 39–55.

76. Gadamer, *Truth and Method,* p. 93.

77. Ibid., p. 94.

78. Ibid., p. 92.

79. This follows from the nature of play: "Play is really limited to presenting itself. Thus its mode of being is self-presentation *(Selbst-Darstellung)*" (ibid., p. 97).

80. Ibid., p. 99.

81. "The audience only completes what the play as such is . . . the play itself is the whole, comprising players and spectators" (ibid., p. 98); "it puts the spectator in the place of the player" (p. 99).

82. On the analogy between the work of art and religious ritual, see ibid., pp. 97–99.

83. Ibid., p. 103.

84. Ibid., p. 101; my italics.

85. Ibid., p. 102. Cf. also p. 101: "Transformation into a structure is not simply transposition into another world." By "structure" Gadamer means the finally achieved work in the wholeness of its being—its lasting *Gestalt.*

86. A letter to the *Missouri Republican* in 1849; cited by Barbara Novak, *Nature and Culture: American Landscape and Painting 1825–1875* (Oxford: Oxford University Press, 1980), p. 27. In the very first known panorama, that built by Robert Barker in London in 1788, spectators

had to climb a tower to a viewing platform, which was "positioned in such a way that the painting's horizon-line coincided with the spectators's eye-level, which meant that spectators experienced a sensation of looking down at the scene. This was only one of several carefully calculated visual effects [e.g., illumination by hidden skylights] . . . All this was done to maximize visual drama" (Alan Wallach, "Making a Picture of the View from Mount Holyoke," in David C. Miller, ed., *American Iconology: New Approaches to Nineteenth-Century Art and Literature* [New Haven: Yale University Press, 1993], pp. 82–83).

87. "In the panorama, the world is presented as a form of totality; nothing seems hidden; the spectator, looking down upon a vast scene from its center, appears to preside over all visibility" (Wallach, "Making a Picture of the View from Mount Holyoke," p. 83). Recall that in Barker's original panorama, spectators had to *mount* the viewing tower, further enhancing the sense of being above the panoramic scene.

88. On the contrast between these two notions of perspective, to which I have alluded earlier in this book, see Svetlana Alpers, *The Art of Describing: Dutch Art in the Seventeenth Century* (Chicago: University of Chicago Press, 1983), pp. 41–59.

89. In particular, the near wall of the stage is experienced as let down so as to include us in this fateful mise-en-scène. This wall is the wall of representation against which we beat our self-contained minds in vain. As Gadamer says, "The closed world of play lets down, as it were, one of its walls" (*Truth and Method,* p. 97). Gadamer borrows this striking image from Rudolf Kassner.

90. Gadamer, *Truth and Method,* p. 111.

91. Arnold Berleant, "The Viewer in the Landscape," in *Living in the Landscape: Toward an Aesthetics of the Environment* (Lawrence: University of Kansas Press, 1997), p. 164.

92. Cited by Alpers, *The Art of Describing,* p. 42, from Alberti's treatise *On Painting and Sculpture,* trans. Cecil Grayson (London: Phaidon, 1972), p. 55. Alpers warns that Dürer's representation is imperfect insofar as it makes purely mechanical what is in essence mathematical in Alberti's original conception.

93. Gadamer, *Truth and Method,* p. 271. Gadamer is here discussing the "fusion of horizons" *(Horizontsverschmelzung)* that occurs between past events and present understandings of these events; but his point can be applied as well to the case of painting in regard to its power of re-implacement.

94. By speaking of "unframing," I am not denying the continuing power of framing effects of the several sorts discussed in the Interlude. In particular, unframing suspends the cloistering action of the external frame; but the role of what I call the "quasi-frame" remains intact in the very midst of such unframing.

95. Concerning the primordial dator intuition, see Husserl, *Ideas,* section 19, p. 136. I say that such perception *claims* to give us the thing itself; whether it does so in fact is another question.

96. John R. Stilgoe, *Common Landscape of America, 1580 to 1845* (New Haven: Yale University Press, 1982), p. 3. But Stilgoe goes too far when he asserts that "a landscape happens not by chance but by contrivance, by premeditation, by design; a forest or swamp or prairie no more constitutes a landscape than does a chain of mountains. Such land forms are only wilderness, the chaos from which landscapes are created by men intent on ordering and shaping space for their own ends" (ibid.). Surely there is considerable overlap at the edges: much of wilderness is perceived by us as landscape, even if the mark of the human is minimal or nonexistent; and wilderness itself can extend into cultivated landscape spaces—as happens in certain remotely located towns in the American West, or in deliberately uncultivated parts of gardens.

97. The passage is in John Ruskin, *The Seven Lamps of Architecture* (London: Deut, 1907; first edition, 1849), pp. 180–81. I have analyzed this passage in Edward S. Casey, *Getting Back into*

Place: Toward a Renewed Understanding of the Place-World (Bloomington: Indiana University Press, 1993), pp. 240ff. I take up the question of the respective contributions of nature and culture to landscape and wilderness at ibid., pp. 229–40.

98. Gadamer, *Truth and Method,* p. 103: "the basic mimic situation that we are discussing not only involves what is presented being there, but also [the fact] that it has in this way come to exist more fully."

99. Kant, *Critique of Judgment,* p. 176.

100. I refer to Arthur C. Danto, *The Transfiguration of the Commonplace: A Philosophy of Art* (Cambridge: Harvard University Press, 1981).

101. I am thinking, for example, of the desolate surrealist landscapes of Tanguy or Max Ernst or Arshile Gorky, and of the many vignettes contained in *The Dictionary of Imaginary Places,* ed. Alberto Manguel and Gianni Guadalupi (New York: Macmillan, 1980).

102. On the idea of "aesthetic surface," a notion first formulated by D. W. Prall, see Edward S. Casey, "Expression and Communication in Art," *Journal of Aesthetics and Art Criticism* 30:2 (1971): 197–207.

103. Stilgoe, *Common Landscape of America, 1580 to 1845,* p. ix; Stilgoe's italics. Stilgoe adds: "Common design is that design understood and agreed upon by all, and passed from one generation to another, usually by example" (ibid.). On "vernacular landscape," see J. B. Jackson, *Discovering the Vernacular Landscape* (New Haven: Yale University Press, 1984), esp. pp. 83–87.

104. Constable wrote to a friend in 1832 that he had fallen into the following conversation with strangers on a coach while he was traveling to Dedham: "In passing through the valley about Dedham, one of them remarked to me—on my saying it was beautifull—'Yes Sir—this is *Constable's* country!'" (cited in Michael Rosenthall, *Constable: The Painter and His Landscape,* p. 5; Rosenthall's italics; from *John Constable's Correspondence,* ed. R. B. Beckett [Ipswich: Suffolk Records Society, 1966], vol. 4, p. 387).

105. Maurice Merleau-Ponty, "Eye and Mind," trans. C. Dallery in *The Primacy of Perception,* ed. J. Edie (Evanston, Ill.: Northwestern University Press, 1964), p. 166, with apparent reference to Cézanne's attitude that a given painting can be regarded as complete at every stage, even if it is never fully finished. On the comparatively recent tradition of *plein-air* (open-air) painting as this originated in the late eighteenth century at and around Rome—and as it was most brilliantly pursued by Corot as a precursor of Cézanne and Monet in the later nineteenth century—see the catalog *In the Light of Italy: Corot and Early Open-Air Painting* (Brooklyn: Brooklyn Museum, 1996). I am not forgetting that Constable occasionally rested his case with an oil sketch that he did not pursue further in a painting. This was not, however, his primary practice.

106. Concerning chorology, see John Sallis, *Chorologies* (Bloomington: Indiana University Press, 1998).

107. These examples are reproduced as Figures 90 and 73, respectively, in Alpers, *The Art of Describing.*

108. For further discussion of Muir's bivalent approach to Glacier Bay, see Casey, *Getting Back into Place,* pp. 231–32.

109. Concerning the close relationship between mapping and writing, especially with regard to certain rhetorical strategies prominent in the French Renaissance, see Conley, *The Self-Made Map,* Introduction, chapter 4, and Conclusion.

110. Stilgoe defines *Landschaft* as "a collection of dwellings and other structures crowded together within a circle of pasture, meadow, and planting fields and surrounded by unimproved forest or marsh" (*Common Landscape of America, 1580 to 1845,* p. 12). Stilgoe opposes *Landschaft* as "the land shaped by men" to wilderness (pp. 12–21). See also the extensive treatment of wilderness

in relation to cultivated landscape in the pioneering works by Keith Thomas, *Man and the Natural World: Changing Attitudes in England, 1500–1800* (London: Penguin, 1984), and Hans Peter Duerr, *Dreamtime: Concerning the Boundary between Wilderness and Civilization,* trans. F. Goodman (Oxford: Blackwell, 1985).

111. On the *roland,* see Stilgoe, *Common Landscape of America, 1580 to 1845,* pp. 18–19.

112. Because *cultus* means both "cultivation (of fields)" and "worship," the *roland* stone, and the *Landschaft* more generally, bring together the natural and the cultural world. The same is true of "landscape" in its modern sense of a cultivated natural scene.

113. Stilgoe, *Common Landscape of America, 1580 to 1845,* p. 24.

114. Alpers, *The Art of Describing,* p. 136.

115. Stilgoe, *Common Landscape of America, 1580 to 1845,* pp. 24–25. Stilgoe points out that *landskip* in its later usage also connoted "large-scale ornamental gardens objectifying ideals of beauty" (p. 25). Here the influence of the French, rather than the Dutch, is evident. Still later, in the development of the English "natural park" in the eighteenth century, the formal garden was itself transformed into a miniature landscape.

116. Gadamer, *Truth and Method,* p. 102.

117. Stilgoe, *Common Landscape of America, 1580 to 1845,* p. 25. This is the final, and still extant, sense of "landscape" in British usage.

118. By "more than one view" I mean that landscape painting after the seventeenth century is no longer confined to the monofocal perspective paradigm of the Renaissance wherein the painter is securely stationed at a fixed point in his outlook on space. After the breakthrough effected by the Dutch, the painter (and thus by implication the viewer) can be situated at several viewing points and yet still produce a perfectly coherent painting of a given landscape.

119. Merleau-Ponty, *Phenomenology of Perception,* p. 329.

120. On *Ideenkleid,* see Edmund Husserl, *The Crisis of European Sciences and Transcendental Phenomenology,* trans. D. Carr (Evanston, Ill.: Northwestern University, 1970), pp. 51–54.

121. Indeed, it is Husserl's thesis that geometry in ancient Greece grew out of geography. See his essay "The Origin of Geometry" in ibid., pp. 353–78, and Jacques Derrida's astute comments on this issue in his *Introduction to Husserl's "Origin of Geometry,"* trans. J. Leavy (New York: Nicolas Hayes, 1972).

122. Berleant, *Living in the Landscape,* p. 181.

123. Ibid., p. 186.

124. *Cosmographia* is in fact the title of Apianus's sixteenth-century adaptation of Ptolemy's *Geographia*; it was a commonly used term in seventeenth-century mapmaking as well and signified the mapping of the entire cosmic realm, that is, heaven and earth alike.

125. Eugene Minkowski, *Lived Time,* trans. Nancy Metzel (Evanston, Ill.: Northwestern University Press, 1970), pp. 277ff.

126. Cited in Chang Chung-yuan, *Creativity and Taoism* (New York: Julian Press, 1963), p. 240.

EPILOGUE

1. Cited in Stephen Owen, *Traditional Chinese Poetry and Poetics: Omen of the World* (Madison: University of Wisconsin Press, 1985), p. 12.

2. Owen argues nevertheless that the reader *assumes* that the poem is of one particular experience or, more exactly, of a set of objects (stars, river) as experienced from the unique point of view of the poet (ibid., pp. 13ff.).

3. See chapter 10 in this volume and, in particular, Joseph Needham, *Science and Civilization in China* (Cambridge: Cambridge University Press, 1959), vol. 3, pp. 543ff.

4. Cited in Owen, *Traditional Chinese Poetry and Poetics*, p. 24, from the text *Wen Fu*.

5. Ibid., p. 25; Owen's italics. Cf. also p. 21: "literature is the entelechy of a previously un-realized pattern . . . the written word *(wen)* is not a sign but a schematization."

6. As Owen observes, even in this extreme Western case of a poem whose exact circum-stance of composition we have no reason to doubt, "it matters not at all whether Wordsworth ever actually stood on Westminster Bridge on September 3, 1802, and gazed at the city of London" (ibid., p. 14).

7. From the entry "Moominpapa's Island," in Albert Manguel and Gianni Guadalupi, *The Dictionary of Imaginary Places* (New York: Macmillan, 1980), p. 251.

8. See Owen, *Traditional Chinese Poetry and Poetics*, p. 23: "the text, the entelechy of one process [= Nature], is only the beginning of another living process in the mind of the reader."

9. Although based in the earlier work of Humboldt, Ritter, Hettner, and Richthofen, the conception itself received its most precise formulation in Carl O. Sauer's classic article "The Morphology of Landscape" (*University of California Publications in Geography* 2 [1925]: 20). For an overall account of Sauer's definition and its origins in earlier German geographers, see Richard Hartshorne, *Perspectives on the Nature of Geography* (Chicago: Rand McNally, 1959), chapter 2.

10. For an assessment of the cultural and social overdetermination of maps, especially in terms of political power and purpose, see Denis Wood, *The Power of Maps* (New York: Guilford Press, 1992), and Mark Monmonier, *Drawing the Line: Tales of Maps and Cartocontroversy* (New York: Holt, 1995). For an account of the sheer variety (and hence at least apparent arbitrariness) of maps, see J. Makower, ed., *The Map Catalogue* (New York: Vintage, 1986).

11. Cited from the lectures on "Physische Geographie" by Richard Hartshorne in *The Nature of Geography* (Chicago: University of Chicago Press, 1939), p. 135. Kant's influence on the course of nineteenth-century geography was considerable. We see this in Hettner's statement that geography is "a science of the filling of space. It is a *spatial* science in the sense that history is a tem-poral science" (cited from Hettner's *Geographie* by V. A. Anuchin, *Theoretical Problems of Geog-raphy*, trans. S. Shabad [Columbus: Ohio State University Press, 1977], p. 109; Anuchin's italics).

12. F. Lukermann, "Geography as a Form of Intellectual Discipline and the Way in Which It Contributes to Human Knowledge," *Canadian Geographer* 8 (1964): 168 and 167, respectively. Lukermann also says that "the geographer studies place. It is the character of the place, the areas of the earth's surface that form the subject matter of geography, not the distribution over the world of a single or multiple class of phenomena" (p. 168).

13. Ibid., p. 169.

14. "The location of a place is not completely defined for the geographer until it is de-scribed in relation to all other interacting places. Involved here are events and places of the past as well as the present" (ibid., p. 170).

15. Fernand Braudel, *The Mediterranean and the Mediterranean World in the Age of Phillip II*, trans. Siân Reynolds (New York: Harper & Row, 1966), vol. 1, p. 23. Cf. also p. 102 for a compa-rable statement.

16. "The description of nature (i.e., the state of nature at the present time) is far from suffi-cient to indicate the basis for explaining the whole variety of its changes. It must be resolved, de-spite all of the very justified hostility to boldly proposed opinions, to create a *history* of nature that would be a separate science" (from Kant's essay "Concerning the Various Races of People" as cited in Anuchin, *Theoretical Problems of Geography*, p. 78; Kant's italics). Similarly, Alexander von Hum-boldt, much influenced by Kant, argued that "it is impossible to separate completely the descrip-tion of nature from the history of nature. A geognost cannot understand the present without the past" (cited from Humboldt's *Cosmos* by Anuchin [ibid., p. 87]).

17. Carl O. Sauer remarks that "human geography" is a "cultural experience of a particular

space" ("The Fourth Dimension of Geography," in *Selected Essays 1963–1975* [Berkeley: Turtle Island Foundation, 1981], p. 281). Similarly, Lukermann says that "place, as an event in human experience, becomes above all a cultural concept" ("Geography as a Form of Intellectual Discipline," p. 169). See also these excellent essays: Denis Cosgrove, "Introduction: Mapping Meaning," in D. Cosgrove, ed., *Mappings* (London: Reaktion Books, 1999), pp. 1–23; and Christian Jacob, "Mapping in the Mind: The Earth from Ancient Alexandria," in ibid., pp. 24–49.

18. Lukermann, "Geography as a Form of Intellectual Discipline," p. 168.

19. Archytas, as cited and translated by S. Sambursky, ed., *The Concept of Place in Late Neoplatonism* (Jerusalem: Israel Academy of Sciences and Humanities, 1982), p. 37.

20. Lukermann, "Geography as a Form of Intellectual Discipline," p. 169; my italics.

21. Ibid., p. 167.

22. Carl O. Sauer, "The Education of the Geographer," *Annals of the American Association of Geographers* 46 (1956): 289.

23. "Geography in the United States was given its academic entry by geologists, who for years remained its sponsors and guides. Some of us [geographers] started in geology and were attracted to the new direction of linking study of the face of the earth to its human occupants" (Sauer, "The Fourth Dimension of Geography," p. 280).

24. *Feng shui* means mind and water. An excellent introduction to the subject is Steven J. Bennett's essay "Patterns of the Sky and Earth: A Chinese Science of Applied Cosmology," *Chinese Science* 3 (1978): 1–26. See also Sarah Rossbach, *Feng Shui: The Chinese Art of Placement* (New York: Dutton, 1983), pp. 21–29, and Lam Kam Chuen, *Feng Shui Handbook* (New York: Holt, 1996).

25. Alexander von Humboldt, *Cosmos,* trans. E. C. Otté (New York: Harper & Brothers, 1850), vol. 1, p. 93.

26. For examples of this genre, see Barbara Novak, *Nature and Culture: American Landscape and Painting, 1825–1875* (Oxford: Oxford University Press, 1980), Figures 34 and 35.

27. Cited in ibid., p. 66.

28. Humboldt, *Cosmos,* vol. 1, p. 23.

29. Ibid., 94.

30. For Humboldt's verbal sketches, see his monumental *Aspects of Nature, in Different Lands and Different Climates* (1849), a book that describes in minute detail Humboldt's elaborate travels in the New World. During these same travels, Humboldt also made a number of maps, especially of Central America. His omnivorous naturalism thus extended from geography to cartography, while promoting landscape painting as a recommended procedure. In his person as in his work, Humboldt embodied an extraordinary set of talents and diverse interests. As Novak remarks, for such contemporaries as Emerson and Church, "Humboldt offered an immensely attractive prospect—merging exploration, exoticism, baroque energy, and pragmatic observation infused with flashes of transcendence" (*Nature and Culture,* p. 67).

31. Sauer, "The Fourth Dimension of Geography," p. 282.

32. Rainer Maria Rilke, *Duino Elegies,* trans. J. B. Leishman and Stephen Spender (New York: Norton, 1963).

33. On earth as self-secluding, see Martin Heidegger, "The Origin of the Work of Art," in *Poetry Language Thought,* trans. Albert Hofstadter (New York: Harper & Row, 1971), pp. 47ff. On ingathering, see Edward S. Casey, *Remembering: A Phenomenological Study,* 2d ed. (Bloomington: Indiana University Press, 2000), pp. 292–95, 301, 303.

34. It is a revealing fact that Joseph Needham makes only five cursory references to geomancy in the third volume of his monumental *Science and Civilization in China,* a study that attempts to restrict itself to logocentrically legitimated scientific disciplines.

35. Joan Blaeu, *Le Grand Atlas* (Amsterdam, 1663), Introduction, p. 1 (cited in Svetlana Alpers, *The Art of Describing: Dutch Art in the Seventeenth Century* [Chicago: University of Chicago Press, 1983], p. 159). I alluded to this remark in chapter 8.

36. John Dewey, *Essays on School and Society* (Carbondale: Southern Illinois University Press, 1976), p. 13.

37. Heidegger, "The Origin of the Work of Art," pp. 48–49, where "self-opening" is also employed. The importance of the distinction between *Welt* and *Erde* is anticipated by Humboldt when he speaks of "a confusion of ideas in the synonymic use of the words 'earth' and 'world'" (*Cosmos,* vol. 1, p. 68).

38. Humboldt, *Cosmos,* vol. 1, p. 79; Humboldt's italics.

39. I borrow this term from Rupert Sheldrake, *The Presence of the Past: Morphic Resonance and the Habits of Nature* (New York: New York Times Books, 1988).

40. "Topography" is here meant in its nineteenth-century painterly sense of an iconic or closely descriptive rendition of an entire landscape, not in the narrow cartographic sense of pictographic insets or landscape vignettes.

41. "The idea gets its objective reality from a cause in which there is at least as much formal reality as there is objective reality contained in the idea" (René Descartes, *Meditations on First Philosophy,* trans. D. A. Cress [Indianapolis: Hackett, 1979], pp. 27–28). Bergson restates the Cartesian principle, but specifies that it applies only to the realm of "unorganized matter": in this realm, "the present contains nothing more than the past, and what is found in the effect was already in the cause" (Henri Bergson, *Creative Evolution,* trans. A. Mitchell [New York: Random House, 1944], p. 17; all in italics). I am claiming that precisely at the level of less than fully organized matter, that is, *chōra,* the Cartesian assumption does not apply.

42. "Since everything that is in motion is moved in some place, it is obvious that one has to grant priority to place . . . Perhaps thus it is the first of all things" (as cited in Sambursky, *The Concept of Place in Late Neoplatonism,* p. 37).

43. Concerning such mapping, see Stephen S. Hall, *Mapping the Next Millennium* (New York: Random House, 1992), chapters 2–6.

44. For further on truth-to the "how" versus truth-to the "that," see Casey, *Remembering,* pp. 281–82.

45. The origin of "tropics" lies in *tropein,* "to turn." Concerning John Muir's writings on glaciated landscapes, see Edward S. Casey, *Getting Back into Place: Toward a Renewed Understanding of the Place-World* (Bloomington: Indiana University Press, 1993), pp. 231–32, 240, 264–65.

46. I refer to Heidegger's notion of *Umriss.* See Heidegger, "The Origin of the Work of Art," p. 63: the rift *(Riss)* between earth and world "does not let the opponents break apart; it brings the opposition of measure and boundary into their common outline *(Umriss).*"

47. For a detailed account of Srin-mo, see Janet Gyatso, "Down with the Demoness: Reflections on a Feminine Ground in Tibet," *Tibet Journal* 12:4 (1987): 34–46. I take the term *points de capiton* from Jacques Lacan, for whom they stand for the condensed nodes of overdetermination, for example, in the case of symptoms or slips of the tongue.

48. This is the epigraph to chapter 9. It is from Wallace Stevens, "Angel Surrounded by Paysans," in *The Collected Poems of Wallace Stevens* (New York: Knopf, 1954), p. 178. It is as if Stevens were answering Rilke's haunting question cited earlier: "Earth . . . is not your dream to be one day invisible?" In its depths, yes; but not as it becomes landscape: earth then attains visibility.

49. This is not to deny that there is landscape on other planets as well as on the moon. But, as Husserl has argued, these other cosmic bodies are themselves the experiential equivalents of earth for those humans who might come to inhabit them. (See Edmund Husserl, "Foundational Investigations of the Phenomenological Origin of the Spatiality of Nature," trans. F. Kersten in

P. McCormick and F. Elliston, eds., *Husserl: Shorter Works* [Notre Dame, Ind.: University of Notre Dame Press, 1981], pp. 222–33.)

50. On ley-lines, see Alfred Watkins, *The Old Straight Track: Its Mounds, Beacons, Moats, Sites, and Mark Stones* (London: Sago Press, 1970), and John Michell, *The Earth Spirit* (New York: Crossroad, 1975). "Ley" is an obsolete form of "lay" as the latter word is employed in an expression such as "lay of the land." I am construing ley-lines not literally or spiritually but metaphorically—as lines laid down on the earth, in effect more than in fact.

51. "*Landscape* thus means *shaped land,* land modified for permanent human occupation, for dwelling, agriculture, manufacturing, government, worship, and for pleasure" (John R. Stilgoe, *Common Landscape of America, 1580 to 1845* [New Haven: Yale University Press, 1982], p. 3; my italics). J. B. Jackson adds that the suffix *scape* of *landscape* is "essentially the same as *shape,* except that it once meant a composition of *similar* objects, as when we speak of a fellowship or a membership" (J. B. Jackson, *Discovering the Vernacular Landscape* [New Haven: Yale University Press, 1984], p. 7; Jackson's italics).

52. For further discussion of "prospect" in this respect, especially in its contrast with "refuge" (i.e., the other fundamental position vis-à-vis landscape), see Jay Appleton, *The Experience of Landscape* (New York: Wiley, 1975), pp. 81–121.

53. On Husserl's notion of the near-sphere, see Ulrich Claesges, *Edmund Husserls Theorie der Raumkonstitution* (The Hague: Nijhof, 1964), pp. 83ff.

54. B. K. Roberts, "Landscape Archeology," in J. M. Wagstaff, ed., *Landscape and Culture: Geographical and Archeological Perspectives* (Oxford: Blackwell, 1987), p. 83.

55. Ibid., p. 79. Roberts's full statement is as follows: landscapes "may be defined as the assemblage of real-world features—natural, semi-natural and wholly artificial—[and] give character and diversity to the earth's surface and form the framework within which human societies exist."

56. William James, "The Place of Affectional Facts in a World of Pure Experience," in *Essays in Radical Empiricism* (New York: Longmans, Green, 1947), p. 153. It is worth noting that Merleau-Ponty, who, more than James himself, pursued the theme of the lived body's intrinsic ambiguity, has frequent recourse to landscape in his seminal discussions of *le corps vécu* in *The Phenomenology of Perception.*

57. Cited by W. G. Hoskins in his introduction to K. J. Allison, *The East Riding of Yorkshire Landscape* (London: Hodder and Stoughton, 1976). In the same introduction, Hoskins characterizes scenery as something to which everyone has access easily as an aesthetic object, while landscape itself can be considered "scenery examined with a trained eye" (this last formulation is B. K. Roberts's paraphrase of Hoskins's point: cf. Roberts, "Landscape Archeology," p. 77).

58. For a systematic exploration of these ambiguities, see D. W. Meinig, "The Beholding Eye: Ten Versions of the Same Scene," in D. W. Meinig, ed., *The Interpretation of Ordinary Landscapes: Geographical Essays* (Oxford: Oxford University Press, 1979), pp. 33–50. J. B. Jackson notes that only the word *land* "rivals in ambiguity the word *landscape*" (Jackson, *Discovering the Vernacular Landscape,* p. 6; Jackson's italics). In my remarks, I have not even touched on the rich semantic density of the word *land,* which Jackson discusses here and to which I have alluded in chapter 9.

59. W. G. Hoskins, *The Making of the English Landscape* (London: Hodder and Stoughton, 1967), p. 8. I do not mean to deny the real possibility of an "abstract" land photography (e.g., as in Terry Evans's photographs of Midwestern landscapes seen from airplanes); but such abstraction is then precisely *of the concrete,* which is not so much transcended as transmuted by the photographic image—just as occurs, *mutatis mutandis,* in landscape painting.

60. Ibid., p. 78.

61. Ibid., p. 234.

62. It is not surprising that there exists a journal titled *Landscape History,* as well as an entire discipline termed "landscape archaeology." On this newly emerging discipline, see Roberts, "Landscape Archaeology," pp. 77–95.

63. Hoskins, *The Making of the English Landscape,* p. 79.

Glossary

Apocalyptic sublime: term applied to early-nineteenth-century American and English landscape painting to denote sublimity of a cataclysmic nature, with an emphasis on the revelation of final things; sometimes termed the "romantic-Gothick" sublime.

Beautiful: aesthetic valorization in terms of formal perfection and various qualities that induce equanimity: for example, symmetry, balance, compensation.

Cadastral map (also **Cadaster**): a plan or survey indicating the borders of subdivided land, created to determine ownership and a basis for taxation.

Cardinal direction: any of the four primary astronomical directions as these are distributed on the surface of the earth: North, South, East, West.

Cartogram: a condensed map that displays quantitative data in pictorial form; it is not usually drawn according to scale.

Cartographer: mapmaker.

Cartographic: mapping that meets rigorous criteria of representation, not only those of isomorphism but also of consistency of scale, regularity of spatial layout, reliability of direction, and so on; measurement of distance is a central feature of cartographic mapping.

Cartographic sign: a vehicle by means of which a map conveys information about a geographic entity (e.g., a place or region); such a sign can be a word, a pictographic image, a conventional pictogram, or a symbol.

Cartography: the making of maps, esp. those purporting to be accurate.

Chorography: representation (typically, in a map) of a region of the earth; the tracing of particular regions, *chōrai,* often with consideration of aesthetic criteria.

Chorology: any enterprise that combines word and image; in particular, a form of mapping that, in addition to writing (e.g., inside the cartouche) includes pictograms of landscape.

Contemplative sublime: term applied to mid-nineteenth-century American and English landscape painting to designate sublimity of an especially tranquil and pensive sort.

Contour-line map: a relief map that indicates comparative height or depth by means of continuous lines whose breadth or comparative closeness specifies the precise configuration of the land.

Cosmography: the descriptive mapping of the heavens and the earth, regarded as a single totality: that is, as the "cosmos"; within its scope is included astronomy, geography, and geology.

Displacement: pushing or pulling an item out of its initial or habitual place into another—often undesirable—place; displacement bears a connotation of degeneration, loss, or alienation, typically in a movement from place to site.

Display: the presentational aspect of the subject matter of a map or painting.

Earth: not merely the physical globe or planet but the cosmic principle of self-secluded origin; the tangible basis of landscape.

Eidetic: having to do with form, structure, pattern; the term derives from *eidos,* Greek for "form."

Eikon: image based on formal resemblance; according to the Platonic conception, the eikon is of diminished status compared with the original term of which it is an image—which is not necessarily true of the "icon."

Extensity versus Extension: whereas extension (in Descartes's sense of *res extensa*) implies an abstract spatial homogeneity, indifference to content, and indefinite divisibility, extensity (in Bergson's sense of the term) connotes a heterogeneous, concrete, and indivisible totality of places.

Frame-working: the action whereby the frame of a painting (or, more rarely, a map) sets the work in motion, creating an "energy map" (in Eve Ingalls's term); represented objects and events take place and find place thanks to the dynamic format provided by the frame.

Geodesy: science of the size and shape of the earth.

Geography: the study of the earth as a whole; its drawing (in lines) and/or describing (in words); its minimal unit is a place.

Geometry: literally, "earth measurement"; science of the properties and relationships between points, lines, angles, surfaces, and solids.

Geomorphology: the scientific study of the form of the earth, especially by attending to the configuration of particular landmasses; by extension, a sense of the earth's multifarious shapes.

Graticule: a network of lines that stand for parallels and meridians traversing the surface of the earth or the face of a map; a two-dimensional projection of the terrestrial sphere.

Grid: a referential system in which points are established by their comparative distance from two perpendicular axes; more loosely, the structuring of a surface by a set of uniformly shaped squares or rectangles.

Hachure-line map: this kind of map indicates the relief of the land by means of comparative distance between continuous parallel lines.

Icon: an image or sign that maintains a transparent isomorphic relation with that which it signifies; unlike a replica, it need not be physical: it can be mental or semiological in status.

Image: mental or physical representation of something with which it has a relationship of likeness but not necessarily of strict isomorphism.

Interplace: a place between places; in particular, a place that serves to connect otherwise disparate places or regions so as to create a more or less continuous single (albeit complex) place.

Isomorphism: resemblance between two or more items in terms of the likeness of certain formal qualities (e.g., shape, size) and, less frequently, certain material qualities (e.g., color, texture).

Land: stretch of the earth's surface that is a unit of habitation (hence having its own history) or that can be taken in by a single comprehensive glance.

Landform: the shape of a portion of the earth's surface.

Landmass: part of the exposed surface of the earh considered with respect to its sheer materiality.

Landscape: literally, "shape of the land"; a word deriving from the Dutch *landschap* that signifies *(a)* a vista or "cut" (hence the *-scape*) of the perceived world, construed as "country" or "land" or "field" set within a horizon; *(b)* the circumambience provided by a particular place; *(c)* by extension, seascape, cityscape, and so on; *(d)* a genre of painting that, in contrast with landskip, is concerned with the material essence of a place or region rather than with its precise topography, and with transplacement rather than with transposition.

Landscape-map: a map-painting that calls upon us to enter it as a landscape scene pictured in a work of art: for example, Kuwagata Keisai's *Nihon Meisho No E* (Plate 16).

Landskip: an early form of "landscape" and, more particularly, a conception of painting as well as decoration that emphasizes topographic exactitude; also applied to maps and verbal descriptions in late-seventeenth-century English.

Large-scale map: a map that offers close-up views of geomorphic landforms and other features of a given region.

Latitude: angular distance north or south of the equator, as measured in degrees along a meridian.

Likeness: resemblance accomplished by similarity of structure; it need not meet the requirement of strict isomorphism in order to reach its goal of verisimilitude.

Liminal place: a zone of open exchange between regions, entities, or elements.

Locale (also **Locality**): place construed as part of an encompassing region.

Location: pinpointed position on a map that represents geographic space; specification of a site.

Locus: the smallest unit of place that is not determined cartographically but experientially; even though it is circumscribed in actual experience, it can be indefinitely large in extent.

Logos: Greek term for word, structure, rational account.

Longitude: angular distance from the prime meridian at Greenwich, England, to a given point on the earth's surface; determined in degrees or (more concretely) in temporal units of hours, minutes, or seconds.

Map-painting (or **Painting-map**): a landscape representation that can be considered equally well as a map or as a painting (e.g., Pieter Saerendam's *The Siege of Haarlem*).

Material essence: in the strict Husserlian sense of the term, the essence of something considered with regard to its content ("material" not as physical but as the subject matter of an idea or thought); by extension, the gist of a region, its landscape sense, as captured by a painting or map of that region.

Meridian: imaginary line that bisects the earth from north to south poles; in practice, half of the great circle that passes through both poles.

Mind: the interior self that cogitates in more or less formal ways; according to Descartes, the realm of *res cogitans* or mental substance that exists in strict separation from *res extensa* or extension in space; for Kant, the inner domain *(das Gemüt)* of imagination and reason, which are co-implicated in the experience of the sublime.

Morphē: Greek word for form, shape.

Morphology: the study of form and (more particularly) shape; formally considered, morphology is most fully realized in the precisely delineated shapes studied in plane and solid geometry; informally, morphology concerns itself with the inherently vague shapes that are at stake in landscape painting.

Natural sublime: the sublime as represented in landscape paintings that focus on the natural world in its refulgent display; this is the sublime as it inheres in the *land* of landscape, that is to say, that part of the earth's surface in which sublime phenomena are experienced (and represented) as arising; but it is also a domain in which mind and matter, spirit and nature coinhere.

Panorama: a sweeping view or extensive vista, whether in the presence of nature or as represented in a painting.

Parallel: imaginary line that encircles the earth parallel to the plane of the equator and that represents degree of latitude.

Periploi: ancient Mediterranean navigational written guides, used mainly for sailing along coastlines between established harbors and ports; termed *portolani* in the medieval period.

Phantasm: image that departs from isomorphic representation (e.g., by altering the proportions of the original term) yet still captures the material essence of something.

Physiographic map: relief map that indicates landscape features by pictograms (e.g., of mountains); characteristic landforms and/or landmasses are seen from an oblique angle and as quasi-three-dimensional, typically from afar.

Pictogram: pictorial sign employed in mapping to designate a given geographical entity; though highly schematic in character, a pictogram is an iconic image that is based on some degree of formal likeness with this entity; broader in scope than a pictograph, a pictogram is contrasted with other cartographic signs such as words and abstract symbols.

Pictograph: a visual image of an identifiable object or place; to some degree isomorphic in the mode of representation; frequently employed in primitive maps (e.g., Neolithic petroglyphic maps).

Pictoriality: the quality of a representation whereby it constitutes a visual image of something (i.e., object or event), with an emphasis on the graphic or linear character of the representation itself.

Picture-map: a map composed exclusively, or at least largely, of pictographic elements.

Picturesque: aesthetic valorization of visual representations whose criteria are the charming and the pleasing; typically, it occurs on a human scale; closely affine with the quaint.

Picture-story: a map considered as simultaneously imagistic and narrative.

Place: main unit of landscape; scene of situatedness; experienced by the entire body; having its own history; not to be confused with space.

Place-by-proxy: a place that stands in for another; for example, an interior psychic place serving as the representative of an exterior natural place.

Place-of-exhibition: room or hall wherein an artwork or a map not only becomes publicly accessible but is such as to interact intimately with the painting or map.

Place-of-identification: the fate of a place that has been internalized into the human psyche, which deeply identifies itself with a place that is initially experienced as outside the self.

Place-of-origin: the specific place that initially inspires a landscape painter, whether or not that place is represented as such in the finished work or figures importantly in the evolution of the work.

Place-of-representation: the place of a landscape scene insofar as this is an integral part of a completed painting; it is the place at stake in the visual representation, belonging to its manifest content—in contrast with the place-of-origin (which is important mainly in the genealogy of the work's production) and the place-of-identification (which matters mostly within inner psychic states); place-of-origin and place-of-representation can coincide in a given case (e.g., Thomas Cole's *The Oxbow* [Figure 4.2, Plate 9]). More generally, place-of-representation is a painting or map regarded as a formatted and framed scene in which various objects, places, or events are re-implaced and endowed with representational status.

Place-of-the-surface: the place afforded by the canvas or paper on which a map or painting is presented; closely related to the size of the work; manifests visual and tactile and kinesthetic properties through the adumbration of the represented object; ties together representing and represented places in a single sensible ground of presentation.

Placescape: a representation of land or sea or city that emphasizes the locus and felt quality of place (above all its shape and scope).

Placial: having to do with place.

Placialization: the formation of place, for example, in landscape paintings and maps, but also in historical narration and prose fiction.

Planar (also **Planiform**) **map:** a map of a given terrritory that lacks express representation of the relief of specific geomorphological features (e.g., mountains, plateaus).

Planiform: flat; specifically, characterizing a planar map that does not attempt to represent depth or relief.

Planisphere: form of map that represents a sphere (or a part of a sphere) on a flat surface by way of projection.

Planitude: flatness or smoothness; the basis of "planigraphic," "planiform," and "planimetric" maps that prize regularity of representation (especially in the context of exact measurement) over the depiction of relief or isolated landscape features.

Plat: originally, place, spot, or locality; more recently, flat surface; also plan or layout of a particular piece of land (i.e., a plot).

Plattmaker: early English for mapmaker.

Portolan charts: navigational maps employed between the fourteenth and sixteenth centuries by European explorers; thought to have evolved from *periploi* and *portolani*; known for their colorful combination of linear, imagistic, and verbal elements; characterized by rhumb lines stemming from wind roses depicted in the map.

Position: an identifiable and definite location which is posited as such and to which other locations are internally related in the overall structuration of a painting or map.

Presentation: the display or exhibition of objects or experiences in a more or less coherent format; it is the "manifest content" (Freud) of any experience insofar as this experience has become accessible in a comparatively unmediated manner and is not the possession of mind alone; this English word corresponds to German *Darstellung*.

Presentational symbol: Susanne Langer's term for a cultural construction (such as a map) that exhibits its constituent parts in spatial coexistence with each other; a presentational symbol stands in contrast with representation, which (in the normal case) signifies items that do not belong to its own medium or intrinsic structure, as well as with discursive (i.e., linguistic) symbols.

Projection: the symbolic representation of three-dimensional space (especially the dimension of depth) in two dimensions, typically on the flat surface of a map, but also on a globe or other cartographic vehicle.

Psyche: the soul that actively imagines and remembers, appreciates and evaluates, recognizes and understands; in contrast with mind, psyche exists in intimate connection with places and things, thanks to processes of identification and internalization.

Psychotopics: genuinely psychical places, whether these belong to the psyche from the start or whether they have been formed by identification with externally situated places.

Quasi-cartographic: descriptive term for a painting that can also be regarded as a map (e.g., Jan Christaensz Micker's *View of Amsterdam* [Plate 12]).

Quasi-place-of-the-frame: the frame regarded as a by-work; it acts to include as well as to exclude; as a paradigmatic interplace, it is both inside and outside, being a determinable zone between natural and represented place—hence its essential ambiguity whereby it combines the difference between the actual place represented and the virtual place-of-representation.

Reconfiguration: alteration of a place and its occupants by means of changing its locus vis-à-vis that of other contents in the place-of-representation; this is a wholly relational concept.

Region: a portion of the earth's surface that has become a significant cartographic or painterly unit; it is constituted by a group of closely concatenated places that are spatially continuous with each other as well as temporally coexistent and thus cohistorical; a region in this sense is at once more constricted and more concrete than the kind of "material region" posited by Husserl as the proper domain of a "regional ontology."

Re-implacement: an effect of representation whereby the place of origin (whether locatable on earth or sea or wholly imaginary) is given a new locus within the painted or mapped work; integral to re-presentation of a place or region.

Reinstatement: the incorporation of something (e.g., a place or a thing) within the representational content of a painting or a map, such that the reinstated item is at once respected in its original form and yet subjected to subtle transformation.

Relief map: a cartographic representation of the major configurations of landmasses taken in relation to each other, and especially with regard to their comparative elevation.

Relocation: the change of locus that occurs when an actually experienced place (real or imagined) finds another location within a painting, that is, in the place-of-representation.

Re-placement: finding a new place for something.

Replacing: substituting one thing (physical object, image, word) for another; taking the place of the other.

Re-placing: a species of relocating; placing again; putting in another place.

Replica: a reproduction of something that keeps the proportions (but not necessarily the size) of the item reproduced; in contrast with an icon, which can be mental or semiological, a replica is physical in status; a replica need not represent that of which it is a replica.

Repoussoir: an item in a painting (e.g., a tree or a wall) that introduces and guides the eye of the viewer as it begins to explore the painting as a whole; in the West, it was first systematically employed by Claude Lorraine in the seventeenth century.

Representability: capacity to be represented in the various senses connoted by such terms as *representation, re-presentation, representation, re-representation,* representation.

Representamen: Charles Sanders Peirce's term for any kind of sign that serves as a representation of something else.

Representation: generic term for all ways of considering, remembering, or otherwise coming to terms with an object or experience in a different medium from that afforded by unassisted perception; the word answers to the German *Vorstellung*.

Representation: representation in Asher B. Durand's emphatic sense of the term, a representation that, though deriving from a model, stands on its own and asks to be judged for its inherent merit: "Paint and repaint until you are *sure* the work *represents* the model—not that it merely represents it" (Durand, "Letters on Landscape Painting," p. 1855).

Re*presentation*: representation that is bound to pictorial exactitude by means of rigorous isomorphic resemblance to a perceived, remembered, or imagined original term; close in meaning to *eikon*.

Re-presentation: representation that is not bound to depictive accuracy but that, instead, reinstates an original term in a transformed modality (e.g., color, linearity, composition, perspective); when taken literally, however, this term signifies exact transposition of an identically same content and is then equivalent to "*re*presentation."

Re-representation: a second representation, often in a medium different from that in which a first representation has appeared, but that carries forward the content of the latter into a new representational format: for example, a photograph of a painting.

Representational: realistically depictive, isomorphically true, achieving verisimilitude; the adjectival equivalent of *re*presentation.

Representative: being exemplary of or standing proxy for; not bound to the strict isomorphic standards of *re*presentation or what is representational.

Scenery: landscape scenes considered as mere backdrop.

Self-presentation: the manifestation of something on its own surface without recourse to support from other material media, symbolic constructions, or systems of interpretation.

Simple location: Whitehead's term for what I designate as "site"; it means specifically that the location of something is determined exclusively ("simply") in terms of its position vis-à-vis other positions in a homogeneous field of space.

Site: place reduced to location and position in space; determined by its relation to other similarly specified sites; at stake in topographically oriented painting such as landskips as well as in scientifically exact cartography and geography.

Small-scale map: a map that provides distant views of items in the mapped landscape.

Space: a totality of extension, unending and limitless; locations or positions within it are sites; as an abstraction from place, it is experienced in disembodied detachment; it is mapped by cartographic rather than by chorographic or topographic means.

Sublate: Hegel's verb for the movement from thesis to antithesis and then to synthesis whereby content is at once negated, preserved, and superseded; in this book, the word implies a movement of painting from topographic to topopoetic concerns, and is linked to the general action of sublimation.

Sublimation of place: the fate of place in landscape painting whereby an initially experienced (or imagined) place is transmogrified by processes of identification and representation to become a place-of-representation in a painting.

Sublime: in contrast with the beautiful, this is the outsize (Kant: the "mathematically sublime") or the emotionally extravagant (the "dynamically sublime").

Subliminal: not just tacit or implicit but (taken literally) up to or under a threshold.

Survey map (also called **Land map**): cartographic representation of survey space for purposes of documentation, development, or determination of ownership (in this latter case, it is also a cadastral map).

Survey space: a conception of space (first fully posited in seventeenth-century Holland) whereby space is considered to be unframed, all-inclusive (i.e., the surveyor is included in the space surveyed), as well as viewed from above.

Symbol: a cartographic sign that is neither pictographic nor verbal in character but that condenses and conveys, in a culturally coded manner, information regarding the geographic item that is mapped: for example, the emblem of the Goddess Roma in the Peutinger Table (Figure 7.8) or the evenly segmented blue lines that designate roads under construction in a contemporary Rand McNally Atlas.

Symbol map: a map that depends on abstract or schematic representations of particular features of a place or landscape, for example, circles that stand for villages.

Terrain: that aspect of a landscape which has to do with the perceptual appearance of its surface features.

Topographic: in general, concerned with the precise representation of landscape features by means of drawing or tracing; in painting, the topographic is the ideal of landskip art and is realized by such mimetic techniques as that of transposition; in mapping, it signifies the representation of landforms by the tracing out of their inherent shapes, sometimes supplemented by icons or symbols of such shapes; but it also may connote representations of certain specific landscape features of the area mapped, for example, in the form of pictorial "vignettes" decorating the margins of a map.

Topography: "the features of a region or locality [taken] collectively" *(O.E.D.)* as these features are represented in landscape paintings or maps; a topographical map or painting is composed of imagistic or pictographic elements, for example, detailed views of a landscape seen close-up.

Topoi: Greek term for "places," with the connotation (emphasized by Aristotle) of being closely confining and precisely bounded.

Topomnesia: remembrance of place, especially the remembering that informs a painter's rendition of a place or region he or she has once experienced in first person; in such memory, the place or topos constitutes the major theme or primary content, though not necessarily in an explicitly topographic manner.

Topopoetic: poetic expression of place, where *poetic* is not limited to verbal representation but includes painterly (or photographic) expression.

Totality: in geography, the collection of all locations in a given portion of the known world; not a mere sum but a whole of parts that is resistant to division or subtraction; as a "detotalized totality" (Sartre), it defies numerical specification, not because it is less than a given sum, but because it cannot be broken into strictly numerable units.

Transformation: change whereby the form or shape or location of a place is altered in the course of its representation in a painting or map, even as its material essence is retained; this often happens by reconfiguration.

Transplacement: the transformation of elemental presences in a landscape (e.g., earth, water, air) into ethereal nonpresences (i.e., silence and the void) in paintings that embody and evoke the contemplative sublime and the transcendental in Emerson's sense; Fitz Hugh Lane's paintings are exemplary instances of this etherealization of the elemental in art.

Transposition: a procedure by which an object's exact position—in a natural scene or in a sketch of that scene—is retained in a finished painting thanks to the use of a quasi-cartographic grid in which this position is clearly indicated, albeit on a different scale; also, the conveyance of factual truths bearing on its subject matter.

True-about: the sense in which a painting or a map is pictographically informative of the detailed structure of its subject matter, typically by means of isomorphic representation.

True-to: the way in which a map or a painting conveys the gist of a landscape, that is, its material essence or sensed presence, by means of a re-presentation of it.

Verisimilitude: the state of representational truth in painting or mapping that is achieved by likeness (but not necessarily by strict isomorphism); the full realization of truth-to; also, the sensuous quality of appearing to be real.

World: the cosmic principle of open-ended but coherent display; often possessing a collective or communal connotation.

Index

EDWARD S. CASEY is leading professor of philosophy at the State University of New York, Stony Brook. His research interests include aesthetics, philosophy of perception, environmental philosophy, philosophy of mind, philosophy of space and time, and psychoanalytic theory. He has written *Imagining* and *Remembering*, closely related studies in the phenomenology of mind, along with *Spirit and Soul: Essays in Phenomenological Psychology*. *Representing Place* is the third in a series of books on place; previously published are *Getting Back into Place* and *The Fate of Place*.